Laboratory organisation and administration

LIBRARY,

NATIONAL INST
FOR H! GH

TELEPEN

120076 3

WITHDRAWN

Laboratory organisation and administration

K. GUY, F.I.S.T

Laboratory Planning Officer,
University of Port Elizabeth

LONDON   BUTTERWORTHS

THE BUTTERWORTH GROUP

ENGLAND
Butterworth & Co (Publishers) Ltd
London: 88 Kingsway, WC2B 6AB

AUSTRALIA
Butterworths Pty Ltd
Sydney: 586 Pacific Highway, NSW 2067
Melbourne: 343 Little Collins Street, 3000
Brisbane: 240 Queen Street, 4000

CANADA
Butterworth & Co (Canada) Ltd
Toronto: 14 Curity Avenue, 374

NEW ZEALAND
Butterworths of New Zealand Ltd
Wellington: 26–28 Waring Taylor Street, 1

SOUTH AFRICA
Butterworth & Co (South Africa) (Pty) Ltd
Durban: 152–154 Gale Street

First published in 1962 by Macmillan & Co Ltd
Second edition published in 1973 by
Butterworth & Co (Publishers) Ltd

© Second edition Butterworth & Co (Publishers) Ltd, 1973

ISBN 0 408 70397 0 standard
     0 408 70398 9 limp

Filmset and printed in England by
Cox & Wyman Ltd
London, Fakenham and Reading

Foreword

Scientific endeavour is today accepted as an essential factor in both agricultural and industrial development and accordingly also for social advancement. The need for more scientists and technologists is fully realised and is being met by increasing both the number and size of the laboratories in schools, colleges of technology, and universities; and the manpower so produced is being rapidly absorbed in the new and enlarged government and industrial research laboratories.

The planning of modern laboratories is becoming increasingly complex. Before the detailed plans for a new laboratory can be prepared the specialist requirements must be enumerated and the resulting specifications are demanding more attention than ever before. The design of laboratories for highly specialised requirements, such as radiochemical operations, the specifications of the type of material for construction ranging from stainless steels to plastic cements, the layout of stores—these and related details cannot be left to an architect and are usually not readily available to a department head. It is doubtful if any laboratory has ever been built which has been found fully satisfactory to those who use it, and this state of affairs must arise from the lack of the necessary information.

The administration of a laboratory is the job of a specialist. The number of persons seeking this channel of employment is not meeting demand and the turnover of staff in junior positions is creating problems in training. It is accordingly important to have available some authoritative book by means of which the beginner can acquaint himself with the many facets of his job, and the experienced technician can have ready reference to specialist aspects.

A book of this nature can never be complete because modern experimental techniques are always changing and new materials of construction constantly being introduced. This book does, however, contain a wealth of information of value accumulated over a number of years and should prove useful to the technical officer as well as

FOREWORD

to the scientific administrator. The different topics which form the subject matter of the various chapters are important aspects of the complex problem of planning and administering scientific laboratories and from this aspect the book makes a real contribution.

The author of this book received his early training in London and was later Chief Technician in the Chemistry Department of the Sir John Cass College. In South Africa he has assisted in the plans for the complete renovation and modernisation of a large chemistry department and of new undergraduate and research laboratories.

It has been a pleasure to have had the services of Mr Guy as Chief Laboratory Technician in this department during which time there has been development unparalleled in the fifty years of our history. It is very satisfying to see that the enormous amount of data which has had to be sifted during the last few years has been assembled and presented in its present form. I feel certain that this book will fulfil the purpose for which it was written, namely to assist all those who are concerned in the organisation and management of scientific laboratories.

FRANK L. WARREN
Professor, Head of Dept. of
Chemistry and Chemical Engineering,
University of Natal, South Africa

September 1961

It is gratifying to me that *Laboratory Organisation and Administration* has been attended with such success that a new edition has been called for.

Since the inception of this book, Mr Guy has been appointed Laboratory Planning Officer to the University of Port Elizabeth and has travelled extensively to study further modern trends in laboratory design and management.

It is my pleasure to note that the value of the additional information collected for the second edition amply reflects the care and enthusiasm with which it has been compiled.

FRANK L. WARREN
Professor Emeritus University of Natal,
UCT/CSIR Research Fellow
Natural Products Research Laboratory,
University of Cape Town, South Africa

Preface

To satisfy the ever-increasing demands of industry for scientists, technologists, and technicians, unprecedented advances in technical education have been made in Britain during the past decade. The functions of the technical colleges have been extended and stronger links with industrial training have been forged. In addition, vast sums of money have been made available by the Government for new buildings, for the creation of colleges of advanced technology, and for expansion of the universities. The national reorganisation of technical education, which has been wisely coupled with special courses and part-time release systems, has given many more students an opportunity to avail themselves of improved study facilities, and as a result there has been an enormous increase in the number of persons undergoing laboratory training.

I hope this book, which is intended for persons who are at present, or who will be in the future, concerned with the organisation, management or design of laboratories, will to some extent contribute to the advances in technical education which are taking place. An attempt has been made to present the material in such a manner that it will be helpful, not only to technicians for whom it was primarily written, but also to science teachers and technologists who in the course of their studies may not have received specific instruction in the management, design, and care of laboratories and their equipment.

In this second edition the chapters are designed to cover the revised syllabuses of the courses in Laboratory Technicians' Work which lead to the 'Certificate of the City and Guilds of London Institute' in conjunction with the Institute of Science Technology. They deal in particular with the sections on general laboratory procedure and laboratory management which occur in the intermediate and final syllabuses respectively.

Two new chapters have been added. The first deals with technical literature and the second with experimental procedure. The whole of the subject matter has been revised, and where necessary brought

PREFACE

up to date. Also, through correspondence with many technicians throughout the world, it has been possible to enlarge the section concerned with the preparation of reagents so that it now includes the majority of those commonly used in laboratories everywhere. Because of the rapid advances made in laboratory design, the chapter dealing with this subject has in particular been revised and considerably enlarged. In some sections of the book the special regulations given may apply only in the United Kingdom, but since very similar regulations are in force in other countries, it is anticipated that technicians overseas will have no difficulty in interpreting them accordingly.

I have made no attempt to concern myself with organisation and management in the highest sense, but have simply endeavoured to present the practical aspects of the subject which are directly related to the laboratories themselves.

The laboratory operates to the best advantage when well designed and efficiently maintained. I hope, therefore, that this book, based on my personal laboratory and teaching experience, presents clearly and factually the ways and means of bringing about these conditions. If the book appears to deal more fully with teaching laboratories, this is due to the fact that my own experience was gained mainly in educational establishments.

I should like to express my deep appreciation to the many technicians and teachers who made suggestions concerning the contents of the book. Any further suggestions for improvement in future editions will be most welcome.

K.G.

Acknowledgements

I wish to acknowledge the following firms for the kind assistance they have given me and for permission to reproduce illustrations which appear in this book.

African Oxygen Ltd, Medical Division, P.O. Box 5404, Johannesburg, South Africa.

Antara, 435 Hudson Street, New York 14, N.Y., U.S.A.

Armstrongs (Hull) Ltd, Terry Street, Hull, Yorks.

Baird and Tatlock, P.O. Box 1, Romford RM1 1HA.

Block and Anderson Ltd, 58–60 Kensington Church Street, London W8.

Donald Brown (Brownall) Ltd, Lower Moss Lane, Chester Road, Manchester M15 4JH.

Colt Ventilation and Heating Ltd, Surbiton, Surrey.

Constructors Ltd, Kingsbury Road, Erdington, Birmingham 24.

Develop Kommanditgesellschaft DR Eisbein & Co, Dieselstrasse 8, Gerlingen 1, Western Germany.

A. B. Dick Company, 5700 West Toughy Avenue, Chicago 48, Illinois, U.S.A.

H. J. Elliott Ltd, E-Mil Works, Treforest Industrial Estate, Pontypridd, Glamorgan.

E.R.D. Engineering Co Ltd, Kelpatrick Road, Cippenham, Slough, Bucks.

A. Gallenkamp & Co Ltd, Technico House, Christopher Street, London EC2P 2ER.

Greiner Scientific Corporation, 22 North Moore Street, New York 10013, N.Y., U.S.A.

Grundy Equipment Ltd, Packet Boat Lane, Cowley Peachey, Uxbridge, Middlesex.

ACKNOWLEDGEMENTS

Hilger and Watts Ltd, 98 St Pancras Way, Camden Road, London NW1.
Kewaunee Scientific Equipment Corporation, Adrian, Michigan, U.S.A.
Kodak (South Africa) (Pty) Ltd, Black River, Parkway, Maitland, Cape Town.
J. Kottermann KG, 3165, Hänigsen Ü. Lehrte, Western Germany.
Ernst Leitz GmbH, Wetzlar, Western Germany.
Landis and Gyr AG, CH–6301, Zug, Switzerland.
W. Markes & Co Ltd, Wolverhampton Road, Cannock, Staffs.
Neumade Products Corporation, 720 White Plains Road, Scarsdale, N.Y. 10583.
Top Rank Television, P.O. Box 70, Great West Road, Brentford, Middlesex.
Rank Film Laboratories, Denham, Uxbridge, Middlesex.
Rank-Xerox Ltd, 33–41 Mortimer Street, London W1.
Roneo Vickers, Roneo Ltd, Lansdowne Road, Croydon CR9 2HA, Surrey.
Savage and Parson Ltd, Watford, Hertfordshire.
Vickers Ltd, Vickers Instruments, Haxby Road, York YO3 7SD.
VEB Carl Zeiss, Jena, Germany.
Vulcathene, Glynwed Plastics Ltd, Bessemer Road, Welwyn Garden City, Herts.

I am also indebted to the many other firms who have supplied helpful literature.

I also wish to acknowledge the assistance of the library staff of the University of Natal and the University of Port Elizabeth and various colleagues and other friends for their criticism and comments on various sections of the draft.

In particular my thanks are due, for his guidance and encouragement, to Professor F. L. Warren who read the text and made so many valuable suggestions.

Finally, I wish to thank my wife who devoted so much of her time to assist me with this book.

Contents

1 DESIGNING THE LABORATORY   1

LABORATORY BENCHES—Dimensions of benches—Reagent shelves—Space between benches—Bench tops—Understructure of the bench—Design of benches—Supporting the bench top—SINKS AND DRAINAGE—Sinks—Drainage waste pipes—CHEMICAL RECEIVERS AND TRAP UNITS—BENCH SERVICES—Cold water supply—Electrical supply—Gas supply—Steam supply—Vacuum—Compressed air—Service outlets—FLOOR SURFACES—Wood—Linoleum—Terrazzo—Concrete—Asphalt mastic—Quarry tiles—Vitreous tiles—Asphalt tiles—P.V.C. tiles—Cork tiles—Plastics varnish—VENTILATION—Natural ventilation—Mechanical ventilation—Central ventilating systems—Local ventilation systems—Fume exhaust systems—General methods of fume cupboard extraction—General design of fume cupboards—Fume cupboard services—LIGHTING IN THE LABORATORY—Measurement of light—The purpose of lighting—Electric lighting—Lighting efficiency—Illuminating power of light sources—Efficiency of light fittings—Distribution of light from fittings—Depreciation in illumination—Number and position of fittings—Colour of walls and furniture—Height of fittings—Lighting values—NATURAL LIGHTING—Measurement of daylight—Window design—Window obstructions—SAFEGUARDS IN LABORATORY DESIGN—Cleanliness—Atmosphere—Ventilation—Overcrowding—Fire hazards—Furniture—Stores—Laboratory workshops—DESIGN OF THE PHOTOGRAPHIC DARKROOM—Basic requirements

2 INSTALLATION OF LABORATORY EQUIPMENT   94

VIBRATION—BALANCES—Balance room—Balance supports—MERCURY BAROMETER—Fortin barometers—Kew barometers—GLASSBLOWING EQUIPMENT—Benches—Burners—Lathes—Ovens—HEAVY EQUIPMENT—Liquid air plant—LABORATORY STILLS—SPECTROGRAPHIC EQUIPMENT

3 STORES MANAGEMENT   109

TYPES OF STORES—Central stores—Main stores—Dispensing stores—DESIGN OF THE STORES—Main stores—FUNCTION OF THE STORES AND STOREKEEPER—Accounting for stores—Economy—Repairs to apparatus—Loans and transfers of apparatus—

CONTENTS

DEPARTMENTAL STOCKBOOK—SIZE AND LAYOUT OF STORES—PURCHASE OF STORES MATERIALS—Amount of materials to be purchased—Purchasing arrangements—Special purchases—CUSTOMS AND EXCISE—Overseas purchases—RECEIPT OF GOODS—Signing for goods received—Checking of goods received—Returned apparatus—DISPATCH OF GOODS—Return of empty containers—Goods dispatched by road—Goods dispatched by rail—Goods dispatched by post—DOCUMENTATION FOR STORES—Order form—Acknowledgement—Advice of dispatch—Delivery note—Invoice—Statement—Credit note—Debit note—RECORDING STOCK—Records—PRESERVATION AND STORAGE OF MATERIALS—Chemicals—Apparatus—MANAGEMENT OF PETTY CASH ACCOUNTS—Petty cash book—Petty cash vouchers—Petty cash systems

4 PREPARATION AND STORAGE OF REAGENTS 166

PREPARATION ROOM—Location—Services—Furniture—Apparatus—Standard solutions—Cleanliness—Distribution of reagents—Record book—Strength of solutions—CHEMICAL REAGENTS—Indicators—BIOLOGICAL REAGENTS—Fixatives—Stains—Mounting media—Agar media—SPECIAL REAGENTS—Special recipes

5 LABORATORY INSPECTION AND MAINTENANCE 197

PERIODIC INSPECTION OF LABORATORIES AND WORKSHOPS—Inspection—OVERHAUL, MAINTENANCE, AND CLEANING OF FURNITURE, EQUIPMENT, AND APPARATUS—Equipment—Furniture—Apparatus—CLEANING OF APPARATUS—Glassware—Silica ware—Transparent silica materials—Platinum ware—Plastics laboratory ware

6 SAFETY IN LABORATORY AND WORKSHOP 236

GENERAL—DANGERS FROM GLASS—Cutting glass tube and rod—Cutting glass sheet—Bending glass tube—Heating glass—Carrying glass—Issue of glassware—Boring corks to take glass tubing—Glassware for vacuum or pressure work—Broken glass—Other dangers from glass—DANGERS FROM GENERAL LABORATORY OPERATIONS—Furniture—Equipment—Acids—Ampoules—Sodium—Pipetting—Sulphuretted hydrogen—Perchloric acid—General precautions and requirements—DANGERS FROM POISONS—Mercury—DANGERS FROM FIRE—Inflammable solvents—Cause and prevention of fire—Firefighting equipment—Firefighting—Gas—Precautions against fire—Hazardous operations—DANGERS FROM EXPLOSION—Laboratories handling explosives—Explosions due to inflammable solvents—Explosions due to liquid air—Explosions due to dust—Explosions due to gases—Explosions due to chemicals—General precautions with

potentially explosive materials—Some common causes of explosions—
Perchloric acid—Pressure vessels—DANGERS IN STORAGE—
Planning chemical stores—General precautions for stores—Acid stores—
Ammonia stores—STORAGE OF CHEMICALS—Hazardous
combinations of chemicals—Volatile liquids—Chemical containers—
DANGERS FROM THE DISPOSAL OF CHEMICALS—DANGERS
FROM GAS CYLINDERS—Storage—Testing for leaks—Colour code
—Transportation—Valves and fittings—Cylinders in the laboratory—
Cylinder fires—DANGERS FROM ELECTRICITY—High tension—
Spectrographic equipment—Types of enclosures—General laboratory
work with H.T.—General installation—Overloading—Experimental
wiring—Earth connections—Precautions with instruments and portable
tools—Specially dangerous circumstances—DANGERS FROM
RADIATION—Types of radiation—Units of radiation—Quantity of
radiation—Sealed and unsealed sources—Effects of radiation—Maximum
permissible dose—Measurement of radiation—Measurement of
contamination—Protective measures against radiation—Personal protection
—Active areas—Safety personnel—Methods of working—Decontamination
—Waste disposal—Storage of radioactive sources—Transportation of
radioactive materials in buildings—Design of laboratories for radioactive
work—DANGERS FROM BIOLOGICAL HAZARDS—SAFETY
IN THE WORKSHOP—Lighting—Ventilation—Heating—Floors—
General finish—Spacing—Storage—Tidiness—Personal cleanliness—
Care of the eyes—Protective clothing—Hand tools—Machinery—
Maintenance—General safety hints—FIRST AID IN LABORATORIES
AND WORKSHOPS—First aid cabinets—First aid room—Types of
injuries and their treatment—Types of poisoning and their treatment—
FACTORIES ACT—Implications of the Act—Applications of the Act—
Interpretation of the Act—Enforcement of the Act—Administration of
the Act—Offences, penalties, and legal proceedings—Miscellaneous—
General provisions of the Act.

7 SPECIAL NEEDS OF TEACHING LABORATORIES 337

DISCIPLINE—Laboratory services—Laboratory clothing—
Maintenance of personal equipment—Bench cleanliness—Prevention of
theft—Continuous experiments—Reporting of damage—Supervision—
Regulations—Suggestion book—GENERAL REQUIREMENTS—
Accommodation—Lighting—Ventilation—Drainage—Services—
Distilled water—Water pressure—FURNITURE—Student lockers—
Other furniture—STORES—EQUIPMENT—Balances—REAGENT
BOTTLES—Labelling—Size of bottles—Pattern of bottles—Canada
balsam bottles—CHEMICALS—RESIDUES BOTTLES—RECOVERY
OF RESIDUES—Recovery of silver—Recovery of iodine—Recovery of
solvents—Recovery of platinum—Purification of mercury

8 OPTICAL PROJECTION 361

PRINCIPAL METHODS—Diascopic projection—Episcopic projection
—OPTICAL COMPONENTS—Condensers—Lens—Lamphouse—
Illuminating sources—The reflector—PROJECTION SCREENS—Plain
white screens—Silver screens—Beaded screens—Translucent screens—
Wall screens—Daylight screens—Portable screens—TYPES OF
PROJECTORS—Slide projector—Episcope—Epidiascope—Overhead

CONTENTS

projector—Filmstrip projector—Micro-projector—Micro-projection with the compound microscope—Specially constructed instruments for micro-projection—Micro-projection for special purposes—Cine film projector—ARRANGEMENTS FOR PROJECTION—Seating—Position of screen—Position of projector—Position of loudspeaker—Lighting—Acoustics—Ventilation—Darkening arrangements—Cine projector procedure—RECORDING AND REPRODUCTION OF SOUND—Types of sound track—Sound head—Magnetic stripe—Basic recording components—Recording on magnetic stripe—Lantern slides

9 LABORATORY RECORDS AND TECHNICAL INFORMATION 416

FILING SYSTEMS—MATERIALS REQUIRED FOR FILING—Primary index guides—Secondary guides—Folders—Miscellaneous folders—FILING EQUIPMENT—Filing cabinets—Visible filing equipment—FILING ARRANGEMENTS—Alphabetical—Numerical—Dewey Decimal Classification—Subject filing—Geographic filing—Memory files—FILING CONTROL—Filing the material—Disposal of old material—Return of borrowed records—FILING SPECIAL MATERIALS—Trade catalogues—Illustrative material—Technical information—Lantern slides—Storing and filing filmstrips—Films—DOCUMENT COPYING PROCESSES—Gelatine duplicators—Stencil duplicators—Stencil duplicating machines—Spirit duplicators—Xerographic process—Offset lithographic process—Camera copying—Non-optical photocopying processes

10 ORGANISATION OF DEMONSTRATIONS AND EXHIBITIONS 452

PERMANENT AND SEMI-PERMANENT EXHIBITIONS—Permanent exhibitions—Semi-permanent displays—Arrangement of exhibits—Display of exhibits—LECTURE DEMONSTRATION—Purpose of lecture demonstration—Preparation for demonstrations—Principles of demonstration—Method of demonstration—Aids to demonstration—Setting up apparatus for demonstration—Preparation room—Lecture theatre—Lecture bench—OPEN-DAY EXHIBITIONS—Preliminary organisation—General requirements—Nature of exhibits—Suggested exhibits—Display of equipment, samples, and specimens—More advanced experiments—EXHIBITIONS ARRANGED BY CHEMICAL SOCIETIES

11 ORGANISATION OF THE WORK AND TRAINING OF LABORATORY TECHNICIANS 469

ENGAGEMENT OF STAFF—Interviewing—ORGANISATION OF LABORATORY WORK—Distribution of work—ORGANISATION OF TRAINING—Training junior technicians—PROFESSIONAL ORGANISATIONS FOR TECHNICAL STAFFS—Institute of Science Technology—Institute of Medical Laboratory Technology—Institute of Animal Technicians—FUTURE OF LABORATORY TECHNICIANS

CONTENTS

12 ELEMENTS OF EXPERIMENTAL PROCEDURE 486
EXPERIMENTAL RECORDS—BASIC STATISTICAL METHODS

13 TECHNICAL LITERATURE 492
LIBRARY SERVICES—PROFESSIONAL TECHNICAL ORGANISATIONS—STANDARD WORKS—Textbooks—Reference books—British Standards—COMMERCIAL PRODUCTS—FILMS

APPENDIX 1 (common names of chemicals)   501

APPENDIX 2 (usual concentrations of some acids)   504

BIBLIOGRAPHY   505

INDEX   517

1
Designing the laboratory

This introduction to laboratory design tries to bring to the attention of the potential laboratory designer some materials and methods in use today so that by applying these he may obtain value for the sum of money at his disposal. No attempt has been made to discuss the complications of larger building programmes and the constitution of planning committees which these might involve, or other problems, such as the selection of sites, which are common to major projects.

The function of the laboratory designer is to assess the requirements of all the staff concerned and to present the information to the architect and builder in such a way that the completed laboratories satisfy all concerned. Before beginning any planning the designer must be in possession of the following information:

1. The amount of money available.
2. The kind of work to be done in the laboratories.
3. The number of persons who will use them at any one time.

First, consider the laboratory in terms of its main features such as the walls, ceilings, floors, and benches; then think about the other requirements which will give it life. These include heating, lighting, ventilation (including fume extraction), drainage, and the supply services to benches and other furniture. The designer must incorporate these component parts of the laboratory into a plan and must indicate clearly what he requires and the amount of money to be spent on the items concerned. From this information, the various craftsmen can advise how best to deal with his particular requirements. Always bear in mind that the cheapest materials are usually the worst for the job.

LABORATORY BENCHES

Laboratory benches fall into three main types and are known as island, peninsular, or wall benches.

Formerly, benches were built into the laboratory in a permanent style, but today unit assembly has become very popular because it enables benches to be dismantled and reassembled in other ways. This allows a measure of flexibility, which is an advantage, particularly in laboratories where the work may change from time to time. Unit furniture, however, may prove initially slightly more expensive, although now it is being mass-produced it is unlikely to remain so. At present, a complete three-metre centre bench with oak movable underbench units (depending on the pattern of units chosen) costs £10–65 more than a fixed integral bench of the same size.

Flexibility in laboratories may be allowed for to the extent that the benches and even the very walls of the laboratory may be moved to new positions. The degree of *essential* flexibility, however, must be most carefully considered, bearing in mind that over-indulgence may result in considerable unnecessary expenditure.

The chief difficulty for the designer considering laboratory furniture is the choice between wood and metal. Whatever material is finally selected, remember that standardisation reduces costs and provides a uniform appearance. Standardisation should, however, be limited, for it is impossible to design a standard bench suitable for all types of laboratories. Nevertheless, a reasonable degree of uniformity should be the aim.

Another very important factor today is the appearance of the units and more especially the use of colour, which has often influenced considerably the final choice of furniture. The advantage formerly given to metal units by the applied colour finishes has been cancelled out by the introduction of wooden units with plastics facings. The final selection, therefore, is governed once more by more practical considerations, many of which are found in the following pages.

DIMENSIONS OF BENCHES

Height

The heights of benches are governed by the average height of the persons using them and whether they will mainly work sitting or standing. Much, too, depends on the nature of the work, but it is generally agreed that in school laboratories a bench height of 0.75 to 0.85 m is suitable for standing work and one of 0.7 to 0.75 m for sitting work. For adult workers in advanced laboratories a standing working height of 0.9 m is best for males and 0.85 m for females. A sitting height of 0.75 m is suitable for both male and female workers.

It is important to adjust the bench height for the difference between young persons and adults where work will be done standing, but not so necessary if the work is done sitting. Adjustable seating may overcome difficulties here, or where the nature of the work may vary. It is sometimes convenient if a part of the bench is made at a different height from that of the remainder. I found that benches designed in this way for use by research workers met with considerable approval. In this particular instance, double-sided peninsular benches 3 m long were extended out from a side wall and designed to accommodate two persons. A 2.25-m length of the top was made 0.9 m high and the remainder (nearest the wall) was reduced to 0.75 m in height. This provided a writing desk with a knee space below. In the centre of the bench top, between the two sitting positions, a low bookshelf was fitted.

Width

The effective working width of benches is reduced by the various service outlets, such as gas taps, and is governed largely by the necessity of having reagents and equipment on shelves or bench scaffolds which must be easily and safely reached. The main laboratory furnishers seem to have reached accord on bench sizes which are generally 0.9 m high (0.75 m for sitting benches) and 0.7 m wide for single-sided and wall benches. Wall benches, which are used for larger items of equipment, such as ovens, incubators, or muffle furnaces and at the back of which wall shelves are not fitted, should be 0.75 m wide. Benches made for work to be done on both sides should be 1.2–1.5 m wide. The actual size is partly determined by the width of the centre shelves, the arrangement of sinks or centre troughs, and service outlets. In most laboratories a bench 1.4 m wide is suitable.

REAGENT SHELVES

Although the design of reagent shelves is influenced mainly by the number and type of bottles to be kept on them, service runs may also have some effect on the details of their construction. Remember that the overall height of the shelf units is of great importance, especially if unrestricted visibility across the laboratory is desirable.

Reagent shelves should be made from hardwood and may be single- or double-sided depending on whether the bench itself is single- or double-sided. The shelves may have one or two tiers but in some cases flat slabs are used which may be placed directly on the bench tops. In some junior laboratories it may be necessary to use slabs for reasons of economy and to provide a single row of bottles to serve persons on opposite sides of the bench. In such cases, to show clearly the contents of the bottles from either side, each bottle should have two labels on its side, one opposite the other.

All upper surfaces of the shelves should be covered with Vitrolite, glass, or other resistant material as a protection against the unseen spillages which run down the sides of bottles. Also, to prevent bottles being knocked off them, the shelves should be protected in the front with hardwood edging, slightly thicker than the shelf and so forming a lip. Similarly, a centre separator on double-sided shelves should also be provided to prevent breakages which sometimes occur when bottles are accidentally bumped together.

The spacings between shelves for bench top and side shelf units are governed by the standard size of the bottles to be used in the laboratory. The sizes of bottles vary according to the preference of the user and the nature of the discipline concerned but the most popular are 250 and 500 ml, which are used for liquid reagents. These sizes, allowing for reasonable clearance, require a shelf spacing of 170 mm and 190 mm respectively, but since the difference is small it is better to allow a common spacing of 190 mm. This spacing accommodates bottles of either size and also overcomes any later difficulties arising from changes in laboratory policy.

It is advisable also to standardise the size of bottles used for solid reagents on bench top units so that these too suit the shelf spacing. They may be refilled as required from stock bottles received from chemical suppliers. It may also be necessary to store Winchester quart bottles on side shelf units, and for this purpose a shelf spacing of 406 mm is necessary. Again, by keeping to the maximum spacing distance, half Winchester quarts and bottles of other intermediate sizes may be accommodated.

Chemicals used infrequently are normally kept on side shelf units in the bottles in which they are supplied, and since the bottles may have various shapes and sizes, adjustable shelving should be used to accommodate them.

SPACE BETWEEN BENCHES

When we consider the layout of laboratory benches it is always difficult to decide the space to be left between them. In order to arrive at a satisfactory conclusion we must know the answers to the following questions:

1. Is the work to be done sitting or standing?
2. What allowance must be made for opening of cupboard doors? (Extra space is required when persons bend down to remove apparatus from bench cupboards.)
3. What passing traffic is involved? (For example, in teaching laboratories provision must be made for supervisors to move among the students.)
4. Are laboratory trolleys to be used?
5. Is the laboratory to be used by adult or junior persons?

To reduce the space between benches to the bare minimum is a sure way of spoiling the laboratory whatever its other good points may be. In the well-planned laboratory ample space is allowed for supervision, servicing, and comfortable working. It is important, too, that auxiliary services such as fume cupboards, side reagent shelves, balance rooms, and shared equipment are well positioned and that their use entails only the least possible movement of the laboratory population.

Whatever the general arrangement of the benches the space between them falls under the headings which appear in *Table 1.1*.

Table 1.1 SPACE ALLOWANCE BETWEEN BENCHES

Conditions	Best space allowance	Minimum space allowance
Aisles with no persons working on either side	1.2 m	0.9 m
Aisles with a fume cupboard or similar facility on one side involving occasional use	1.2 m	1.0 m
Aisles with one worker on one side (e.g. single-sided benches)	1.2 m	1.1 m
Aisles with one worker on either side working back to back	1.7 m	1.4 m

The measurements shown in the column headed *Best space allowance* provide for sitting work, the passing of occasional persons, and intermittent trolley traffic.

DESIGNING THE LABORATORY
BENCH TOPS

In selecting materials for bench tops the availability of the material, its length of life, resistance to attack, and cost must be considered. The cost must be estimated on the installed price, for although certain materials may be initially cheap they are hard to work and the labour costs in making them up may be very high indeed.

The choice of material is determined by the nature of the work to be undertaken. The bench top might be called upon to withstand moisture, heat, attack by chemicals or pests, and contamination. Certain other properties of the material may also be important such as its resilience, electrical properties, ease of cleaning, and general appearance. The designer is required to consider all these properties before making his final decision.

Timber

Teak has for many years been the foremost material used for bench tops and today, in competition with hosts of new materials, because of its general suitability it still occupies first place. It has many advantages, for it is extremely durable, resists attack by chemicals and heat, and has a pleasant appearance. On the other hand, teak is costly and not so readily available as it once was. Consequently, other hardwoods with comparable qualities are now being used, such as afrormosia, mahogany, iroko, European oak, European beech, makore, and afzelia. Other suitable timbers, not widely used in Britain, are opepe, guarea, New Zealand kauri, cypress, and Pacific maple. The choice of wood for bench tops is limited by local availability and price but the best substitute for teak is undoubtedly afrormosia, which if properly maintained will easily last the normal lifetime of a laboratory. It is cheaper than teak and very similar in appearance. Iroko is also suitable but is a less predictable material. Mahogany is unsuitable for heavy wear because of its lower impact resistance.

Bench tops of manufactured board are also gaining popularity and give good service and do not distort or shrink. This type of bench top is veneered and lipped with a suitable hardwood such as teak. The minimum thickness of the veneer should be 3 mm.

Whatever timber is chosen, it must be properly seasoned and free from pests. The moisture content should not exceed 10–12% in heated laboratory conditions. The thickness of bench tops varies, according to the user's requirements, from 30 to 40 mm. Adequate drying out time in well-ventilated conditions must be allowed when the benches are first installed and the building itself should be allowed to dry out prior to their installation. Bench manufacturers recommend an acclimatising period of four weeks *in situ* during which time the central heating system of the building should be carefully controlled.

Bench tops may be constructed from either wide or narrow boards. Wide boards have the greater tendency to warp, and for this reason the narrow boards, usually 230 mm wide, are often used. Good service from bench tops constructed from narrow boards, however, depends very much on the way the boards are jointed and glued. I have found, too, that jointed tops tend to buckle in damp conditions at positions where they are cut to accommodate sinks.

After installation, the bench top may be treated with several applications of raw linseed oil, and thereafter wax bench polish should be regularly applied. Waxing the bench top, however, gives it better protection, and this treatment is undoubtedly best for benches used in chemical or other laboratories where the tops are subjected to chemical or acid attack. The wax may be ironed into the bench top but the easiest method is to dissolve high-melting-point wax in heated commercial xylene. Two coats of the hot mixture are applied with a paint brush. The wax is allowed to harden for forty-eight hours and then rubbed over very lightly with steel wool; during the next few days, before the bench is used, several liberal coats of wax polish should be applied. The further treatment of the bench top is to treat it regularly with wax polish. Some users prefer to acid-proof their wooden bench tops and use, in a certain sequence, solutions of potassium chlorate, copper sulphate, hydrochloric acid, and aniline. The result of this treatment is an almost black bench top.

More recently, many manufactured brands of protective finishes for bench tops have become available but up to the present time, for conditions of severe wear, they do not appear to have superseded the oil or wax finishes. Among the best of the protective finishes is polyurethane, a modern base-plus-hardener material which dries to give a hard and smooth finish. This material has a reasonable length of life and is resistant to water and a large number of solvents, but is affected by concentrated acids and alkalis[1].

Metal

A number of metals are used for bench tops including stainless steel, Monel metal, nickel, aluminium, zinc, lead, galvanised iron,

8 DESIGNING THE LABORATORY

and steel, but these have limited applications. To avoid expense thin gauges of metal may be mounted on non-warping wooden bases. Some of the advantages or disadvantages of the metals, according to the conditions under which they may be used, are listed in *Table 1.2*.

Table 1.2 RELATIVE MERITS OF METAL BENCH TOPS

Metal	Appearance	Resistance to attack	Cost	Some of its uses
Stainless steel	good	resistant but is attacked by sulphuric and hydrochloric acids to some extent	high	food laboratories, darkroom wet benches, radioactive laboratories, biochemical laboratories
Monel metal	fairly good	fairly resistant to chemical attack but tends to darken	high	mainly as for stainless steel
Lead (chemical purity)	poor	tends to buckle with heat; attacked by mercury	high	battery charging rooms, darkroom wet benches, chemical preparation benches, washing-up rooms
Aluminium (pure) and its alloys	fairly good	fairly resistant but attacked by alkalis and mercury; in alloy form very heat-resistant	low	explosives laboratories
Zinc	poor	attacked by chemicals	low	physical and textile laboratories

The usual metal bench top for general laboratory use is made of specially coated sheet steel about 1.25 mm thick. One method of coating the steel consists of treating it with an anti-rust material and a paint-bonding process. It is then sprayed with a red oxide, undercoated twice, and again sprayed with a resistant enamel.

Metal bench tops and metal bench units have several advantages over wooden tops, since they cannot be attacked by pests and are fireproof and free from shrinkage, warping, and distortion. In addition, they have greater mechanical strength and a greater storage capacity in terms of material–space ratio. The appearance of coated sheet-steel bench tops and units is good and they are easily cleaned. On the other hand, their disadvantages must be well considered by a designer, for although rubber is used to deaden sound, they are considerably noisier, they have no natural warmth, and if chipped they may rust. Metal bench tops cannot be so easily reconditioned as wooden ones after years of service.

Other bench tops

For specific purposes other bench tops have their uses. They include rubber, linoleum, quarry tile, ceramic tile, glass, slate, soapstone, asbestos cement composition, cement, and soft asbestos. Possibly the most important of these is the asbestos cement composition, which is finding increasing use in the laboratory. In chemical laboratories it is used mainly in the construction of fume cupboards and combustion benches because of its ability to withstand high temperature and its resistance to fire. Its natural colour is a light grey which is not unpleasing in appearance, and it can also be obtained in other colours and in ebony black grades in several finishes. It may also be purchased specially treated against acid attack and in thicknesses which vary from 3 to 75 mm. It can be machined, and some grades are particularly suitable for electrical work. Because it is porous in its natural state, asbestos cement for bench tops and similar purposes is impregnated under pressure with resins and acid-resisting paints and lacquers.

Slate

Slate was once used extensively in laboratories for supporting delicate apparatus. It is now found mainly in balance rooms, and being solid and heavy it has many advantages for this purpose. It is fairly resistant but may be affected by some acids such as chromic and sulphuric. It must be carefully selected because its surface tends to flake. Care must be exercised in the installation of the material and this work should be carried out by a stone setter. It is not recommended for use in fume cupboards.

Rubber and linoleum

Materials such as rubber and linoleum are limited in their general application and are used mostly in physical and similar laboratories. They are both kind to glassware. Linoleum, which is obtainable in many thicknesses, has been used successfully in darkrooms. It is quiet, warm, and fairly resistant, but is attacked by strong acids and alkalis.

Soft asbestos

Soft asbestos is used almost exclusively for glassblowing benches and it is well suited for this purpose. It tends to cling to clothing and can be annoying for this reason, and is usually treated with waterglass to harden the surface and make it less porous. The use of compressed asbestos cement bench tops is nowadays favoured by glassblowers.

Quarry tiles

Provided the jointing and bedding material is good and resistant to chemical attack, quarry tiles are eminently suitable for combustion benches. Great care is necessary to ensure a level surface with this type of bench top, since an uneven surface can be a nuisance. This is particularly so if heavy equipment has to be moved, for it is more convenient to be able to slide it to one side than to lift it bodily to a new position.

Glass

A bench top covered with glass is easily kept clean and for this reason is well suited to laboratories where perfect cleanliness is essential. Toughened glass or opal glass are the best for general laboratory purposes. Toughened glass may be used for the working top in fume cupboards, but since it must be thick to withstand mechanical shock, it is very expensive. The opal variety is extensively used as a surface for filtration benches. It is also used to protect reagent bottle shelves, and for this purpose should not be less than 6 mm thick or it will be easily broken. It should be held in position by a beading affixed to the front of the shelf which also protects the edge of the glass. The height of the beading should be such that its top edge is about 3 mm higher than that of the upper surface of the glass. This prevents bottles being accidentally pushed off the shelves. The only substance likely to affect glass to any extent in the laboratory is hydrofluoric acid. The disadvantage of glass is that it becomes so easily damaged by heat or sharp impact.

Glazed earthenware tiles

Glazed white tiles are also used to a great extent for filtration benches and have also been used for the working tops and backs of fume cupboards. They are set in cement or in acid-proof cement, but it is difficult to set them level. Their main disadvantage is the tendency of the glazed surface to craze; when this happens chemicals penetrate the surface and the appearance of the bench rapidly deteriorates. Tiles such as these are also easily damaged by sharp impact.

Soapstone

Soapstone is quarried and is inert and non-absorbent. It can be machined, and jointed by a tongue and groove, and the joints are filled with glycerine and red lead. It is not very widely used in laboratories at the present time.

Fabricated bench tops

In the U.S.A., fabricated bench tops of outstanding appearance and resistance are made. A variety of them are produced; one important type is the 'Kemresin' top made from a mixture of modified epoxy resin, curing resins, and inert fillers, which produces a thermosetting material. Since the finished bench top consists of a uniform mixture throughout its thickness of 32 mm it does not depend on a surface coating for protection. The top is jet black in colour and has a most attractive appearance.

Plastics

Plastics bench tops are used increasingly in radioactive and other laboratories where cleanliness or sterility is of paramount importance. Their use as bench tops in laboratories where more general work is carried out is limited by their mechanical wearing properties or by their resistance to solvents and corrosive reagents. They may also be damaged by heavy instruments, especially if the equipment is mounted on point supports.

Suitable plastics materials are expensive and are therefore generally used in the form of thin veneers which may be glued or otherwise affixed to cheaper base materials such as blockboard. The edges of the veneers should be protected by lipping them with teak.

The most important plastics materials are unplasticised P.V.C., polythene, and laminated plastics. An outstanding newcomer, polytetrafluroethylene (P.T.F.E.), offers better properties of resistance than any of the foregoing. It is durable and withstands temperatures up to 250°C. It is unattacked by chemicals, except fluorine and molten alkali metals, oils and waxes, or solvents and has first-class electrical properties. Unfortunately although this material seems the obvious choice for plastics bench tops, it is very expensive, which prevents its extensive use at present. With increased production the price of this material will undoubtedly decrease.

Polythene and P.V.C. are both weldable materials. They soften at temperatures above 70 °C but are resistant to a wide range of chemical substances.

Laminated plastics are homogeneous boards built up of layers of resin-impregnated core sheets, a coloured pattern sheet, and on top a transparent impregnated sheet. When used as a veneer their normal thickness is 1.6 mm. These plastics are easily kept clean and are very suitable for bench tops if the wear is moderate. They are unsuitable for heavy mechanical wear. They do not deteriorate with age, and can withstand temperatures up to 150 °C.

Table 1.3 VARIOUS SOLVENTS WHICH ATTACK PLASTICS BENCH TOPS

Polythene	P.V.C.	P.T.F.E.	Laminated plastics
Acetone	Amyl acetate	unattacked	conc. alkalis (prolonged exposure)
Dibutylphthalate	Bromine		conc. hydrochloric acid (prolonged exposure)
Ethyl acetate	Carbon disulphide		conc. sulphuric acid (prolonged exposure)
Hydrogen peroxide (100 vol.)	Carbon tetra-chloride		conc. nitric acid (prolonged exposure)
Hydrofluoric acid	Chlorsulphonic acid		
Iodine in KI	Dibutylphthalate		
Mineral oils	Ethyl alcohol		
Nitric acid	Lactic acid		
Conc. sulphuric acid at 60 °C	Conc. nitric acid		
Turpentine			

The common chemicals or solvents which attack various plastics tops are given in *Table 1.3* below the name of the plastics material concerned.

UNDERSTRUCTURE OF THE BENCH

The material for the understructure of benches must receive just as much care in its selection as the bench tops. A good-quality hardwood is advised and it should, like the bench top, have a moisture content such that in its new environment it is not subjected to conditions likely to cause it to warp through shrinkage. The construction is of paramount importance for it will generally house the services of the bench and it must do so in a way that allows accessibility for repairs. This point cannot be too strongly emphasised. Those who have worked in laboratories with faulty plumbing in inaccessible places have indeed had a bitter experience. Nothing is more infuriating than to have to contort oneself in the confined space of a cupboard to get at the seat of the trouble. Plumbing and other services may be brought to benches in several ways and these will be discussed later. For the moment it is assumed that these are available at the bench and it must be decided how they can be made accessible from any point along its length. One of the following methods may be applied:

(a) All services above bench level

Although this method of servicing benches is cheap and allows the pipe runs to be easily repainted and maintained, the appearance of the laboratory suffers considerably. This is especially true of those laboratories in which service pipes are fully disclosed and are brought down from the ceiling or carried at suspended levels to feed island benches below. This method of servicing benches also has the disadvantage that it restricts the use of some other fittings above bench level and hampers work in sink areas.

Another way to carry the services above the bench level is to enclose the pipes in a box. Since the box may be made from the same material as the bench top, it does not detract from the appearance of the bench and may also be neatly incorporated in the reagent rack design. Similar boxes may also be used for wall bench service runs and may be attached either to the bench or to a wall. In the latter case bench units may be slid in beneath them.

(b) No cabinet work below the bench

Benches of this description have a very limited use, for it is usually essential to have cupboard and drawer space adjacent to working places.

14 DESIGNING THE LABORATORY

(c) Part cabinet work below the bench with knee space sections and removable rear panels

This type of bench has great merit, for access is very easy. There is, however, an objection to its use in educational laboratories where maximum cupboard space is required. Note that island benches have been designed where the bench reagent racks are removable to give access to services from above. This allows for a restricted manoeuvre only, however, where the use of long tools is necessary.

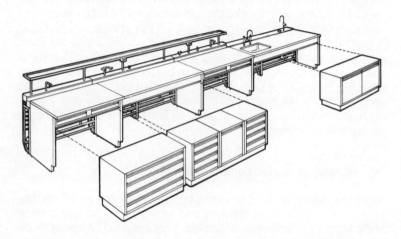

Figure 1.1. Wall bench with withdrawable units (Courtesy Grundy Equipment Ltd)

(d) (Figure 1.1) Removable cabinet work below bench tops

The removal of the entire cupboard completely discloses the services and allows free use of tools. This cuts down the maintenance time and in these days of high labour costs represents a considerable economy.

(e) Fixed cabinet work below benches and removable backs in the cupboards

The physical difficulties of entering cupboards, plus the necessity for first removing the apparatus within them, make this design decidedly undesirable.

DESIGN OF BENCHES

Benches resolve themselves into two main patterns:

1. Fixed.
2. Unit-constructed.

Note here that the so-called 'fixed' bench may incorporate removable units in strategic positions. Its construction thus allows limited maintenance facilities and is worthy of consideration by those who prefer 'fixed' benches.

Fixed benches

These may be designed especially for the work to be carried out, and since they are a permanent fixture may be suited to the general shape of the laboratory. Even though it may be decided to have fixed benches in preference to flexible units it is not essential that benches be specially designed, since certain basic laboratory units, including complete laboratory benches, are available from some laboratory furnishers. These have the merit that many years of thought and design have been incorporated in their construction.

Unit-constructed benches

Unit construction is particularly suitable to cope with the modern practice of changing laboratory layouts when the progress of the work demands new bench arrangements. It has obvious advantages and can save enormously on costs when such rearrangements become necessary. Benches of any type, width, or length can be quickly made up provided they are based on multiples of the unit size and wooden or metal units may be used.

Laboratories built up this way are flexible and the furniture and fittings can be quickly installed since so little assembly is required *in situ*. It may be true that in teaching laboratories there is less need for flexibility in bench layouts. Nevertheless, the fact that cupboards and other units may be simply slid out from under the bench top, thus allowing complete access for regular inspection and repairs to service and drainage pipes, is in itself sufficient to commend this design (*Figure 1.1*). Service units are prefabricated in lengths which are also multiples of the unit size and these may carry all the service pipework for the particular bench services required (*Figure 1.2*).

16 DESIGNING THE LABORATORY

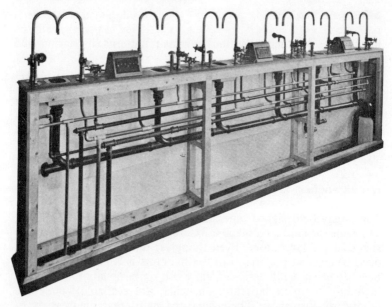

Figure 1.2. Prefabricated service unit (Courtesy Armstrongs Ltd)

If cost is an important factor when the type of underbench unit to be used is considered, the useful comparison of costs given by Dobson[2] serves as a guide. It is summarised in *Table 1.4*.

Table 1.4 SUMMARY OF COMPARATIVE PRICE LEVELS

Mahogany	1.00
Oak	1.07
Steel	1.25
Aluminium	1.32

SUPPORTING THE BENCH TOP

The bench top may be supported in one of three ways:

1. As an integral part of the bench itself.
2. On legs, brackets or cantilevers.

Special cantilevers called bench standards make the fitting-up of bench tops a relatively easy matter and bench units may be slid into position below them. These standards may also be suitably shaped to accommodate drainage troughs in the bench. By using cantilever frames much greater flexibility in the positioning of bench units is possible. The wall-type frames are also designed to permit, with the aid of special steel sections and pipe clips, the suspension and adjustment of service pipework at any level with ample fall. They are also drilled so that cover panels can be readily attached to cover up any visible pipework.

Note also that braced cantilevers too are commonly used for laboratory bench construction. In this case underbench units must fit between the confines of the leg spacings and suspended wooden or metal clip-on type underbench units may be used this way. These units may have a floor clearance of 190 or 90 mm *in situ*,

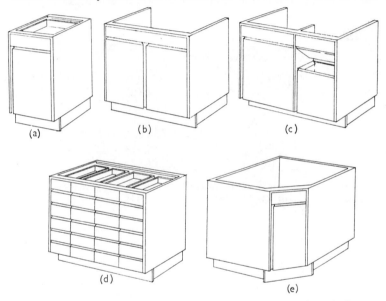

Figure 1.3. Various types of wooden units. (a) Cupboard unit with single drawers. (b) Cupboard with sink recess. (c) Cupboard and waste box. (d) Drawer unit. (e) Corner cupboard (Courtesy A. Gallenkamp & Co.)

depending on the size of units used. This clearance assists cleaning, especially in laboratories where it is necessary to hose down the floor. The units are also fitted with castors, which allows them to be quickly moved to new positions when necessary. Yet another interesting feature of this particular bench design is the adjustable feet fitted to the leg supports. The adjusting screws are fully concealed by plastics sleeves to maintain a neat appearance. A large variety of metal bench units is available.

3. Mounted on carcase units.

Some of the various types of carcase units are illustrated in *Figure 1.3*. In some cases the bench top is built up in sections, each bench having its own integral and separate working top. The success of such a bench, however, depends on the accurate setting up of the units and the adequate sealing of the joints at positions where the sections butt together. To ensure this the underbench units are normally fitted with height adjusters to take up variations in floor level and sealing strips are used at each joint (*Figure 1.4*).

Plumbing lines may be housed in service boxes or may be fixed to the wall in a recess left behind the bench units. The kind of work to

Figure 1.4. Shows how a sectional bench top may be made up from bench units with separate working tops. The bench top is quite separate from the service box, which may be fixed to a supporting frame assembly attached to the wall (Courtesy A. Gallenkamp & Co. Ltd)

be done on the bench, and the nature of the apparatus to be stored in cupboards and drawers, determine the size of the various units required to make up the bench.

Rubbish box units, too, should not be overlooked. In island and peninsular benches it is preferable that these be installed as a centre unit on each side of the bench rather than at the ends. The position of the shelves inside the cupboard needs some thought, and remember that some manufacturers make cupboard units with adjustable shelves which prove very useful. The handles on cupboard doors, if made of metal, soon corrode in the presence of fumes. In chemical laboratories, provided they are securely fixed with brass screws inserted through the back of the door, wooden handles are best.

Modern-styled laboratory furniture has recessed drawer and door pulls and these allow a completely flush exterior and also eliminate the hazards associated with protruding handles (*Figure 1.4*).

SINKS AND DRAINAGE

Remember always that the services provided on benches represent a large proportion of the bench costs and give a great deal of thought to this matter when a decision, in respect of the services to be provided, is being made. The service almost all benches require is a water supply and this entails the provision of sinks and a drainage system.

SINKS

The location of sinks always provides a problem for the laboratory designer and they should if possible be positioned in accordance with the sequence of work to be done on the bench. In teaching laboratories of university level, it is best to allow one sink for each student. They may be placed in openings in the working surface close to the working positions or at the ends of benches. Although sinks placed in the bench top take up bench space, this is compensated for by the convenience of this arrangement. Sinks may also be placed in the centre of the bench top to be shared between persons working on opposite sides of the bench. Although economical this arrangement is inconvenient.

It is difficult to choose between troughs and sinks and the arguments put forward for or against have been summed up by saying that benches with troughs usually also require a sink, whereas in benches fitted with sinks a trough is merely a convenience. Although troughs occupy a great deal of bench space there is no doubt that they do serve an excellent purpose in organic chemistry laboratories. If the trough is situated in the dead space below the centre reagent shelves, it does not interfere with the effective working space. The reasons for fitting at least one large sink to benches fitted with troughs is that it provides for large waste disposal, it can be used as a cooling bath, and it provides for local washing up of apparatus. Most sinks are fitted into working benches and are made of glazed fireclay, which is often erroneously referred to as porcelain. These are available in a range of sizes and shapes and can be referred to in any laboratory furniture catalogue. They have various rim finishes, known as: (a) external, (b) rimless. Another type of sink, now no longer manufactured as a standard item, has an external and internal rim; the function of the internal rim is to reduce splashing. An important feature of all laboratory sinks is the provision of an adequate weir-type overflow outlet.

Fitting of sinks

The actual fitting of sinks is a matter of considerable importance as neglect may result in flooding later on. Sinks with rims are fitted flush to bench tops, which may be recessed to accommodate them. Sometimes sinks are fitted below the bench top, and in this case the top overhangs the sink. Rimless sinks are usually fitted so that a part of the top of the sink is showing and the bench is shaped to fit snugly to the sink and a beading may be provided. In all cases it is most important that no gap exists between the sink and the bench top, and in any event a sealing compound must be used. Where bench tops overhang the sink, and particularly where the sink has no internal rim, the bench itself prevents splashing. In this case the overhanging portion is grooved underneath to prevent water from coming into constant contact with the sealing compound. Sinks fitted in this way have the advantage that water spilled on to the bench may be swept into them with a bench squeegee.

Materials for sinks and troughs

Sinks receive very harsh treatment in most laboratories and are the eventual dumping ground for almost all the chemicals which enter the laboratory. Much of this material is corrosive and is left to lie in the sink and do its damage. It therefore follows that a careful choice of material for sinks must be made. To avoid heavy wear on sinks, reversible plugs may be fitted in waste outlets and this allows a thin layer of water to remain in the sink which dilutes any corrosive material poured in. At the same time, it also prevents splashing, which is a common and disturbing occurrence in laboratories, especially where large numbers of students work in close proximity. Faucets which are placed too high also aggravate this situation.

Glazed fireclay

A common and inexpensive material used extensively for sinks is glazed fireclay. Glazed porcelain covers the porous material and hence should be of good quality and of adequate thickness. These sinks are made in one piece, which is a decided advantage over other types with joints or seams. They can be broken by heavy physical shock and by heat but they are quite resistant to moderate use, and are lasting and easily kept clean. This type of sink is undoubtedly the best for teaching and other laboratories where the sink is subjected to heavy wear.

Porcelain on metal

The base metal for porcelain-on-metal sinks is iron or steel. The main fault with sinks of this type is the tendency of porcelain to chip off. Wherever chipping occurs, they are liable to be attacked by corrosive substances, and for this reason they are not suitable for heavy wear. They are also easily scratched, and after a period of time may become stained.

Metal or metal-lined sinks

Sinks of metal, or metal-lined, can withstand physical damage and resist temperature changes. Because most of them are easily cleaned and have a pleasant appearance, they are well favoured in certain types of laboratories where cleanliness is of paramount importance. Stainless steel, which takes a satin polish, is the most widely used metal. If it is kept well flushed, it is not subject to attack to any great extent by chemicals, but is unsuitable for sinks in working benches in chemical laboratories. The pressed units, with the draining boards forming an integral part of the whole, are particularly good. Stainless steel is not unkind to glassware since it has a natural resilience. Metal units can be sound-deadened but they still tend to be noisy and are also expensive.

Polythene

Polythene sinks are now obtainable as one-piece mouldings. In the past these sinks were fabricated or consisted of a metal base lined with polythene. The moulded variety, which is kind to glassware, is suitable for laboratories in which quantities of solvents harmful to polythene are not used.

Drip cups

Drip cups are used in positions where a sink is unnecessary or would take up valuable bench space but where a drainage outlet is essential. For this reason they are often used in fume cupboards and on benches where working wastes are required. Drip cups are circular or oval in shape and may be bench- or wall-mounted and are available in polythene or glazed fireclay or glass. Polythene or glass drip cups are the most popular and the final selection depends on whether polythene or glass pipes have been used for the drainage lines. The circular cups are available in diameters ranging from 76 mm up to 165 mm and the oval-shaped fittings have top openings up to 114 mm wide by 184 mm long.

Sink outlets

Sink outlet sizes are standardised. It is wise to insist on a reasonable size of outlet, for small ones constantly choke and as a result users contrive to break or remove the outlet grilles in an effort to improve matters. This causes troubles to arise in the drainage system at points where the debris collects, which may be far removed from the sink concerned. The drop pipe should be glass or polythene and the waste pipe should also be of acid-resisting material. Since polythene usually gives trouble if sinks are used for hot water or for steam condensates, it is best to use resistance glass which is both acid- and heat-resistant. However, if some provision is made for cooling the hot effluents a polythene drainage system can be used. Standing overflow tubes are useful in sinks since the sink may be used as a cooling bath and for other similar purposes. Fixed grilles in wastes are preferable to the loose ones, which are invariably removed and lost.

Sinks for washing glassware

Large sinks for washing glassware are a decided asset at selected positions in the laboratory and this excellent facility is often overlooked in the design of laboratories. Where much washing up is to be done a double sink is a distinct advantage and allows for adequate rinsing of glassware. The wall behind the sinks should be protected by an adequate splashback. The water taps should also have a larger discharge outlet than is normally provided for those on bench sinks, and if a supply of hot water is also provided the water mixer type is most useful. A special tap with a jet nozzle outlet should be set quite high on the wall for washing burettes. A reasonable size of draining board must also be provided and in many cases this can be made of the same material as the sink. The remarks made about the wearing qualities of sink materials also apply to draining boards of the same material. Where a considerable amount of washing up is done wooden draining boards deteriorate quickly and pressed sink units with the integral board are a distinct advantage. Stainless steel pressed units leave little to be desired in this connection provided not much hydrochloric acid or sulphuric acid is spilled upon them and they are kept flushed with water. Draining boards should not slope steeply and those with the draining grooves deepening towards the sink but with flat lands are the most suitable. A peg board, with inclined hardwood or plastics pegs, fixed to the wall is handy for draining glassware. The front of the board should be covered with sheet rubber about 3 mm thick to prevent the lips of glass vessels from becoming chipped. The pegs should vary in diameter so as to accommodate the various neck sizes of the apparatus to be drained. Cupboards for storing cleaning materials should be provided below the wash-up unit.

DRAINAGE WASTE PIPES

Waste pipes suffer heavy wear and for this reason must be well sloped. The nature and temperature of the effluent affects the choice of material. The main materials used for the pipes are chemical stoneware, cast-iron, high-silicon cast-iron, cast-iron with glass or vitreous enamel lining, lead, polythene, polypropylene, rigid P.V.C., and glass. Copper, Monel metal, stainless steel, Fibreglass, and rubber are sometimes used for special purposes.

Chemical stoneware

Chemical stoneware open channels, once widely used for drainage purposes, are nowadays seldom employed and closed polythene pipes are used in preference. In cases where chemical stoneware is used, the glazed variety is the most suitable. It is strong, unattacked by corrosive liquids (except hydrofluoric acid), easily cleaned, and because of its smooth surface does not tend to retain solid suspended matter. It should not be used in exposed places, as it may be broken by impact. Drainage channels are made in half- or three-quarter-section. The latter are used where there is a risk of splashing and are more hygienic. The half-section has the advantage that it can be more easily cleared, particularly in bench runs. Both channels and pipes are usually joined by spigot-and-socket connections and asbestos cord and cement is used for joining. Depending on the effluent, cements should be either acid-resisting or alkali-resisting. Other materials used for joining have a pitch or asphalt base, but these may give trouble when hot water, such as the overflow from stills, continuously passes down the channels.

Polythene

Polythene is widely used nowadays for waste lines. It is usually black in colour owing to the addition of a pigment which prevents it from being attacked by strong light. A complete range of bends, elbows, and joints in the same material is also available.

Polythene is light and flexible, is easily handled, and can be cut with conventional tools. Pipes and fittings may be hot-welded together using a filler rod, and the process is similar to metal welding. The Polyfusion method of joining pipes and fittings is the most suitable, and homogeneous joins may be made in a few seconds by this process. A simple tool, shaped to receive pipes and sockets of various diameters, is used. The tool is heated electrically or by other means and the pipe and the socket are held on to it until the surfaces have melted. They are then simultaneously withdrawn and the melted end of the pipe is pressed into the socket. The surfaces amalgamate in contact giving a perfect join of great strength. If the pipe has later to be dismantled, screwed fittings at selected points should be used. This is done by a couple and a loose nut. Flanged connections are also available if desired. Polythene can easily be threaded in the conventional manner and has many advantages to offer for laboratory plumbing. At room temperatures it is resistant to acids, alkalis, and most solvents, especially when they are diluted by the waste water which normally flows in the pipes. Polythene is also unaffected by soft water and mercury. Chlorine attacks it at the surface and bromine and iodine are absorbed, which causes the material to become somewhat brittle. It has a natural elasticity, can be used in exposed conditions without fear of damage from frost, and is suitable for hot water up to 70 °C. Polythene tubes can be bent quite easily, and immersion in hot water assists this, but sharp bends or strains must be avoided. To prevent sagging, the pipe must be very carefully supported either continuously on boards or metal channels or by clip supports.

If the pipe is 25 mm diameter or less, place clips at a distance apart equal to twelve times the outside diameter of the pipe. For pipes exceeding this size the distance apart should be equal to eight times the diameter. The clips should not grip the pipe too firmly, for its thermal movement is seven times that of copper. To allow for this expansion, thermal stress-relief joints may be used.

Polythene can be used throughout an installation; this avoids the use of dissimilar materials—an important advantage for laboratory plumbing. At temperatures above 60°C it is liable to attack by some oxidising agents and aromatic and aliphatic hydrocarbons.

A high-density polythene known as polypropylene has even more resistant properties than low-density polythene. It is somewhat less flexible than low-density polythene and more expensive but is being increasingly used especially for laboratory plumbing.

P.V.C.

Because it is resistant to chemical attack, rigid P.V.C. compares favourably with low-density polythene but its extra rigidity makes it less convenient to use.

Lead

Only chemical lead, which is of high purity, should be used for laboratory drainage purposes[3]. In the past, lead proved itself reasonably satisfactory, but as already mentioned is attacked by mercury. It is also attacked by nitric acid at high temperatures and gives trouble in the presence of acids when joined to brass wastes. Lead pipe is heavy and unless it is well supported along its length it sags. It is expensive and for laboratory drainage purposes has now been almost entirely superseded by other materials.

Cast-iron

If cast-iron is used for drainage pipes the more expensive high-silicon variety (Duriron or Tantiron) is best. It is a risky material in laboratories where acids are used and in any event should not be concealed in solid floors. It should not be used directly under benches or in the immediate vicinity of sinks, but is suitable for main drainage runs in which the effluent has been already considerably diluted. Its main function is for vertical stacks where liquids do not lie over long periods and used in this way may have quite a long life. It is cheap and is obtainable lined with glass or vitreous enamel, and provided such pipes are used as complete lengths they are very good. At places where the pipes are cut or joined and at bends, however, damage to the lining may occur and corrosion sets in.

Glass

Borosilicate glass, which is virtually uncorrodible, is an extremely versatile material for laboratory waste lines and is now very popular. It has the advantage that blockages can be easily seen and it is more robust than its name suggests, though reasonable care is required during installation. The high cost of this material previously prevented its wider use, but owing to cheaper joining methods and the cutting and forming of glass pipes on site, installation costs have now been considerably reduced.

CHEMICAL RECEIVERS AND TRAP UNITS

Effluent in chemical laboratories should be diluted before it is allowed to enter the main drains. The effluent is normally discharged into chemical receivers where dilution takes place and any solid matter is retained. In some cases the use of receivers inside the laboratory is not recommended, since they may accumulate hazardous material, and provision should be made for more effective dilution outside the laboratory.

Until recently waste material was commonly discharged from bench sinks through drop pipes directly into open channels. Such channels, centrally placed, extend the whole length of the bench and discharge into the chemical receiver. These open channels have the advantage of being quickly inspected and easily cleaned out, but the fumes from them corrode neighbouring service pipes and they are unhygienic. A closed pipe is better and should be made of that material most likely to withstand the effects of the particular bench effluent. Remember that at this point the drainage system is subject to the effects of undiluted waste material and is therefore likely to suffer the most damage, so it is essential that the most suitable material should be employed for this part of the drainage, whatever the cost. For most conditions polythene or glass is suitable and the diameter of the pipe should be fairly large so that the various objects which inexplicably find their way down sink waste pipes do not easily block it. A pipe of reasonable diameter can more easily be cleaned of such obstructions, and should have blanking caps fitted at each end. This allows the complete length of the run to be rodded through when necessary.

After dilution in the receiver, the effluent must be led away to the main drains by the shortest possible route. In this respect, wall and peninsular benches may not present a problem, since they may be adjacent to outside walls and the effluent may be discharged directly into down pipes. With island benches, drainage may be effected by open drainage channels laid in floor chases and covered by removable panels in the floor. For laboratories situated above ground level, closed pipes are safer and lessen the possibility of flooding. Sometimes chases cannot be employed in floors above ground level and provision may have to be made for the pipes to be concealed above false ceilings in the floor below or else suspended from the ceiling itself. Unfortunately, these drainage methods often subject the occupants of the lower floor to the discomforts associated with faulty laboratory plumbing and this is a state of affairs to be avoided.

Chemical receivers are available in several shapes but two types are in general use and these are the siphon trap and the settling tank. Both of these act as mercury and silt traps and as dilution tanks. The siphon type has the additional advantage of providing a seal to the drain but, since the bottom of the siphon is only about 100 mm above the bottom of the receiver, regular attention is necessary to prevent the accumulation of excess silt.

The free outlet type of receiver is sometimes modified by fitting it with a plastics tube about 20 mm in diameter. One end of the tube rests about 25 mm above the bottom of the receiver and the other end is taken out through the normal receiver outlet. This modification is reputed to act efficiently as a siphon and to assist the rate of discharge from the receiver. Whatever type of receiver is used, it must discharge at a rate commensurate with the effluent input and several sizes of receivers are available. The cleaning of receivers, which is a distasteful task, can be made considerably more pleasant if a thick pair of gauntlet-length rubber gloves are worn.

It should be mentioned that although these types of receivers act as mercury traps, the recovery of mercury from them is not an easy matter.

The dilution of effluents is now usually effected by the use of a catch pot, which completely encloses the effluent and prevents the escape of fumes and smells associated with the open-type receiver (*Figure 1.5*). Depending on the drainage arrangements they are designed to serve, catch pots may have one to three inlets of variable sizes and may be of one- or two-gallon size. To clean them out the liquid contents are first removed via the drain plug and the complete catch pot removed. The lower half is then unscrewed and emptied.

Figure 1.5. Catch pot recovery trap and diluting receiver (Courtesy Vulcathene Glynwed Plastics Ltd)

Other similar types are available in the same capacities, but have a dilution chamber made of borosilicate glass. Catch pots of this type are particularly suitable for waste systems where quantities of solid matter are involved and the amount of collected solids contained in the easily removable glass portion may be easily seen.

The normal lead 'S' bend traps are not recommended for laboratory use. Below single sinks, individual traps made of polythene or similar material are suitable for use in laboratories where sinks are subjected to normal use. The design of these traps allows for the easy recovery of spilled mercury and the type with the visual glass base is particularly good (*Figure 1.6*).

Figure 1.6. Visual (glass) base trap (Courtesy Vulcathene Glynwed Plastics Ltd)

All traps required regular inspection and cleaning.

BENCH SERVICES

The way services are brought to the benches varies with the circumstances involved. They may be laid in floor chases fastened to the underside of floor joists in suspended floors, brought overhead, or, in the case of wall benches, brought along the walls behind the bench units. Whichever method is chosen, all piping should be concealed for the sake of appearance. If complete concealment is not possible, or proves too expensive, the exposed plumbing should be neatly installed so that the appearance of the laboratory is not adversely affected.

If open channels are used in the floors to carry away the effluent from the bench receivers, the pipes carrying the bench services, if also in the floor, must be so arranged that they do not pass over the open channels at any point. The feed runs should be kept as far away from the drainage channels as possible and access panels to both the pipe runs and drainage runs should be provided.

The kinds of services which can be laid on to benches are legion and are governed by the requirements of the work to be done. The following, however, are commonly used: cold water, hot water, steam, gas, compressed air, vacuum, and electricity. Of these, in most laboratories, cold water, gas, vacuum, and electricity are the most essential. The distribution of the services to various points along the bench can be from a central natural tunnel formed between the movable units, so that all services are easily accessible. This method is possibly the best. Sometimes services are run along the bench and placed just below the reagent racks. In this case, the water supply is set the lowest so that taps hang downwards pointing directly into troughs. The water and electricity service pipes are kept well apart. Concealed services in fixed-type benches have obvious disadvantages from the point of view of servicing.

Control valves to all the various services must be provided at accessible positions in all the main laboratory benches, under fume cupboards or groups of fume cupboards, or at appropriate positions in side benches. In this way particular sections of the laboratory can be isolated for plumbing repairs, which is a tremendous advantage, particularly in teaching laboratories where large classes are involved.

The various services are identified by colours in accordance with BS 1710[4]. The full length of the pipe should be covered with the appropriate ground colour. Alternatively, the pipe may be identified by painting it with the ground colour along only part of its length or by means of colour panels, labels, or coloured adhesive tapes. In all cases, however, where the pipes are only locally identified, the means of identification should be applied near to valves, junctions, or walls, etc.

Painting along the full length of the pipe is best, not only for identification but also because of the protection it gives the pipe itself. Painting considerably prolongs the life of the pipes and this is very noticeable in chemical laboratories if the conditions of painted and unpainted galvanised pipes are compared after a relatively short period of service.

In order to identify the pipe contents more precisely the actual

DESIGNING THE LABORATORY

state of the contents may be distinguished by means of a colour band superimposed on the ground colour (e.g. compressed air up to 13.8 bar has a white ground colour and compressed air above 13.8 bar has a red band superimposed on the white ground colour).

Under certain lighting conditions some shades of colour may not be easily distinguishable. BS 1710 recommends, therefore, that the contents of the pipe be shown on the ground colour in either black or white lettering and that this be done near to valves, junctions, and walls, etc. Lettering may also be used where necessary to indicate flow or return pipes. This is done by writing FLOW or F on the one pipe and RETURN or R on the other.

BS 1710 also gives the colour codes for piped supplies of industrial and medical gases and for the recognition of radioactive content hazards in pipes.

Lettering may also be used on electrical supplies to indicate whether the circuit is intended for lighting or heating purposes and to indicate the voltage.

Table 1.5 gives some of the colours used for the identification of piped services most likely to be used in laboratories and materials recommended for service piping are shown in *Table 1.6* (page 36).

Table 1.5 COLOUR CODE FOR LABORATORY SERVICE PIPES

Service	Ground colour	Colour band
Air compressed up to 13.8 bar	white	—
Air compressed over 13.8 bar	white	Post Office red
Chemical gases*	dark grey	various
Electrical services	light orange	—
Drainage	black	—
Fire installations	signal red	—
Town gas	canary yellow	—
Vacuum	white	black
Water		
cooling (primary)	sea green	—
condensate	sky blue	—
drinking	aircraft blue	—
cold (down service from storage tanks)	brilliant green	—
hot domestic	eau-de-nil	—

* For identifying colour bands for particular gases see BS 1710[4].

Another way to distinguish services is by the shapes and colours of wheels and levers attached to taps or valves, thus allowing the services to be identified by touch and sight. The shapes of the control handles are usually in accordance with BS 3202[5] and the inner ring on the handle has a letter code and is coloured as recommended in BS 1710.

COLD WATER SUPPLY

The materials commonly used in the manufacture of pipes for cold water supply are copper tube and galvanised iron tube; which is chosen depends on the conditions in the laboratory. To ensure the correct pipe sizes the consumption of water must be assessed. Bear in mind when estimating pipe sizes that even at peak periods all the taps will not be in use at the same time. An accurate assessment is difficult since the number of taps regularly used, as compared with the number actually provided, may be different in the various laboratories. From time to time, too, the number of taps in use varies according to the particular class exercises being undertaken. A general figure of 40% is, therefore, normally allowed and is known as the diversity factor. The amount of water the laboratory taps pass also varies in accordance with the supply pressure. If the average water pressure is known, however, then the amount of water which a particular size of tap outlet passes can be obtained from manufacturers' data. The size of water pipes required to supply a given number of outlets at a given pressure can then be found. This is done by multiplying the amount of water each outlet passes by the diversity factor. The figure obtained is divided into the carrying capacity of the pipe, which can be obtained from tables.

The total number of outlets required in particular laboratories is determined by the level of the students using it and the nature of the work. At university level, for instance, second- and third-year students in the inorganic laboratory require three taps each (including a swanneck outlet for washing burettes) and for organic work they require four.

A good pressure of water in the laboratory is important, particularly where there may be special requirements. The water pressure varies according to the district and the height of the draw-off point above the mains. If overhead storage tanks are used the pressure depends on the distance of the point below the storage tank. To obtain water at high pressure a booster pump may be used in connection with the low-pressure supply.

ELECTRICAL SUPPLY

Various means are used to protect electrical supply cables. Provided it is well protected from corrosion or can be regularly repainted, screwed conduit is the safest for laboratory purposes. The electrical sockets should be positioned well away from water taps, steam outlets, and possible wet spillages. This poses a problem on centre benches where multi-services are provided. If the sockets are placed at the front or ends of benches to meet the requirements, the leads from them to apparatus themselves constitute a hazard. A great deal obviously depends on the design of the bench itself. The author has found the best position on benches with centre reagent shelves to be inside the upright, supporting the shelves, and close to the underside of the lowermost shelf. This position provides maximum protection for the socket and keeps leads off the front working portion of the bench. For installation purposes the total possible load must be calculated and a diversity factor of 20% is normally allowed. Fleming-Williams[6] has said that the main switchboard, or at least the main cable in the building, should have several times the capacity of the maximum estimated demand. Without doubt, an adequate margin should be allowed and switchboards, and conduits too, which carry supplies to the laboratory should allow for any possible future additional requirements. In consultation with the electrical contractor a satisfactory margin should therefore be agreed upon.

In some rooms, such as microchemical laboratories, variations in the supply voltage should be controlled by a voltage control regulator or experimental work will be affected.

GAS SUPPLY

Town gas is widely used in laboratories and is supplied at a pressure of 28 cm w.g. For workshops and glassblowing rooms where higher pressures may be required gas boosters are used.

In some localities, and particularly in certain overseas countries where town gas is not available, liquefied gases of high calorific value are common. These propane or butane gases are supplied in light-gauge metal cylinders. The cylinders may be fitted with individual regulators. The pressure for the normal supply to laboratory bench outlets is 28 cm w.g. but special regulators are available for use where high-pressure delivery is required. Individual cylinders fitted with regulators may, subject to local safety regulations, be used inside buildings to serve a particular laboratory or even single burners. It is much safer, and more customary in larger establishments, to situate the cylinders outside the building where they may be arranged in two banks and connected to a manifold. The banks are linked by an automatic changeover valve which also indicates when one bank is empty. The gas is piped to various laboratories where pressure regulators are situated outside the building in the immediate vicinity of the areas being serviced.

STEAM SUPPLY

Although modern heating devices have greatly reduced the necessity of a steam supply on the laboratory bench, a steam supply is still preferred by some workers, especially in organic chemistry laboratories. The installation becomes a much cheaper proposition in buildings served by a central plant for central heating purposes. For laboratories which do not merit a central heating system, or where climatic conditions are such that it is unnecessary, an electrically heated steam generator is remarkably efficient. The boiler, along with other local generating plant, should be housed in a central room which is so constructed as to reduce noise to a minimum. If steam is produced in a large central plant it is economical to distribute it at high pressure in small-diameter pipe and to use local reducing valves at places where lower pressures are required. Adequate drainage points in steam mains should be provided. All steam piping must be efficiently lagged, and due allowance made for the increased diameter due to the lagging when the pipe runs are determined.

VACUUM

Vacuum may be provided on laboratory benches by means of central pumping units connected to a pipeline system, by individual rotary vacuum pumps, or by water jet pumps. Because of the large initial outlay it is not usually economical to install a central system unless twenty-four points or more are involved, but where requirements merit a large installation the pumping unit chosen must have sufficient capacity to deal with the number of points to be serviced. Such units may be automatically controlled. These systems are best suited to conditions where a moderate vacuum of 710 mm of mercury is required although, provided the installation is well designed, a vacuum of 10 mm can be obtained. As the system grows in size, however, so the difficulty of maintaining the vacuum increases and great care is necessary in its installation and maintenance. Also, take adequate precautions to ensure that corrosive liquids and vapours do not corrode the line or harm the pump itself; such precautions may include the use of traps in the line or the provision of gas ballast pumps. As a further safeguard individual workers on the bench should also use traps and this also prevents corrosion of the vacuum taps. This kind of system demands good discipline on the part of the users. Another type of pump suitable for central systems has a continuous flow of water passing through it and the rotor itself enclosed in an elliptical casing continually flushed by the water flow. Air and vapours sucked in by the pump are discarded along with the water outflow. Such pumps do not suffer damage for vapours and the waste water may be passed to a tank and recirculated.

A moderate vacuum can also be provided by the use of inexpensive individual vacuum pumps, and these are ideal if only a few vacuum points are required. Some models are designed for use as both vacuum pump and compressor and the larger ones of this type produce a vacuum of 710 mm of mercury.

For pressures below 10 mm of mercury individual rotary pumps are essential and the type of pump must be selected in accordance with the requirements of the work.

Water jet pumps, or filter pumps as they are sometimes called, which may be of glass, metal, or plastics, are widely employed for rough vacuum work, especially in teaching laboratories. They may be attached to the water tap by rubber pressure tubing and both the metal and some plastics varieties can also be screwed directly on to water taps threaded to accommodate them. To obtain the best results from these pumps, which produce a vacuum of 12–15 mm, a water pressure of between 1.7 and 3.4 bar is required.

Although the installation of water jet pumps is initially much cheaper than piped systems the cost of water used in operating them is high. Because of the wastage of water, too, their use is frowned upon by most water authorities, and water recirculation systems are often employed to overcome their objections and save expense. Nevertheless, because this type of pump is reliable and convenient and cannot be easily abused it is extremely popular. Modern water jet pumps incorporate a device to prevent suckback when sudden reductions in the water pressure occur or when the pump is turned off. Unfortunately, these devices are not always completely efficient and bench traps are usually used in conjunction with the pumps. An attachment to overcome the splashing associated with filter pumps has been described by Edwards[7]. Another suitable method is to attach a length of plastics garden hose to the water outlet. The hose, which should be large enough in bore not to impede the water flow and to prevent whipping, can be clipped to the edge of the bench top where it overhangs the sink.

COMPRESSED AIR

Compressed air can be made available in the laboratory by means of cylinders, portable air compressors, or a piped supply from a centrally placed compressor unit. Pressures of 0.35–0.70 bar are normally required for bench use, although higher pressures may be required for special purposes.

If only a few outlets are to be served a piped supply is uneconomical and cylinders may be used, but since these are cumbersome small portable compressors mounted on trolleys are better. The dual-purpose vacuum pump–compressor unit is particularly suitable.

For piped systems various types of compressors are used and the size of the unit required can be determined with reasonable accuracy by taking into account the number of outlet cocks likely to be in use at the same time. It may be assumed that each cock will discharge 0.7 l/s at 0.35 bar pressure.

Table 1.6 MATERIALS RECOMMENDED FOR SERVICE PIPING (FROM BS 3202[5])

Service	Materials
Cold water	lead, copper; polythene to BS 1973[8]; P.V.C. medium or heavy galvanised steel tube to BS 1387[9]
Hot water	copper, or medium galvanised steel to BS 1387[9] for normal domestic or industrial purposes; heavy galvanised steel tubes to BS 1387[9] for special cases (underground)
Heating (hot water or steam)	medium black or galvanised steel tube to BS 1387[9] for normal installations, heavy for special cases; copper
Steam	medium or heavy black steel tube to BS 1387[9]; copper
Condensate	copper
Gas (town or petrol)	light or medium black steel tube to BS 1387[9]; copper or aluminium
Liquefied petroleum gas	copper
Compressed air	medium or heavy black or galvanised steel tube to BS 1387[9]; copper or aluminium
Vacuum	copper or aluminium; medium steel to BS 1387[9]; chemical lead to BS 334[3] for corrosive conditions
Distilled water	stainless steel; glass to BS 2598[10]; P.V.C.; polythene to BS 1973[8]; aluminium, pure tin

If clean air is required a suitable means of removing oil, water, and dust must be provided. Oil may be eliminated by using water-sealed or graphite-lubricated pumps, instead of oil-lubricated types, and water can be removed by using after-coolers or separators and absorption filters.

Air entering the compressor should also be cool and filtered to exclude dust. The compressor should be connected to a stabilising tank to smooth out the supply. The pressure of the air fed to the tank should be higher than that in the line and a pressure regulator should be fitted to ensure a constant pressure in the line. The receiver should also have a pressure gauge and a pressure-controlled switch to cut out the compressor when not required. In order to comply with safety regulations the tank should also be fitted with a relief valve and a drain cock. To ensure quiet operation the compressor should be fitted with a silencer and housed in a special services room.

SERVICE OUTLETS

The taps, cocks, and valves used to control laboratory services require careful selection. They should be pleasing to the eye and easy to clean. By far the most important property of any fitting is its resistance to the particular conditions in which it is to be used. Black bronze or black oxidised and chromium-plated finishes, although much favoured in the past, are now seldom used because they deteriorate rapidly in the presence of acid fumes. Such finishes are suitable only for physical and other fume-free laboratories. Epoxy resin and similar stoved enamel finishes, on the other hand, are suitable for any open laboratory conditions. They have a long life, and are available in various colours. Even these durable finishes deteriorate if subjected to severe conditions. In confined spaces such as fume cupboards they are quickly attacked, especially in the presence of nitric acid or other strong oxidising agents, and for such conditions special finishes are necessary.

The importance of the careful installation of all service outlets cannot be overemphasised. Damage to the finish through the use of unprotected tools allows corrosive agents to penetrate the protective coating and thus considerably shortens the life of the fittings. Fittings will last much longer if cleaned with warm soapy water followed by a rub with a lightly oiled rag and this treatment will keep epoxy resin finishes looking like new.

The design of laboratory service outlets is constantly being changed and two recent major improvements have been described by Tongue[11]. The first is a water tap incorporating a non-rising handwheel and a completely enclosed spindle. The sealed headwork

38   DESIGNING THE LABORATORY

prevents water escaping past the gland and eliminates the well-known laboratory nuisance of water dripping on to bench tops.

The second improvement concerns a new thimble valve which has been designed to control a number of different kinds of services, including mains water, more easily and is capable of giving a metered flow control.

Water outlets

Although some special-purpose water taps have fixed corrugated nozzles, most modern water fittings now have a standard threaded outlet to allow nozzles of all types, vacuum pumps, and other apparatus to be fitted to them and changed at will.

Water taps may be classified roughly into three main classes: pipe-mounting, bench-mounting, and vertical-surface-mounting. Within the three classes the various types used are bib taps, drop taps, angle taps, and swannecks. In many cases combinations of these various types of taps may be incorporated in one fitting.

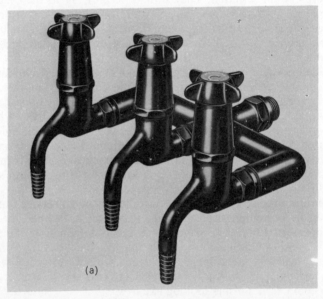

Figure 1.7. (a) Vertical-mounting bib-tap fitting for walls, service ledges, or pedestal units. (b) Five drop taps on a loop fitting. This multi-tap fitting fits over a 6-in (152-mm) diameter drip waste. (c) Water standard with two bib valves and one swanneck outlet. (d) Combined water and gas fitting (Courtesy (a) W. Markes & Co. Ltd, (b) Brownall Ltd, (c) Vulcathene Glynwed Plastics Ltd, (d) W. Markes & Co. Ltd)

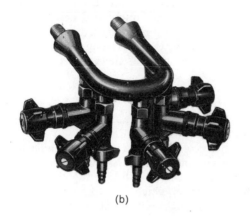
(b)

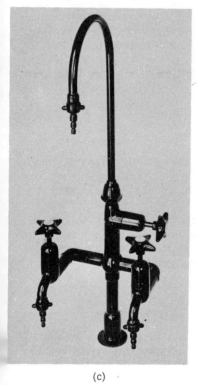
(c)

(b)

Composite units known as multi-tap fittings are also available and are used for areas where an unimpeded bench top is considered essential. Other special-purpose taps such as mixing taps are in use and combined gas and water units designed to save space are also available. Examples of some of the various types of fittings are shown in *Figure 1.7*.

Gas taps

The various patterns of gas service outlets available may be used for pipe, wall, or bench mounting. They may have from one to four outlet nozzles and the choice of fitting is governed by the requirements of the location in which it is to be used (*Figure 1.8*). The outlets on two-way fittings are set at 90° or 180° apart and on three-way fittings 90° or 120°. Four-way fittings have outlets 90° apart.

Modern gas fittings have a yellow lever control and a plug cock. The cock is usually spring loaded and for bottled gases this is essential. The outlet nozzles on most modern fittings are angled slightly upwards and this makes the fitting of tubing to them easier. Some gas fittings have cocks with short fold-down levers and are known as drop-lever cocks.

Most modern plug cock gas outlets are suitable for both town gas and low-pressure bottled gases.

Steam outlets

Steam valves must be suitable for the pressure of steam used. Although pressures of up to 1.38 bar are normal for laboratory benches, the standard valves supplied by many manufacturers are tested and suitable for pressures up to 6.8 bar. Screw-down valves are used for steam control and should have heat-insulated handles. Steam fittings may be of the pipe, angle, vertical-mounting, or bench-mounting variety, but in all cases the outlets must be directed downwards over sinks or other drainage fixtures.

Vacuum taps

Vacuum taps for low-vacuum lines on laboratory benches usually have needle or diaphragm (or thimble diaphragm) type valves. The fittings may be of pipe, angle, vertical-mounting, or bench-mounting type.

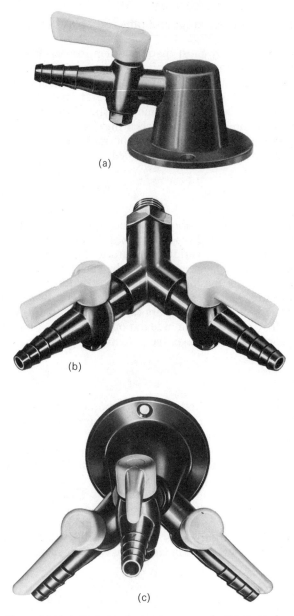

Figure 1.8. (a) Single bench-mounting gas cock. (b) Vertical-mounting 2-way gas cock suitable for pipe fixing. (c) Vertical-mounting 3-way gas cock suitable for wall fixing (Courtesy Brownall Ltd)

Valves for compressed air and other gases

The valves made by most manufacturers for bench vacuum lines are equally suitable for controlling compressed air and other gases such as oxygen, nitrogen, hydrogen, and petroleum gases. Some internal components of the valves, however, may differ according to the supply to be controlled and the type of service the valves are required for should be stated when they are ordered. The control knobs on the valves are coloured and marked in accordance with the service for which they are intended.

FLOOR SURFACES

The materials used for laboratory floor surfaces should be selected in accordance with the type of work to be done in the laboratory. The materials used in various laboratories throughout the building may not, therefore, necessarily be the same. When selecting the flooring, consider the following points: (a) safety, (b) resistance to substances likely to be spilled on it, (c) resistance to wear, (d) comfort, (e) ease of cleaning, (f) appearance, (g) noise, (h) warmth (to the touch), (i) resistance to indentation. The order of importance of these properties depends on the particular application.

WOOD

Hardwood floors are well suited for laboratory purposes and withstand harsh treatment. Teak and other hardwoods such as iroko, oak, and beech wear well and are little affected by substances accidentally spilled upon them. Their good natural appearance is enhanced by the light application of wax polish, which should be regularly applied to preserve them. Wood is not slippery or noisy and is warm and comfortable to stand on. The most suitable hardwood for laboratory floors is strip or strip parquet. Wood blocks are recommended for use in laboratories and rooms which are dry and this type of flooring has been successfully used in such conditions. Wood floors are kinder to glassware than hard floors and reduce glass breakages.

LINOLEUM

Linoleum may be purchased in roll form and is used a great deal in radiochemical laboratories and in rooms where mercury is likely to be spilled. The joints may be filled with a sealing compound to prevent penetration by harmful substances. This material is also used in balance rooms and in other similar locations where dry conditions exist. It should not be subjected to any great extent to the effects of spilled chemicals. It is quiet, comfortable, and available in attractive colours. The lighter gauges are suitable for offices, but in laboratories the heavier grades such as 4.5 mm and 6 mm are best. Linoleum must be laid on level floors, and if necessary levelling sub-base materials may be used beneath it. It can be cemented to wood or cement floors, which checks its stretching tendencies and prevents deterioration due to water seeping beneath it. Linoleum is also available in the form of tiles, and in rooms where jointless flooring is not required and in which the floor is not subjected to harsh conditions these provide an attractive and satisfactory floor.

TERRAZZO

Terrazzo floors consist of small marble chips mixed with cement. By using tinted cement, floors of various colours can be effected. This is an attractive, durable, and hygienic material, easily cleaned by washing. It is hard, noisy, cold, and is attacked by acids. Its main use is in corridors and hallways where the traffic is heavy, but it is also used in laboratories where very hygienic conditions are necessary.

A similar floor, also suitable for heavy traffic and for heavy working conditions such as are met with in engineering laboratories and workshops, is known as granolithic paving. This is laid in the same way as terrazzo but granite chips instead of marble ones are mixed with the cement.

CONCRETE

Concrete floors are dusty, cold, uncomfortable to stand on, and have a poor appearance. They withstand the action of substances spilled upon them and are resistant to acids if cement with a high silicon content is used. To further assist their resistance and to give them a more attractive appearance they may be painted with resistant paint. Pieces of floor covering may be placed on concrete floors at standing positions by benches.

ASPHALT MASTIC

Asphalt mastic floors are fairly resistant to acids and alkalis but are attacked by solvents. They are water-resistant and for this reason are sometimes used in darkrooms and in other places where damp conditions prevail. This flooring material is easily indented by heavy equipment but is quiet and fairly warm.

QUARRY TILES

Quarry tiles vary in thickness from 16 to 50 mm and may be laid on a solid concrete foundation. Although the colours of these hard-wearing tiles are restricted to red, buff, and brown they can when laid be made to look attractive by using design effects. Quarry tiles are resistant to chemicals and acids and are not slippery, but they are noisy and non-resilient. They may be repeatedly washed and coved tiles are available for skirtings.

VITREOUS TILES

The remarks made about quarry tiles apply generally to tessellated clay vitreous tiles. These tiles provide an impervious and resistant floor of good appearance and are available in a range of colours and in several sizes. They are very hardwearing and should be laid with narrow joints. In acid conditions the joints should be made with acid-proof cement.

ASPHALT TILES

Asphalt floor tiles, also known as thermoplastic tiles, have been found generally suitable for laboratory conditions. They should not, however, be laid in laboratories where solvents are likely to be spilled or where the floors may become wet. They are attractive, easily cleaned, fairly resistant to attack, and quiet, but may be indented if heavy objects are placed on them.

P.V.C. TILES

P.V.C. tiles are similar in appearance to asphalt tiles and have the same general properties. They have a greater resistance to solvents and to acids but these substances if spilled should be immediately wiped up and not allowed to lie on them. They are fairly resistant to indentation. P.V.C. is also obtainable in roll form.

CORK TILES

Cork tiles are warm and comfortable to walk and stand on. They are very quiet and hence are used to a great extent in libraries and offices. They are affected by grease, acids, alkalis, and solvents and are not generally suitable for use in chemical laboratories.

PLASTICS VARNISH

Plastics varnish floors such as polyurethane or acrylic types are applied over an epoxy base coat. They have good wear resistance and are impermeable to many chemicals. They are attractive in appearance, available in a number of colours, and, having no joints, do not present a cleaning problem.

They may be made even more decorative if vinyl decorative chips or texture coatings are incorporated when they are being laid. The life of these floors is said to be greater than that for conventional thermoplastic tiles.

VENTILATION

Good general ventilation is essential in laboratories to rid the atmosphere of fumes given off from normal bench processes, laboratory reagents, and gas burners, furnaces, and hotplates. It is also necessary to remove other air contaminants such as body heat, body odours, and excess carbon dioxide and to reduce the risk of infection from airborne diseases. In addition other air contaminants peculiar to laboratories which, in the form of fumes or dust constitute a serious hazard to health, must be removed. This is sometimes done at the source of contamination by localised ventilating devices.

Since human beings can tolerate inadequate ventilation better than they can extremes of temperature, the ultimate success of any method of ventilation is associated with a comfortable working temperature. Various people react differently to the same temperature, but a reasonable working temperature for winter conditions lies between 16 and 20 °C.

The human body loses heat to the atmosphere by radiation, convection, and evaporation. It is therefore necessary that the temperature, velocity, and humidity of the air be maintained at suitable levels. If laboratories are overheated, stuffy atmospheres result; this effect is more noticeable when insufficient variable air movement is produced. A greater degree of air movement is necessary in summer than in winter.

Movement of the air may be effected by natural means of ventilation or by mechanical means.

NATURAL VENTILATION

Natural methods of ventilation are cheaper than mechanical means and in some cases where problems of fume disposal do not arise opening windows is sufficient. Much depends on the proper positioning of the windows so that they do not create draughts or affect

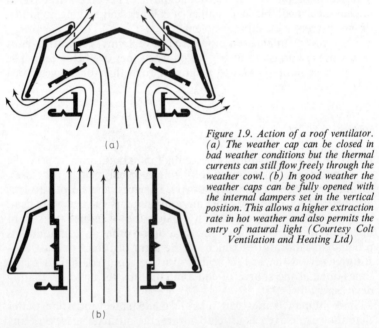

Figure 1.9. Action of a roof ventilator. (a) The weather cap can be closed in bad weather conditions but the thermal currents can still flow freely through the weather cowl. (b) In good weather the weather caps can be fully opened with the internal dampers set in the vertical position. This allows a higher extraction rate in hot weather and also permits the entry of natural light (Courtesy Colt Ventilation and Heating Ltd)

burners, and the window design should be capable of deflecting incoming air upwards. Other simple methods include the use of lever-controlled louvred openings in walls or ceilings. The louvres may be adjusted to avoid sun glare and when closed are rainproof.

By placing the ventilators behind heating surfaces the incoming air may be warmed in winter.

Other methods which depend on the power of the wind or air temperature differences may also be employed and it is claimed by manufacturers that these have powers of extraction equal to mechanical methods. Roof ventilators which create a suction and draw air from the room, irrespective of the direction of the wind as it passes over them, may be fitted to buildings (*Figure 1.9*). These have damping devices so that the degree of ventilation may be adjusted to suit the season. Ventilation by means of roof ventilators is further assisted by the natural displacement of air due to the ascent of warm air inside the building.

MECHANICAL VENTILATION

Mechanical ventilation may be effected by the use of either a central ventilation system or a local ventilation system.

When making a choice between these two methods consider the ventilation problems applicable to the particular laboratories. Whereas a central system might be suitable for blocks consisting of laboratories presenting no special fume or dust hazards, difficulties could be encountered when a number of laboratories with varying degrees of hazard are involved. For this reason laboratories with similar ventilation or heating problems are sometimes grouped together, but where this is not possible both central and local ventilation systems may be necessary. Whichever system is employed certain standards of ventilation must be adhered to, as follows:

Type of laboratory	*Air changes per hour*
Chemical	4–15
Chemical stores	5–10
Physical	3–5
Rooms used for obnoxious fumes	15–30
Radioactive	15–30
Biological	4–6
Animal	4–20

To effect the required number of air changes fresh air is introduced through fresh air inlets situated at low level, window sill level, or in some instances even at ceiling level. In most cases, however, it is best to introduce the air at a low level and extract it at a high level. It must be appreciated, however, that if dangerous heavy fumes are involved, it may be necessary to reverse the position of the inlet and outlet vents. The positions of the inlet and outlet vents relative to one another must be such that the whole working area is cleansed by fresh air and draughts are avoided. It is also important for obnoxious fumes to be contained within the laboratory they are generated in until they are removed by the ventilation system, which means that a negative pressure should exist in the laboratory. In other laboratories where sterile or other special conditions necessitate the exclusion of the outside atmosphere, a positive pressure should exist. This prevents the entry of contaminants through cracks and crevices and in special circumstances may involve the use of windowless rooms or rooms with sealed windows. Airlocks may also have to be provided between the positive area and adjacent rooms or corridors.

Mechanical ventilation methods involve fans which may force air in by a plenum or propulsion system or draw air out by an extraction system. Plenum ventilation creates a positive pressure and extraction ventilation a negative pressure. Sometimes both systems are employed simultaneously, one being complementary to the other.

Propeller fans can move large volumes of air against slight resistance and centrifugal fans are used when considerable resistance is to be overcome. It follows that the propeller fan is unsuitable for use with ducts unless they are short and have no sharp bends, whereas the centrifugal type can be used with long ducts and is particularly suitable for removing dusts or fumes.

CENTRAL VENTILATING SYSTEMS

Central ventilating systems involve large ventilating plants. They are relatively simple to maintain but make heavy demands on space. If such a plant is envisaged, provision for its installation must therefore be made at an early stage in the planning. Normally such plants incorporate a large extract fan serving all laboratories and rooms. In this case the arrangement of the required ducting may complicate the building structure. The fan used in this system usually has one exhaust, which must be situated at a safe level and may involve a high stack.

Arrangements for heating the incoming air may be incorporated in the central system, although it may be additional to the other heating arrangements provided in the building. Because of the diversity of the temperature requirements in various laboratories, local thermostatic control may also be involved.

LOCAL VENTILATION SYSTEMS

Local ventilation systems involve the use of fans usually situated in individual laboratories and often used in conjunction with fume cupboards or fume hoods. This method is employed particularly in laboratories where dangerous fumes or dust hazards exist.

Propeller fans may be used to change the air in the laboratory by placing them in the wall opposite windows or suitably positioned air inlets. The inlets must be large enough not to increase the velocity

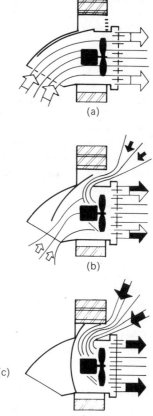

Figure 1.10. Wall inflow unit. To heat the incoming air, steam, hot water, or electric heaters can be incorporated; filters can also be fitted if necessary. Similar units are also used for wall extraction and are almost identical in appearance and construction. Both types of units may be employed for ventilation schemes involving powered inlet and extraction through the walls. For use in rooms where there is no danger from fumes inflow units may be fitted with recirculation devices. (a) How the dampers are set in hot weather so that air is drawn in from outside. (b) In cool weather the dampers are set in the mixed position and the outside air is mixed with warm air drawn from the building. (c) In cold weather the dampers are used to seal off the units from the outside and warm air is recirculated (Courtesy Colt Ventilation and Heating Ltd)

of the incoming air and so cause draughts. Fan openings in the walls should be protected from the wind and rain by hoods. The fan may be made to discharge the foul air clear of the buildings by vertical ducts, in which case both the fan and the open end of the duct should also be well protected. The extraction effect is assisted by the aspirating effect of the wind.

Propeller fans are also used in reversible air flow units, which can be used as a propulsion or extraction system by reversing their action. The speed of the fan can also be varied to suit various conditions. This type of unit is also fitted with recirculating dampers and in winter the warm air can be conserved and recirculated. By adjusting the setting of the damper a mixture of fresh air and warm air can also be circulated. Heater batteries, which may be connected to the steam or hot water system, may be incorporated in these units and the incoming air may be warmed when partial recirculation is insufficient, or, owing to the nature of the atmosphere, impermissible (*Figure 1.10*).

Air filtration

To clean incoming air and in some instances to prevent the extraction of harmful dust or particles, it may be necessary to install air filters in ventilating systems. Various types of filters are employed and may be of a disposable nature. In the case of radioactive laboratories, the disposal arrangements may be subject to rigorous precautions. To permit the easy removal of the filters they must be readily accessible.

Air conditioning

Air conditioning is an excellent if somewhat expensive means of controlling the temperature and humidity conditions in laboratories. This system is independent of outside weather conditions and can be used in conjunction with a general ventilation scheme. If the cost of a general scheme is too great, then air conditioning can be made locally available with small units situated in the rooms or laboratories which require it most. The main advantage of the locally controlled air-conditioning units is that they are adjustable to meet the widely different temperatures and the specific requirements of various laboratories. Air conditioning not only fulfils the special needs of, for instance, microanalytical laboratories, but also permits delicate instruments, which would otherwise need to be housed in special rooms, to be used in working laboratories. To save cost the air from conditioned rooms may be recirculated but this may not be possible in conditions where fumes or airborne hazards exist.

FUME EXHAUST SYSTEMS

The removal of noxious fumes by local exhaust systems is of primary importance in laboratories where dangerous fumes and dusts are produced. In accordance with good laboratory practice, all dangerous fumes should be generated in a fume cupboard or under a fume hood, and should be removed by an extraction system.

It is still common practice to admit air from the laboratory into the fume cupboards to carry away the fumes and at the same time to regard the cupboards as an integral part of the general ventilation system. Even in air-conditioned laboratories, it may still be necessary to effect sufficient air changes to remove fumes which are generated in the laboratory itself. In some instances, however, the fume cupboards may remove unnecessarily large amounts of heated or cooled air, which tends to make this kind of arrangement uneconomical.

However, a sufficient number of fume cupboards should always be provided and the cost of conditioned air losses must be regarded as a necessary item of laboratory expenditure. Alternatively, specially designed fume cupboards which partially prevent conditioned air losses should be installed.

Another important aspect of designing efficient fume cupboards is to prevent the escape of fumes due to local disturbances, and the essential function of any fume cupboard is to remove all the fumes generated within it.

Various conditions such as cross-draughts outside the cupboards, air disturbances caused by pedestrian traffic, and thermal spillage due to convection currents set up by burners or hotplates inside the cupboards may all give rise to the escape of fumes. Because of these possibilities fume cupboards are for most purposes better than fume hoods, since by lowering the cupboard sash spillages can be prevented. When the sash is lowered, however, the rate of air flow through the reduced front opening increases and this produces draughts which affect burners and give rise to other experimental control difficulties. The need to overcome these adverse effects has also affected fume cupboard design, so that new models have been produced which allow a constant rate of air flow irrespective of the sash position. Attempts have also been made to prevent spillages by introducing large air flows through the cupboard front but this has proved uneconomical and may have undesirable effects in the laboratory. It is more important to make air flow adjustable to suit any particular work undertaken in the cupboard and to some extent this has been made possible by variable-speed fans and, more effectively and cheaply, by dampers. Other fume cupboards are

52 DESIGNING THE LABORATORY

designed to allow air to be introduced into the cupboard from outside the building.

Whatever means of control is used, the size of the fan is important and its capacity must be in keeping with the size of the cupboard and the purposes for which the cupboard was designed. Recommended air velocities through the fully open fronts of fume cupboards are 6–12 lineal metres per minute for school laboratories, 12–30 for academic and general industrial laboratories, and up to 61 lineal metres per minute for special requirements and radioactive laboratories. The flow must also be uniform and steady and the volume of air admitted to the laboratory during both winter and summer conditions must be sufficient to meet the cupboards' inflow.

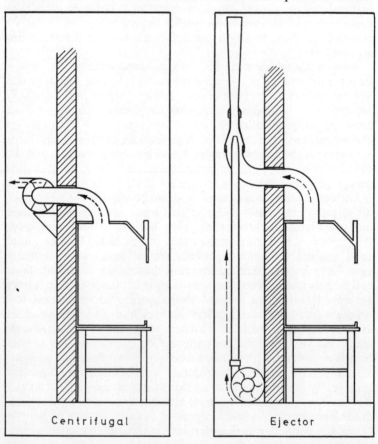

Figure 1.11. Centrifugal and ejector fume extraction systems (Courtesy Baird & Tatlock Ltd)

A reduced pressure within the laboratory does not allow the fans to function satisfactorily.

The design of fume cupboards may be based on the nature of the dangerous fumes or dusts produced in them and the fumes may have to be washed or neutralised to make them safe prior to discharge. In some instances the fumes must be extracted at the source of liberation, which necessitates the provision of special types of cupboards.

GENERAL METHODS OF FUME CUPBOARD EXTRACTION

The two general methods of fume cupboard extraction are (a) positive system, (b) ejector system. *Figure 1.11* shows the working principles of these.

Positive system

In the positive system of extraction the fumes pass through an extractor fan the choice of which depends upon the number of cubic metres per minute of air it is required to remove through the fume cupboard. This may be calculated by multiplying the area of the face opening of the cupboard by the required face velocity. In making the calculation any resistance offered by the ducting must also be taken into account.

Centrifugal fans may be used for positive systems and all modern fans are constructed so that the motor is completely sealed off from the impeller housing; fumes are thus prevented from coming into contact with the motor (*Figure 1.12*). Nevertheless, when selecting a fan the effect of the fumes on the impeller and its housing must also be considered if the fan is to give good service. In this respect centrifugal fans appear to fall into three categories: (a) metal ones with impeller and housing protected with resistant finishes such as enamel; (b) impellers and housings made entirely of resistant materials. (Common materials for housings are rigid P.V.C. and glass fibre bonded with polyester resin; for impellers epoxy resins, Tufnol, and stainless steel may be used.) (c) special-duty fans with housings of cast-iron coated with P.T.F.E. or rigid P.V.C. If hot fumes are involved, steel fans lined with other special materials may be required.

When ordering fans it is important to specify the electrical supply as well as the arrangement of the position of the driving motor and the direction of the outlet. If explosive fumes are likely to be extracted special explosion-proof motors may have to be purchased. Centrifugal fans tend to be noisy and they should if possible be situated on a wall on the outside of a building, well protected from the weather.

(a)

Figure 1.12. (a) Centrifugal fan for positive system fume extraction. The motor is protected from harmful fumes by sealing it off from the impeller housing. (b, opposite) Designation of direction of rotation and discharge of one type of centrifugal fume cupboard fan (Courtesy (a) J. Kotterman KG, (b) Kewaunee Manufacturing Co.)

Axial or propeller fans are also used for the positive extraction system and have the advantage of being cheap and easily installed. They are not designed to work against considerable resistance and the ducting runs should be kept as short as possible and should be of large diameter. Since the motor and fan blades are in the direct path of the fumes, they should be suitably protected. Modern axial fans have casings made entirely of rigid P.V.C. and impellers

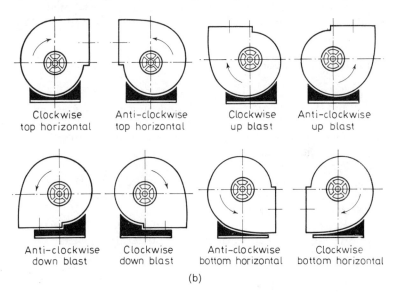

(b)

moulded in phenolic resin and permit gases at temperatures of 32 to 120 °F to be handled. The motor is protected by a polyurethane compound.

Ejector system

The ejector system of extraction, now seldom employed, is very suitable where fumes are exceptionally corrosive. The fan and the motor are situated at floor level and are not in contact with any fumes. Air at high velocity discharged through a venturi system induces a flow of air from the cupboard. Although the fan and motor have a much longer life the system is not so efficient as the positive type and hence, to obtain the same effect, a fan of greater output capacity is required. In addition, the system is noisier and space for housing the components must be provided within the laboratory.

The fumes from fume cupboards may also be extracted by a central extract system, for which purpose a single fan may be situated in a convenient position in or above the building. Alternatively, each fume cupboard, or small groups of adjacent fume cupboards, may be provided with individual fans. Each of the two systems has its particular merits and a choice must be made. The main advantages of the two methods of extraction are given below.

Individual extraction

1. An operator has complete control of his own fume extraction and the cupboard can be independently operated even outside normal working hours.
2. The operator has better control over the fume extraction and can adjust the cupboard to suit the hazardous nature of the experiment.
3. The possibility of dangerous hazards due to the mixing of fumes drawn from several cupboards into the same exhaust ducting is avoided.
4. Repairs to working parts or the adjustment of the structure of the cupboard can be carried out independently of the system as a whole.

Central extraction

1. Since only one central discharge point is involved, it can be sited in the best position with regard to neighbouring buildings.
2. A saving on the cost of multiple fan installations can be made.
3. The large ducts involved lend themselves to easier cleaning procedures.
4. If two fans are installed, the extraction service can be continued in the event of a breakdown.

Whichever of the two general methods of extraction is employed for any particular fume exhaust system, two fundamental conditions must be met. The method of extraction must be such that:

1. It does not allow fumes or dusts to escape back into the laboratory and is adequate to deal with the most hazardous experiment likely to be conducted in the fume cupboard.
2. It should not cause uncomfortable working conditions in the laboratory and when necessary should allow for the conservation of conditioned air.

DESIGNING THE LABORATORY

In advanced laboratories, several types of cupboards are necessary to deal efficiently and economically with all the hazards likely to be encountered. Extract systems used in normal-purpose fume cupboards differ, therefore, from those used in special-purpose fume cupboards.

Extract systems for normal purpose fume cupboards

For normal purposes a fume cupboard must (a) extract the fumes efficiently whatever the position of the sash, (b) have an air intake the velocity of which does not vary to any appreciable extent whatever the position of the sash, and (c) permit a uniform and smooth flow of air.

Single top exhaust

Fume cupboards of the simple type are normally fitted with a vertical sliding sash and have a slow-speed, heavy-duty fan connected to a vent pipe at the top of the cupboard. The vent pipe may be fitted with a damper to provide an extract control. The size and speed of the fan used should be more than adequate to ensure that fumes evolved in the cupboard are suitably extracted. Nevertheless, simple, inexpensive cupboards of this type are usually quite suitable for school laboratories where low extract rates are sufficient and where only a few fume cupboards are normally provided. For more advanced laboratories better extraction methods are necessary.

Multi-level extract

In the multi-level method of extraction the fumes are removed through slots situated at the top and bottom of a baffle plate which is secured to the back of the cupboard. Hot or light gases are extracted at the top and heavy gases at the bottom. Independent manual or automatic dampers, which may be adjusted to suit the work being done, are often provided so that the extraction rate at the top or bottom of the baffle plate may be controlled. Slots may also be provided at intermediate levels in the baffle to ensure an even air flow across the cupboard. In this type of extraction the duct opening is situated at the top of the cupboard (*Figure 1.13*).

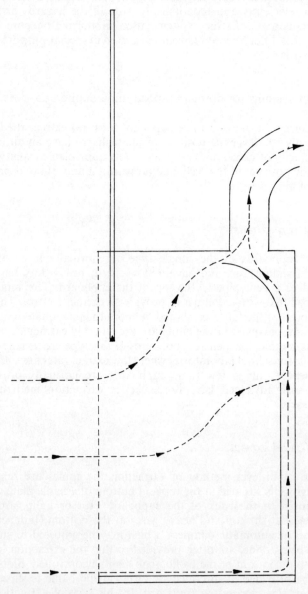

Figure 1.13. Multi-level extract fume cupboard. The baffle situated at the back of the cupboard is adjustable. Extract slots are provided at the top or bottom or at any other desired level and by adjusting the baffle the air flow distribution is controlled

Centre back exhaust

The centre back exhaust is similar to the multi-level extract, but as its name implies the exhaust duct is situated behind the baffle at its exact centre. The extraction slots are situated at the top, bottom, and both sides of the plate.

Compensated extract

The compensated extract is designed to overcome the nuisance of an increased velocity through the fume cupboard front, whatever the working position of the sash. It has an air intake at the top of the cupboard which is covered when the sash is raised and proportionately uncovered as the sash is lowered. The air introduced by the action of the fan compensates for the reduced sash opening. This method may be used to induce air either from the laboratory or from outside the building. The latter method ensures the conservation of conditioned air in the laboratory (*Figure 1.14*).

Balanced extract

In the balanced extract system up to 70% of the air used in purging the fume cupboard may be introduced from outside the building by means of an input fan. The remaining 30% is drawn from the laboratory by an exhaust fan in the usual manner. Depending on the design of the particular fume cupboard, the outside air may be introduced in various ways, but the entry point is the subject of considerable controversy. The system may be further elaborated by providing the input system with controls so that the incoming air may be preconditioned to suit any particular conditions obtaining in the laboratory. Such cupboards are primarily designed for use in fully air-conditioned laboratories. Those designed to introduce the input air inside the cupboard suffer from the disadvantage that their efficiency is reduced when the sash is not closed (*Figure 1.15*).

Extract systems for special-purpose fume cupboards

Direct extract

If highly corrosive or very dangerous fumes are evolved it may be necessary to conduct away the fumes directly from the point of

(a)

Figure 1.14. (a) Compensated extract fume cupboard. (b) Section through the cupboard shows that when the sash is in the raised position all the air enters through the front of the cupboard. (c) When the sash is closed the bulk of the air enters through the air intake at the top of the cupboard. A restricted amount of air enters the small 25-mm space between the sash and the working top to sweep clean the working top surface (Courtesy Grundy Equipment Ltd)

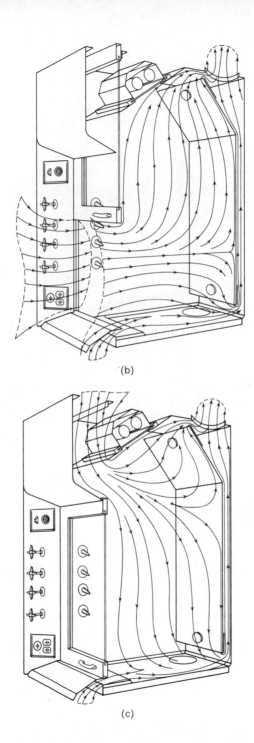

(b)

(c)

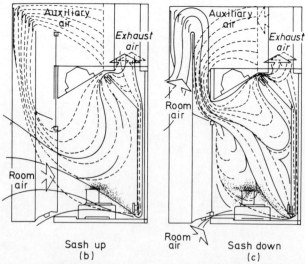

Figure 1.15. (a) Balanced extract fume cupboard. This particular cupboard supplies 70% of the air from outside the laboratory. (b) A section through the cupboard shows that when the sash is in the raised position the air by-pass is closed and the outside (auxiliary) air enters the laboratory via the grille above the face of the cupboard. The auxiliary air along with some air from the laboratory is then drawn uniformly through the face of the cupboard by the action of the fume cupboard fan. (c) When the sash is down the air drawn from outside the building does not enter the laboratory but is deflected via the by-pass directly into the front of the fume cupboard. Some air from the laboratory is drawn into the cupboard through the grille and also through a 25-mm space between the bottom of the sash and the fume cupboard working top (Courtesy Grundy Equipment Ltd)

liberation. The classic example in this case is the Kjeldahl fume cupboard, in which the fume duct is extended into the cupboard. The necks of the flasks are directly supported in holes cut in the duct itself (*Figure 1.16*).

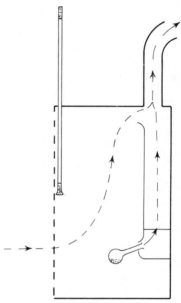

Figure 1.16. Direct extract: the fumes are conducted away directly from the open ends of the containing vessels (e.g. as in the Kjeldahl process). The mouths of the vessels lie in inlet holes in a branch piece connecting with the main exhaust duct

Washed extract

The dangers associated with perchloric acid are generally well known and work with this acid necessitates a special type of fume cupboard (*Figure 1.17*). Since it is essential to prevent liberated fumes from coming into contact with organic substances including the wooden portion of the cupboard, the interior, including the baffle plate, should be constructed of nonabsorbent materials. A water spray through which the fumes must pass is arranged to wash down the lower part of the duct and the baffle plate behind which the fumes pass. The working top must therefore be sloped and drained and all joints in the cupboard must be liquid- and gas-tight. The baffle plate should be removable so that the back of the cupboard may be cleaned. The fume duct should have no sharp bends and should be designed to allow for inspection and cleaning. It is also best if the fan and duct are not linked with other extract systems. Note that for work involving dangerous and corrosive fumes special cupboards completely fabricated from rigid P.V.C. are now obtainable and their advantages should be carefully considered.

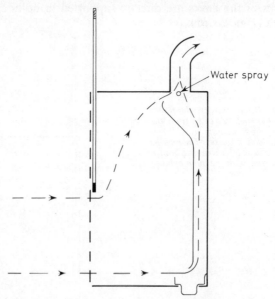

Figure 1.17. *Fume cupboard suitable for work with perchloric acid. The fumes are washed by a spray of water which drains away via a channel set in the working top of the cupboard*

Ducting

After extraction from fume cupboards, fumes are carried away through ducts. The size of the duct is determined by the number of cupboards it serves and the air velocities through their fully open fronts. In determining the duct size, an allowance must also be made for additional cupboards which may be installed in the future.

Generally, the shorter the duct runs the better the fume extraction and, for the same reason, bends in the ducts should be avoided as much as possible. Another reason for avoiding bends and abrupt changes in the size of the duct is the possibility of the formation of dangerous chemical deposits at these points. If long horizontal duct runs are employed they should have a fall and be provided with drainage points and some means of access for inspection and cleaning. It is advantageous if the fan is situated as close to the point of exhaust as possible to avoid any build-up of pressure in the duct and possible leaks in the system. Some fans, however, are designed to be installed in the ducting itself and in this case the fan should be situated vertically above the cupboard but never less than about 1 m from it. The actual point of discharge of the duct must be considered carefully and with especial regard to height of discharge, the disposition and proximity of neighbouring buildings, and the direction of the prevailing winds.

Fumes extracted from fume cupboards are generally highly corrosive and a choice from a variety of different kinds of available ducting materials must be made. In making a choice, both cost and the weight of the material are important considerations. The effects of the fumes on the ducting is the most important factor, however, and the final selection of the material to be used should be made with this in mind.

Commonly used ducting materials are asbestos cement, chemical stoneware, rigid P.V.C., polythene, and polypropylene. The use of the plastics materials is becoming much more popular because they are highly resistant to corrosion, lighter, and of good appearance. The use of plastics should, however, be tempered with caution, especially if the fumes are likely to contain large quantities of solvent vapours or are of high temperature. Asbestos cement is still a very much favoured material, and providing it is acid-treated or painted internally with epoxy or bitumastic paint, it is suitable. The protective paints themselves may be unsuitable if large quantities of organic solvent vapours are used.

Types of joints

All ducts must be well supported and if the material of construction is rigid in nature it should be free from vibration. The joints in the ducts must also be resilient and flanges or spigot-and-socket type joints are used.

Jointing compounds

Details concerning joints and jointing compounds are given in BS 3202[5].

GENERAL DESIGN OF FUME CUPBOARDS

Modern fume cupboards bear little resemblance to the unsightly boxes which formerly constituted these important items of laboratory equipment.

By using smaller-section framings, rails, and guides, a much larger area of the cupboard may now be composed of transparent panels which provide better vision and greater safety for the operator. The use of coloured linings and working tops has also done much to improve the appearance of fume cupboards and has introduced a greater element of cleanliness.

For practical reasons the working top, in contrast to the light superstructure, remains relatively substantial and the understructure must therefore be strong enough to support it. In addition it must withstand the weight of heavy items of apparatus placed upon it, especially in radioactive laboratories where heavy lead shields may be used. To support the superstructure, therefore, brick piers, angle iron brackets, standard units, or table frames of wood or metal are used. For particular purposes the supports of understructure components may have to be specially strengthened.

The recess formed below the working top provides a convenient storage space and by enclosing it with doors it can be easily converted into cupboards. It is common practice to use these for the storage of noxious or fuming chemicals which if kept in the laboratory or in the fume cupboard itself would create undesirable conditions. To dispose of the fumes a small vent may be led from the storage cupboard and connected to the fume cupboard exhaust system. Depending on the material used in the general construction of the cupboard, the doors of the understructure may be of hardwood or treated metal. Sometimes they are of toughened glass, which is particularly suitable if the fume cupboard is of the convertible type. Convertible fume cupboards have withdrawable working tops and when these are removed the cupboard can be used for walk-in purposes.

When the sashes and doors are closed the operator still has full vision into the cupboard over its full height. It is not possible to give exact sizes for fume cupboards since they vary with the type of work to be carried out. Also, their size tends to increase in accordance with the level of the work. An indication of sizes, however, is as follows. The height of the working surface above floor level is normally 0.9 m, and the width of opening should be from 0.75 to 1.2 m (a common size is 0.9 m). The height of the window openings is usually about 0.8 m and the height between the working surface and the top of the cupboard, depending upon the work to be done, varies between 1.0 and 1.8 m. A reasonable depth is 0.75 m; depths greater than this are hazardous because of the dangers involved in reaching into the cupboard.

Framings and panels

Metals such as mild steel and aluminium treated with protective coatings are commonly used for fume cupboard framings. Stainless steel and Monel metal are also used and although they may require no special coating the extra cost involved normally restricts their use to radioactive laboratories. In many cases the whole cupboard is manufactured in treated metal and may be lined with sheets of compressed asbestos. More recently, inert material such as P.V.C. and fibreglass have been used for the superstructure.

Hardwoods, particularly teak, oak, and mahogany, have also been used for many years and possess certain wearing qualities and an elegance hard to surpass. Hardwood treated with oil and finally wax-polished lasts indefinitely. Even so, manufacturers sometimes prefer to coat all woodwork with chlorinated rubber or epoxy resin paints for extra protection. The use of painted softwoods for fume cupboards is not recommended. The wood deteriorates quickly and staining inevitably occurs, giving a poor appearance. Softwoods also burn more easily than hardwoods if hot objects or Bunsen burners come into contact with them. In any case, since only a little timber is used in the superstructure, the use of softwoods means very little saving overall.

In order better to observe the behaviour of experiments the fume cupboard superstructure consists mainly of transparent panels. Transparent plastics have been used for this purpose, but are not the best material since they are easily scratched, are distorted by heat, and may be affected by solvents. The best of these are unplasticised P.V.C. and nonflammable cellulose acetate. Glass undoubtedly remains the most popular material and is available in sheet, wired, laminated, and toughened form.

Sheet glass is a particularly hazardous material but may be used with greater safety in cupboards where there is no risk of explosion. The dangers of mechanical breakage remain, however, and glass of sufficient weight should always be fitted. Wired glass, although mechanically strong, is subject to mild thermal shocks but is sometimes used for top glazing. Laminated glass, which is expensive but mechanically strong, is also subject to thermal shock but can be safely used where any risk of explosion exists. Toughened glass is also expensive, but for fume cupboards is well worth the cost. It is strong, resists thermal shock, and is resistant to fumes except those given off by hydrofluoric acid. The most suitable thickness is 5 mm but for conditions which demand an extra degree of safety thicknesses of 6 mm and 9 mm are available. It is advisable that all cupboards be fitted with an explosion vent consisting of a hinged flap let into the top of the cupboard.

Whenever possible, advantage should be taken of the fact that if the backs of the fume cupboards are suitably glazed they can provide an extra medium for natural lighting in laboratories. The cupboards may in certain cases be allowed to protrude out from the walls so as to allow more space in the laboratory. In situations where the backs of the cupboards face the sun, and especially in hot climates, consideration should be given to the use of Antisun glass, which has the effect of reducing the amount of solar heat. It is sound practice not to extend the side and back glass panels down to the level of the working top of the cupboard but to use some heat-resisting material such as Sindanyo to a height of about 200 mm above the working top. This prevents hotplates and other apparatus likely to cause damage from being placed accidentally in contact with the glass.

In certain cases opaque panels are a desirable feature in fume cupboards, and acid-treated asbestos cement sheet, laminated plastics sheet, mild steel sheets, or aluminium coated with resistant materials may also be used. Suitable coatings include priming paints followed by zinc plating or stoved enamel. Asbestos cement sheet, which may be purchased under a variety of trade names, is the most suitable but untreated varieties of this material tend to stain. The above remarks concerning opaque panels also apply when these materials are used as liners in fume cupboards.

Working top

The working top of the fume cupboard must submit to the hardest wear and here the choice of material is paramount. Bearing in mind the work to be carried out, the following materials, the comparative merits of which are discussed above under 'Bench tops' (pages 6ff), may be considered.

1. Acid-treated asbestos cement sheets.
2. Chemical lead.
3. Plastics sheet (rigid P.V.C., polypropylene, laminated plastics sheet).
4. Slate.
5. Stainless steel.
6. Toughened glass.
7. Resistant metal.
8. Encaustic tiles.

DESIGNING THE LABORATORY

Of these, encaustic tiles have proved best. They withstand hard wear and do not deteriorate, are little affected by heat, and do not suffer as lead does from attack by mercury. They retain a pleasant appearance and do not buckle. The tiles should be laid on the base slab in acid-proof cement and should be pointed with the same material. Cement for this purpose may be purchased or may be made by mixing together 1 part fine asbestos, 1 part fine sand, 2 parts sodium silicate (waterglass). The sodium silicate is added just before the mixture is used. The best colour for the tiles is black, which does not show marks and can be kept attractive by the regular application of ordinary wax polish. A further advantage of tiles is that they allow the working top to be dished and to be provided with a raised front edge. This minimises the dangers from accidental spillages. (Lead, plastics sheet, stainless steel, and other resistant metals also allow dishing to retain any spillages which may occur.) By using tiles, too, the size of the fume cupboard is not determined, as some are, by the maximum available size of the working top material. Some modern fume cupboards are designed to allow the interchange of working tops and a choice from various surface materials may be made according to the nature of the work in hand. The use of removable tops, however, usually entails the fitting of a flexible waste pipe.

Sashes

Fume cupboard sashes may be framed with hardwood, stainless steel, mild steel, or aluminium and the tracks may also be of metal. In many modern fume cupboards, however, toughened glass is used without framings and it is claimed that there is less tendency for the glass to stick in the guides. Another advantage of frameless sashes is improved all-round vision into the cupboard. If vertically opening sashes are used it is essential they be carefully balanced so that they may be brought to rest in any desired position. The suspension for vertical types is usually sash cords and weights, but other types of suspension, such as spring tape balances, are available and these occupy less space. The most reliable materials for suspension cords are nylon, Terylene, and stainless steel cords coated with plastics material; being unattacked by fumes they have a long life. For ease of movement they should be supported on ball-bearing pulleys with nylon sheaves. Some means of preventing the windows dropping right down in the event of a suspension failure is a wise safeguard, and safety catches or wood or rubber stops should be fitted. Laterally opening sashes are also used but are more suited to wide fume cupboards. Sashes may be fitted with small sliding panes in the main windows to allow access to the cupboard through the reduced opening without the necessity of lifting the whole window. Small openings are often provided, too, in the side windows to admit electric cords or rubber tubing. In the U.S.A., fume cupboards are commonly fitted with glass safety shields which can be moved horizontally across the fume cupboard face. The shields provide extra protection and are used when adjustments are made to equipment inside the cupboard. If several fume cupboards are adjoining it is useful if they have common division ends. By removing the division a bank of cupboards may easily be converted into an open continuous working bench unit with a fume hood above.

For safety reasons fume cupboards should always be well lit and it is preferable that fittings be accessible from outside the cupboards. This arrangement, apart from being more convenient, also prevents corrosion of the fittings. For this purpose the tops of the fume cupboards should be fitted with a glazed lighting panel and reflectors. If it is necessary to situate lighting fittings inside cupboards they should always be of a flameproof pattern.

Walk-in fume cupboards

Walk-in type fume cupboards are designed to accommodate large or tall pieces of apparatus such as distillation columns. They may also be used to house the large solvent containers used for paper chromatography thus avoiding the need for specially ventilated rooms. Their height is such that laboratory personnel may work comfortably inside them when setting up apparatus. For working convenience, they are often fitted with service controls inside the cupboard and a duplicate set outside.

FUME CUPBOARD SERVICES

All services situated within fume cupboards should be operated by frontal controls which should be colour-coded or distinguished by shape. These are best arranged along a recessed or inclined front rail. In some cases they are positioned on front pilasters, but on free-standing fume cupboards are sometimes placed at the side. It is essential for ease of control that the pattern of frontal control units used incorporate a universal coupler or that they be of the pattern, recently introduced, in which the front control valve is connected to the outlet by means of a nylon tube. This arrangement permits complete flexibility (*Figure 1.18*). If the fume cupboard incorporates a baffle plate the control to this may also be situated on the front rail. Electrical outlet sockets should be positioned as far away from water fittings as possible.

The relatively small area of the cupboard makes the arrangement of fittings and sinks a problem for the designer. Wherever the sink or other waste outlet is situated, it rarely seems to be in just the right position for a particular operation. One reasonable solution

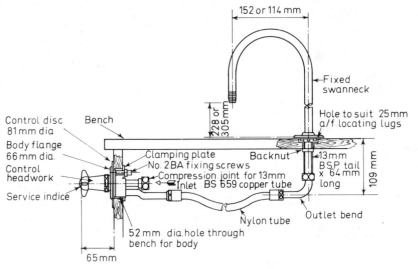

Figure 1.18. A completely flexible type of fume cupboard control (Courtesy W. Markes & Co. Ltd)

to the problem is to set a trough of either rigid P.V.C. or polythene about 100 mm from the back of the cupboard and running parallel to it. The P.V.C. and polythene channel is readily available in a number of standard sizes. The trough extends almost the complete width of the cupboard but sufficient space may be left at each side of it for service outlets. Water faucets are set behind the trough. This arrangement keeps the base of the cupboard free from obstructions and affords drainage facilities at any point. When ordering swanneck water faucets it is important to obtain those with the correct projection. The height of the outlet and the depth of the trough should be such that splashing does not occur. Smucker and Weaver[12] give details of a splashless trough they made. It has one straight side while the other is curved to overcome splashing. Its design might well be profitably studied by manufacturers.

The ordinary laboratory finishes on metal service outlets do not withstand the intense corrosion inside fume cupboards and fittings made of nylon or coated with special plastics finishes should be used.

LIGHTING IN THE LABORATORY

It is difficult for the occasional laboratory designer to estimate the artificial lighting requirements in the laboratory. Provided, however, that the general principles of good lighting are understood, that the technical details of the type of visual tasks to be undertaken are known, and that consultations are held with the electrical contractor concerned, a suitable design is assured.

MEASUREMENT OF LIGHT

Illumination

In everyday conversation we use the terms 'illumination' and 'brightness' without perhaps fully realising the technical significance of these terms. So it helps to consider the measurement of light. Energy released from light sources is received by the eye as light and the continuous flow of the energy, known as 'luminous flux', is measured in lumens. The light is emitted from the source in all directions but the illumination of any surface is considered in terms of the luminous flux falling upon it per unit area. The lux, a unit of illumination, is equal to one lumen per square metre (lm/m^2). When one lumen is evenly distributed over an area of one square metre the illumination on the surface is one lux (1 lx), so that the illumination on a surface is flux/(area distributed).

Brightness

All surfaces upon which light falls reflect it in some measure and the amount of reflectance or brightness of the surface depends upon the reflectance factor of the particular material. The amount of reflected light is measured in apparent lux and is the product of the lux times the reflectance factor.

THE PURPOSE OF LIGHTING

It is essential to regard the laboratory as a working area in which the lighting must be sufficient for the occupants to see easily and comfortably without strain and provide an agreeable working environment. It must assist the personnel in the various operations which they perform. The ultimate choice of the fittings should also enhance the appearance of the laboratory. Vision improves with equal proportional changes in light and generally speaking the more difficult the seeing task the more illumination is required. For good vision other factors must be considered and these are dealt with.

ELECTRIC LIGHTING

The electric lighting used in the laboratory is either an incandescent or a discharge source. With similar illumination values incandescent sources, usually tungsten filament lamps, provide illumination equal to that of the discharge source. Nevertheless, the incandescent type has now been replaced almost entirely by fluorescent discharge tubes. There are several reasons for this; probably the main ones are that the discharge sources give a more diffuse light and the operating cost is considerably lower. In addition, they are nearer to daylight, which is an important feature for colorimetric and other work. Tungsten filament lamps are still used for special purposes and the installation costs are lower.

Incandescent sources are small considering their intensity and are usually shielded to avoid glare. Frosted light bulbs help to reduce glare and have a small efficiency loss of about 1%. A variety of sizes of bulbs of different wattages are marketed and to obtain the maximum life and efficiency from these they must be operated at the correct voltage. Changes of 1% in input volts may introduce 3% change in the light output. A 1% over-voltage gives more light but a life reduction of 8% and a 1% under-voltage less light with a 20% longer life. Tungsten particles are given off from the filament and as a result the glass bulb darkens with age and less light is emitted. It is therefore often economical to change the bulbs before they burn out. The light from the incandescent lamp is composed of all the wavelengths in the visible spectrum but the proportion of the colours varies, and although the light emitted approximates to daylight it is in fact predominantly yellow.

There are a variety of gas discharge lamps and the colour emitted by them depends on the particular gas with which they are filled. A fluorescent type which contains mercury vapour and has the inside wall of the tube coated with fluorescent materials, such as barium salts, is commonly used and is available in a range of colours. Fluorescent lamps are mainly used for direct lighting and are not so dependent on the cleanliness, and consequently the upkeep, of walls and ceilings.

Cold-cathode tubes may be fitted with iron or tungsten cathodes. The warm-cathode type has a tungsten cathode which is preheated before the tube strikes and these incorporate a starter mechanism. Both the cold- and warm-cathode tubes have certain advantages, but for laboratory use the cold-cathode type is extremely popular. Cold-cathode tubes start instantly and their life, since it is not directly proportional to the number of starts, is longer. Fluorescent lighting fittings may require auxiliary equipment such as starters, transformers, or ballasts for their operation and the price of the tubes is also quite high. The metal housing of fluorescent fittings is usually made to contain the auxiliary equipment and in many cases also acts as a reflector. Although metal housings and metal shades may have a stove enamel finish, they tend to deteriorate in the corrosive atmosphere of the laboratory at ceiling level. Serious consideration should be given to the concealment of fittings and it is possible to achieve this by enclosing them behind glass or plastics in ceiling troughs. If this method or similar ones are adopted, make provision for the removal or hinging of the covers to allow for maintenance, cleaning, and tube replacement.

LIGHTING EFFICIENCY

The amount of illumination resulting from a particular lighting design depends on a number of factors:

1. Illuminating power of the source.
2. Efficiency of the fitting.
3. Distribution of light from the fitting.
4. Depreciation.
5. Size of the laboratory.
6. Colour of ceilings, walls and furnishings.
7. Height of the fitting.

ILLUMINATING POWER OF LIGHT SOURCES

From the manufacturers' published data the light output of the source may be elicited but the rated output is reduced by the type of shade or masking device, if any, used with it. In most cases some form of control is necessary to reduce glare or to direct the light.

EFFICIENCY OF LIGHT FITTINGS

If the light from a source is controlled by a reflecting or transmitting medium the degree of control is determined by the reflectance or transmission or absorption factor of the material employed. Light is reflected or transmitted in various ways. Perfect mirror surfaces reflect light regularly, which means the incident light angle is equal to the reflected angle. Light reflected and diffused by a surface is reflected uniformly in all directions. At other surfaces such as porcelain enamel, mixed reflection occurs whereupon part of the light is diffused and some regularly reflected. Materials such as rippled glass cause scattered reflection because the rippling is in effect a number of small mirrors.

Light is regularly transmitted through clear glass but diffuse transmission occurs with milk glass. Sandblasted glass produces a spread transmission and ribbed glass a scattered effect. It is evident that the choice of material used to control light is of primary importance. Owing to its many desirable properties, glass is an excellent material for this purpose since it is easily cleaned, resistant to heat, non-corrosive, and obtainable in many varieties. Among other materials widely used are sheet steel, stainless steel, aluminium, and plastics. Various combinations of these materials are also used. The reflectance values of the metals are quite high and are given in *Table 1.7*.

Table 1.7 REFLECTANCE VALUES FOR METALS

Polished aluminium	85%
Polished chromium	60%
Porcelain enamel	60–80%
Stainless steel	42–47%

Plastics materials are generally very suitable for laboratory use but must be carefully selected since they may tend to warp, discolour, and become brittle.

DESIGNING THE LABORATORY

DISTRIBUTION OF LIGHT FROM FITTINGS

In accordance with the way they distribute light from the source, fittings may be classified into five types (*Table 1.8*).

Table 1.8 THE DISTRIBUTION OF LIGHT FROM DIFFERENT TYPES OF FITTINGS

Type of fitting	% of light in upward direction	% of light in downward direction
Direct	0–10	100–90
Semi-direct	10–40	90–60
General diffusing	40–60	60–40
Semi-indirect	60–90	40–10
Indirect	90–100	10–0

General designs of light fittings and diagrammatic examples of the distribution of light from these are shown in *Figure 1.19*. The choice of fittings is determined by the local conditions.

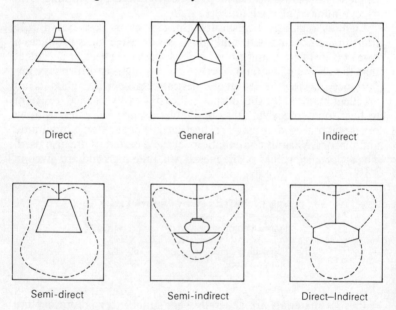

Figure 1.19. How the light from various types of fittings is distributed

Direct lighting

Direct lighting is used in circumstances where good light is required at the work plane as, for instance, in workshops. In general it tends to produce shadows and a dark ceiling effect which may be oppressive. This effect may be minimised by selecting furniture and floors with high reflectance values.

Indirect lighting

With indirect lighting, little shadow effect is experienced, but light-coloured non-glare ceilings and upper side walls are essential. It is to be recommended for office use and for exacting visual tasks, and although somewhat monotonous can be used to good aesthetic effect. A combination of indirect and direct lighting is excellent since shapes become interesting and clear. Contrasts which are too strong are avoided.

Diffused lighting

Shadows may be avoided by using diffused lighting and an even illumination results. It tends to produce glare and for this reason is not recommended for visual tasks and prolonged eye work. Walls and ceilings must have high reflectance values.

Semi-direct and semi-indirect lighting

An adroit use of direct and indirect lighting combines the characteristics of the types of lighting already mentioned.

DEPRECIATION IN ILLUMINATION

To avoid depreciation in the light output the choice and placing of fittings must allow adequate facilities for cleaning and servicing. The amount of depreciation of the initial illumination due to the lack of maintenance and other contributory factors is perhaps not generally realised. Lighting experts usually allow a depreciation of 30% in estimating lighting requirements but depreciations of up to 50% may occur through neglect. Such depreciation may be due to:

1. Ageing of lamps.
2. Dirty fittings.
3. Variation between recommended and actual voltage.

A routine inspection by a qualified person, at intervals not exceeding four months, is to be recommended and for full efficiency walls and ceilings must be regularly cleaned or redecorated.

NUMBER AND POSITION OF FITTINGS

The size of the room must be considered in relation to the number and position of fittings. The amount of reflected light reaching any position in the laboratory is a function of the overall size.

COLOUR OF WALLS AND FURNITURE

The colour of the walls and furniture is important and the use of light pastel shades to make working conditions pleasant and interesting is now common practice. By the application of colours in suitable places laboratories may be made to appear to alter their dimensions and temperatures. Colours may also be affected or enhanced by a wise selection from the range of fluorescent tubes. Walls and ceilings have a tremendous bearing upon the utilisation of light. It illustrates the relative effects and efficiencies of the many available wall finishes to discuss a few of these. Unpainted (white) plaster diffuses the light and has a reflectance value of about 92%. Gloss paint, which has a value of 85%, is easily kept clean and is therefore less likely to depreciate. It introduces glare and shows up irregularities in the wall surface. Matt paint has a reflectance factor of 85% and diffuses more light than the gloss.

Successful laboratory lighting also depends upon uniformity, which means avoidance of glare and the correct amount of contrast. Glare causes eye discomfort and is due to large brightness differences. Direct glare may be caused by fittings which are too bright for the background. The eyes are particularly susceptible to direct glare in the angle between the horizontal line of vision and 45° above it, and to eliminate glare within this angle the fitting should be shaded. Colour contrasts which are not too strong are kind to the eyes and are not tiring. An excess of light from one direction, provided it is partly balanced by light not so intense from another direction, produces a good general diffusion. Laboratory furniture and particularly writing and working tops should have the correct reflectance factor. If they have a matt finish reflected glare is avoided. Some recommended reflectance values are given in Table 1.9.

Table 1.9 REFLECTANCE VALUES FOR ROOM INTERIOR SURFACES

Ceilings	85%
Walls	60%
Working tops	35%
Furniture	35%
Floors	30%

HEIGHT OF FITTINGS

The amount of direct light reaching the working area is a function of the mounting height of the fitting, and to ensure even illumination the spacing between the fittings should not greatly exceed the mounting height.

LIGHTING VALUES

It is not possible to say precisely what the lighting value should be for laboratories. The variety of tasks performed in different kinds of conditions rules out this possibility, but a minimum value of 269 lx is suitable for general lighting in laboratories, at working surfaces 0.9 m above floor level. Values of up to 538 lx are used in laboratories for difficult seeing tasks. It should be noted that a good general lighting may make supplementary lighting unnecessary.

Tables may be consulted which enable the illumination to be calculated in lux. One such table takes into account the size of the laboratory and the mounting height of the fitting and a room index figure is obtained. Another table relates the index to the efficiency of the fitting with allowances for the colour of walls and ceilings, etc. The coefficient of utilisation thus calculated is the percentage of the lumen output of the fittings spread over the working area. The equation

$$\text{illumination (lx)} = \frac{(\text{lamp lumens}) \times (\text{coefficient of utilisation})}{(\text{area (m}^2)) \times (\text{depreciation factor})}$$

gives the average illumination over the area.

NATURAL LIGHTING

In recent times more attention has been paid to the important aspect of natural light in buildings. Earlier, the value of good window design had been ignored to some extent and this resulted in an era during which a maximum window area was the main consideration.

The lack of understanding of the technical terms concerned with lighting is not confined to artificial lighting alone and the term 'daylight', as applied to illumination, is also little understood. So let us also consider how this may be measured.

MEASUREMENT OF DAYLIGHT

For the purpose of measuring daylight the sky is considered to be the light source. The brightness of the sky varies constantly owing to cloud conditions and the position of the sun. The amount of light at any given point inside a building is measured as the percentage of the total light outdoors, and for this purpose the sky is considered as a hemisphere which is equally bright in all parts. The unit of measurement is called the 'daylight factor' (d.f.), which is convenient to use since it is independent of the outside conditions at the time when the measurement is made. The d.f. is measured at the same height of working plane as artificial lighting, which is 0.84 m. The relationship between the new unit and lux power may be illustrated by saying that at noon in the south of England the outdoor illumination is above 10 760 lx for 272 days in the year, and at noon on such days a daylight factor of 1% would represent a value of about 108 lx indoors.

WINDOW DESIGN

It is common practice today to employ large window sections extending the complete length of one side of the laboratory. Provided the light so obtained is balanced by additional light from the opposite direction so as to give even illumination, this provides excellent natural lighting, and good working conditions result when such large window areas are used. Other factors, such as the avoidance of heat losses and the correct diffusion of light, must be considered. Light may be diffused by employing coloured, obscured, or tinted glasses, but these are more costly than the clear glass. Clear glass should be used for windows at eye level, not only to allow the laboratory occupants the pleasure of being able to see out, but also to provide them with an opportunity to focus their eyes on distant objects. This gives considerable relief to the eyes and avoids eye strain.

The main purpose of a window is to admit light, and when choosing windows this must be the primary consideration. The amount of light which penetrates into the laboratory depends upon: (a) size, (b) shape, and (c) height of the windows. If the window is dirty, or glazed with obscure or coloured glass, or obstructed in any other way, the amount of light which enters is reduced. Inside the room and in the immediate vicinity of the window the light intensity will be high but this will steadily fall in value as the centre of the room is approached. In the centre of rooms d.f. values of less than 1% are not uncommon.

The size of windows affects the amount of light admitted and the area of illumination. The contours of the illuminated area are determined by the shape of the window. Whereas windows of the long low type will, if the areas are the same, admit an amount of light equal to that from a tall window, the taller window allows a greater penetration of light. For most laboratories good light penetration is essential to provide illumination on the centre working benches. The lower portion of windows allows the least light penetration and the upper portion the most, and for this reason it is better to position windows as high as possible. If the sill of the window is made too high, the distant light values improve but the amount of light at the working plane is reduced. Clerestory lighting plays a useful role in day lighting and in laboratories often provides excellent top ventilation. Where horizontally placed windows can be used these admit twice as much light as those of the same area in a vertical plane and the light is distributed symmetrically below the window.

The geographical situation of the building influences window design. In hot climates, in addition to their main function of admitting light, windows are required to permit the maximum ventilation and exclude solar heat. Some means of preventing the entry of flying insects when windows are open at night may also be required. In colder climates ventilation without draughts and the prevention of heat losses are the main functions of the window. Double-glazed windows provide excellent insulation but are expensive. Whatever style of window is chosen, it should also be in keeping with the architectural style of the building.

Direct sunlight will also determine to a large extent the selection and siting of windows. Generally speaking, in locations where little sunshine is to be found it is customary to orient the windows so as to allow the maximum amount of sunlight to fall upon them for as long as possible throughout the day. In hotter climates for the greater part of the day, and particularly at peak temperature periods, it is necessary to avoid direct sunlight on the windows. To accomplish this various types of anti-sun obstructions, such as louvres and sun canopies, are often placed outside the windows. It is evident that to suit local conditions the orientation of windows is quite important. In the case of new buildings, the orientation of the building itself should ensure that windows are placed at angles which will use the available sunlight to the best advantage.

Sun blinds which may be raised or lowered are also used inside windows and the choice of window should allow free passage to the blind when the window is open. Similarly, lecture rooms fitted with blinds for darkening purposes may require windows which allow ventilation when the blinds are drawn. If the windows cannot be opened then an alternative means of ventilation is necessary. If venetian blinds are fitted at windows in rooms in or adjacent to chemical laboratories, they should be constructed of noncorrodible materials. Those with plastics slats and nylon cords are suitable.

The installation of windows also necessitates a study of local weather conditions. The windows must ventilate without draughts and should exclude rain even when partly open. The exclusion of draughts and direct sunlight is particularly important in balance rooms and in laboratories where delicate experimental manipulations are performed or control of temperature is necessary. The effect of the weather and laboratory conditions on window frames raises the question as to which type is best for laboratory purposes. Provided steel or wood frames are well maintained either is satisfactory. Steel is termite-proof but if untreated is subject to rusting. Rustproof coated steel, although more expensive, may prove to be an economic consideration for chemical laboratory conditions.

WINDOW OBSTRUCTIONS

Obstructions such as trees, balconies, or neighbouring buildings, restrict the amount of light entering the laboratory windows. The closer the obstruction the more light is cut off. In conditions where the obstruction must be considered permanent, light losses may be minimised by positioning windows to the best advantage and the obstruction should be carefully studied from this point of view.

The shape of the windows should also assist the entry of the maximum possible light. It is, for example, better to use tall vertical windows where the obstruction is long and horizontal. When the obstruction is tall and vertical, long horizontal windows would provide more even illumination.

Windows situated in inside walls, or the use of glass partitions, may materially assist natural lighting in the interior of buildings. Glazed entrance doors with side lights will help to admit light to dim corridors and entrance halls. Inside doors may also be glazed to borrow extra light. Fanlights also serve the same purpose and at the same time provide extra ventilation.

SAFEGUARDS IN LABORATORY DESIGN

When a laboratory is being designed all the possible hazards must be foreseen and steps taken to apply the necessary safeguards. It is inevitable that in implementing these precautionary measures building costs will rise, but these expenses must be met if the laboratory is to be a safe working area. The word 'accident' immediately brings to mind a sudden occurrence which results in injuries to the person or persons concerned. Due regard must also be paid, however, to 'accidents' which are in effect the deterioration of the mental or physical health of persons performing their duties in unsuitable conditions.

CLEANLINESS

The degree of ease with which a laboratory can be cleaned is an important design feature. For perfect safety strict cleanliness and orderliness is essential and it is important that every laboratory worker have his own locker.

ATMOSPHERE

For efficient work and good health, the atmosphere in the laboratory should be cool, dry, and moving. A reasonably diverse temperature is preferable to one which is monotonously uniform. Extreme temperatures must be avoided since very hot and very cold conditions are dangerous to health and result in poor work. Under certain conditions, a close watch should be kept on the inflammability of the atmosphere.

VENTILATION

Good general ventilation is essential and in chemical laboratories low ceilings are unsuitable. The presence of toxic gases, vapours, and dusts constitutes a serious risk to health and must be eliminated by an exhaust system. This is achieved in chemical laboratories by means of local ventilation such as fume cupboards or fume hoods. The construction of fume cupboards is discussed on pages 65–72; suffice it here to emphasise that fume cupboards must be large enough and numerous enough to deal effectively with the toxic and other fumes released in the laboratory.

If certain hazardous chemicals are to be kept ready for use in fume cupboards, provision should be made to prevent these being knocked over.

In areas where the grinding of materials is carried out good ventilation is essential and a local method of dust extraction should be provided above the grinding machine. In some circumstances dust may be subdued by means of a water spray.

In certain laboratories mercury is used in large quantities and jointless flooring should be laid for safety. Porous floors or floors with unsealed joints may have to be completely relaid if mercury vapour from an accumulation of spilled mercury assumes dangerous proportions, for it is extremely toxic.

OVERCROWDING

Many accidents may be attributed to overcrowding of laboratories, and in teaching establishments overcrowding is prevalent. The space between benches should never be less than 1.2 m and 1.5 m should be allowed in the main aisles. A working space of at least 11.3 m^3 per person should be allowed.

FIRE HAZARDS

Since the occurrence of fires is a constant hazard, it is advisable, where possible, to situate chemical laboratories on ground floors. In all laboratories suitable exits must be provided to ensure that a safe way exists of entering or leaving the area under normal and emergency conditions. The District Fire Officer should be consulted about fire precautions when the laboratory is in the design stage. In teaching and other laboratories where large populations are involved, special attention must be paid to the exit facilities. Areas in which quantities of inflammable substances are in use require special consideration. At all times gangways and corridors must be quite free from obstructions and at least two exits from every laboratory should be allowed. All doors should open outwards and should be made so that when locked or fastened they can still be opened easily from the inside. All other exits which afford an escape route in case of fire should be clearly marked and a means of giving warning should be installed. Every part of the building should be within a reasonable distance of a fire exit and this distance depends upon the degree of the fire hazard.

If the work involves the use of large quantities of inflammable liquids it may be advisable to subdivide large spaces by fire-resistant partitions to confine outbreaks of fire to a small area. For very hazardous operations small separate rooms should be provided. If lofts are used for storage purposes, accumulations of combustible material must not be allowed, since in the undivided loft space a fire may quickly spread throughout the building.

Suitable firefighting equipment should also be ready for immediate use, including an adequate number of fire extinguishers suitable for the type of fires most likely to occur. In addition to extinguishers buckets of sand should be provided as an added precaution. For major fires the installation of a sprinkler system commends itself, although certain sections of the laboratories may have to be excluded or otherwise protected from damage by water. Fire hose points with a sufficient footage of hose to reach any part of the department are a desirable safeguard.

Although the fighting of a fire may take precedence over the control of contamination in areas where radioactive materials are being used or stored, the undue spread of contamination must be avoided at all costs. Should a fire occur, inform the chief officer of the fire brigade immediately of the presence of radioactive hazards.

At all times a list of laboratories or stores in which radioactive materials are present should be kept in a prominent and accessible place. The list should be kept up to date by a responsible person assigned to this task. Safety-glass sight panels should be let into the doors to the laboratory. These are not only useful for looking into the laboratory without disturbing the occupants, but also prevent collisions at the doorway.

FURNITURE

The height of benches is generally determined by whether the work is done standing or sitting. The bench width should be restricted because controls at the back of deep benches are dangerous, particularly if reagent shelves are also situated at the back of the bench. For similar reasons it is advisable not to make fume cupboards deeper than necessary. If wooden bench structures are installed then the tops of those used for hazardous purposes should be covered with an appropriate material. For most purposes Sindanyo is an excellent top.

STORES

Because of the mixed types of material kept in them chemical stores are particularly hazardous and considerable thought should be given to their location and design. If quantities of inflammable liquids have to be kept in the stores for immediate use then the stores should have doors which open directly outside the building. A suitable means of ventilation should ensure frequent changes of air. The stores should be constructed of nonflammable materials and metal shelving is better than wood. The electric lighting should be enclosed in vapour-proof fittings and a fireproof door is desirable. Adequate space should be allowed in the aisles and congestion generally should be avoided. The floor should be resistant to corrosive chemicals and easily washed down. A small separate room within the precincts of the store should provide for the dispensing of liquids from carboys. This room must have its own exhaust fan and the floor should be sloped to a drainage outlet hole so that it may be hosed down if necessary. If gas cylinders are stored, another separate room should be provided. For combustible gases this room, too, should have vapour-proof light fittings and the light switch should be placed outside the room. The room must be cool and situated so that direct sunlight cannot enter. Proper storage racks for empty and full cylinders are necessary.

Other hazards arise from the storage of radioactive materials. A special room which should be kept locked should be provided for these materials, especially high-activity sources. The door of the room should bear the radiation symbol and the level of radiation within the room should not exceed safe levels. The materials stored should be adequately shielded where necessary and the room should be mechanically ventilated by fans to the outside air.

For the transportation of materials to and from the storeroom special containers should be used. Further storage problems and details of the hazards associated with radioactive materials may be found in Chapter 6.

Inflammable stores

Separate storage facilities situated outside the main building are required for the bulk storage of inflammable liquids. Inside storage of bulk inflammable substances places the whole building in danger. The storage building must be constructed of solid nonflammable materials such as brick and concrete. The floor should be lower than the door sill and sloped to a drainage outlet. A water tap may be provided for hosing down the floor. Shelving, if fitted, should also be made from nonflammable material such as thick cement, asbestos sheet, or precast cement slabs. To make the store absolutely safe a dry chemical or carbon dioxide automatic fire control installation should be fitted to give constant protection.

LABORATORY WORKSHOPS

The laboratory workshop should be designed so that ample space is left between the machines. All machinery should be adequately guarded. The type of lighting installed must not produce stroboscopic effects.

General

Gas and water pressures must be adequate and constant. An extra draw-off at any point must not diminish the pressure elsewhere. Such variations in pressure may have dangerous results where the local supply to experiments in progress, such as distillations, have been carefully adjusted. This may result in possible explosions or fires through overheating and for other reasons. Gas and water valves should be fitted so that the supply to any section or to the whole of the laboratory can be shut off in an emergency.

Laboratory floors must be suited to the conditions in the laboratory and under these conditions should not be naturally slippery. In addition, floors should not become slippery if spillages occur, and neither should the cleaning materials used on them produce such conditions.

Coat racks should be provided outside laboratory areas to prevent clothing from being left on bench tops or hung around the laboratories in hazardous positions.

DESIGN OF THE PHOTOGRAPHIC DARKROOM

The darkroom is now recognised as an important auxiliary facility for laboratory work and has recently been transformed from a damp and dismal room to a pleasant workplace in which techniques involving great skill may be practised.

Like that of a well-designed laboratory, the layout of a darkroom depends on the nature of the work. It is proposed in this chapter to deal with the general basic requirements for the small darkroom.

BASIC REQUIREMENTS

The darkroom must be (a) of suitable size, (b) lightproof, (c) well ventilated, (d) adequately lit.

The size of the room

The size of the room must suit the number of people who use it at any one time. In estimating this, the cubic capacity must be considered and a minimum of 11.3 m^3 per person should be allowed. In relating the size to the cubic capacity, remember that low ceilings are unsuited to darkrooms. Since all light must be excluded, it is advisable to select a naturally darkened room—if possible, one without windows which is situated in the interior of the building. Such a room will have the advantage of remaining at an even temperature, since its walls are not exposed to the weather. If blinds are necessary, it is a wise precaution to install good ones. A little more money spent on the initial installation prevents the frustration and further expense which are invariably associated with the deterioration of the cheaper variety.

Lighting

More time is spent in the darkroom in lighted conditions than is generally realised and provision for good general incandescent lighting should be made. At the working bench level 108 lx is sufficient. It is also convenient if, by suitable switching arrangements, the amount of light can be reduced when processes which require subdued white light are carried out. The general rules for lighting should be observed, the ideal being good diffused light without shadow at the working surfaces. Safe-lighting will also be required, and a good general safe-light applicable to the work being done should be fitted. This may be attached to or suspended from the ceiling, depending on the ceiling height. If suspended it may be necessary to fit a reflector above it so as to provide a good general diffused light effect at bench level. In addition, local safe-lights should be arranged at suitable lower positions above the appropriate working benches. The arrangements of the light switches for the general lighting must ensure that these cannot be accidentally operated. This may be done in any convenient way such as by using different types of switches, or by placing them above the normal switch height so as to involve a conscious effort when the switches are operated. Foot switches for local lights are often employed in wet areas. A conspicuous warning light should be provided above the darkroom door with the switch positioned just inside the light lock and adjacent to the entry door. Other switches serving the general safe-light and the general white light may also be positioned inside the light lock and should be wired on two-way circuits to alternative switches in the darkroom proper. The switches may be of the luminous variety.

Sufficient electrical sockets should be arranged around the walls to allow for the use of enlargers, dryers, and other equipment. Sockets should also be provided at suitable heights and positions above the wet bench for local safe-lights and also for an electric clock.

Light lock entry

A number of different light lock arrangements are used to suit particular requirements. A satisfactory one is the maze type shown in *Figure 1.20*, but the room must be of a suitable size to allow it to be used. The maze can be constructed from brick or any adequate partitioning material. Since darkroom ceilings are high the maze does not usually extend up to ceiling level, and in this case it may be covered in so as to provide additional storage space above it.

Furniture

In general the furniture in a darkroom may be divided into wet benches and dry benches and these should not be immediately adjacent to each other. For wet benches wood, including teak, is unsuitable. The caustic solutions used tend to lift the wax protection, which results in first of all a poor appearance and finally a general deterioration. Lead wears well and lasts indefinitely but has a poor appearance and is not particularly pleasant to work on. Stainless steel, Monel metal, and P.V.C. are possibly the best materials.

For dry benches hardwoods are quite suitable, and since the wear on these benches is not heavy, other materials such as Formica, thick linoleum, and even hardboard on a cheap wood bench top

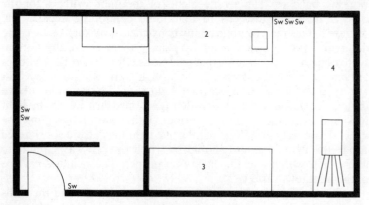

Figure 1.20. Plan of a small darkroom. Units: 1, Retouching desk. 2, Dry bench. 3, Instrument bench. 4, Wet bench

may be used. With these materials a protective beading to the bench top is necessary to prevent the material lifting at the edges. A translucent glass viewing panel illuminated from below and let into the top of the dry bench provides a convenient means of inspecting negatives.

Other useful items of furniture include a simple desk with drawers for writing, retouching purposes, and other work best done sitting down. Also, in some instances, it may be necessary to use the darkroom for experimental work which requires darkness or semi-darkness. In this case, an instrument bench dimensioned to suit the equipment may also be necessary. A towel rail, preferably electrically heated, and a coat rack are also necessary darkroom items.

Plumbing

Conditions in the darkroom are not so severe as those in the laboratory as far as plumbing is concerned. For trouble-free service and simplicity in installation, however, polythene or similar types of materials are to be recommended for the drainage system. Considerable solid matter finds its way into the waste pipes and for this reason, and to ensure the efficient dilution of the effluent, a laboratory-type receiver should be used. If more than one sink is involved, and a common drainage pipe is used, then it should be of such diameter that blockages will not occur. In any case, suitable inspection eyes must ensure that it can be easily rodded through. Strainers should be fitted to sink waste outlets and standing wastes should also be provided so that the sinks can be used for photographic washing purposes. Open gullies are not recommended for the short runs necessary for darkrooms. Gas points should not be installed.

Fire precautions

A portable extinguisher suitable for use in confined spaces must be kept in the room. Rubbish bins, preferably the type with a closed lid and a pedal attachment, should be used and regularly emptied. A separate bin for broken glass is advisable.

Decoration

To assist the lighting the ceiling should have a smooth non-flaking finish such as matt enamel paint and should be white. It should be free of pipes and other encumbrances which may collect dust or corrode and cause small particles to fall into open dishes. The walls should be a light colour with a smooth hard finish so that they may be washed at frequent intervals. It is customary to paint a white line around the matt black walls inside the light lock, which is of assistance when entering from the light.

Ventilation

Adequate ventilation may be provided by the non-mechanical type of darkroom ventilator but special propeller-type darkroom fans provide better air circulation and should be positioned high up for full efficiency. They should be light-tight and must not be strong enough to create draughts.

Temperature control

Independent heating arrangements should be provided. Steam or hot-water radiators are suitable for heating, but electric fires should not be used. A thermostatically controlled concealed tubular heater would be suitable. Local means of heating solutions such as hotplates and immersion heaters are useful and sufficient electrical outlets should be provided to allow the use of these.

Floors

The type of flooring selected should be unaffected by dampness, free from dust, and have no open joints. Terrazzo or hard tiled floors are suitable but uncomfortable. Asphalt, composition tiles set in mastic, or the polyester-aggregate-type floors are recommended. Battleship lino has been used successfully provided it has been well laid so that dampness cannot penetrate underneath. With due care darkroom floors do not become very wet, but they may require frequent washing to prevent the accumulation of dust. It is desirable that the flooring material should be coved at the wall–floor junction. Cement floors are not recommended.

Sinks

Sinks should be deep and have a good overflow. Adequate clearance must be allowed below the water tap for filling tall vessels. For the protection of glassware a rubber sink mat is an advantage in 'hard' sinks. Hot water at the sink is desirable. The bench top at the sink area should be well drained and in wooden tops grooves should be cut. Alternatively, a loose draining board should be provided on the bench top. Stainless steel, glazed fireclay, or polythene sinks are preferable to other kinds.

Storage units

Cupboards below the wet bench are best reserved for apparatus unaffected by damp such as glassware, developing dishes, tanks, and bottles of stock solutions. One of these should be so designed that it will accept developing trays stored on end. This design may take the form of slotted shelves with vertical dividers. All cupboard units under the bench should be removable to allow easy access to plumbing.

Other cupboards below the dry benches may be used for stocks of paper, plates, film, dry chemicals, and apparatus requiring dry storage. Some drawer units should also be included for the storage of packets of paper, plates, and other light-sensitive materials and it is advisable to provide these with light-proof sliding plywood covers. Shelves should be provided above the dry bench for the storage of in-use materials so as to keep the bench clear for working purposes.

REFERENCES

1 A Technical Review Panel Report, 'Polyurethane Coating for Laboratory Benches', *J. Sci. Technol.*, **7**, No. 3, 98 (1961)
2 Dobson, E. W., 'Fume Cupboards', *Lab. Pract.*, **12**, 908 (1963)
3 BS 334, *Chemical Lead (Types A and B)*, British Standards Institution, London (1934)
4 BS 1710, *Identification of Pipelines*, British Standards Institution, London (1960)
5 BS 3202, *Recommendations on Laboratory Furniture and Fittings*, British Standards Institution, London (1959)
6 Fleming-Williams, B. C., *The Design of Physics Research Laboratories*, Institute of Physics, Chapman & Hall, London (1959)
7 Edwards, J. A., *Laboratory Management and Techniques*, Butterworths, London (1960)
8 BS 1973, *Polythene Pipe (Type 32) for General Purposes, Including Chemical and Food Industry Uses*, British Standards Institution, London (1970)
9 BS 1387, *Steel Tubes and Tubulars Suitable for Screwing to BS 21 Pipe Threads*, British Standards Institution, London (1967)
10 BS 2598, *Glass Pipelines and Fittings*, British Standards Institution, London (1955)
11 Tongue, E. W., 'Design and Maintenance of Laboratory Service Outlets', *Lab. Pract.*, **12**, 1088 (1963)
12 Smucker, A. A., and Weaver, H. D., 'A Splashless Table Drain', *J. chem. Educ.*, **36**, 301 (1959)

2
Installation of laboratory equipment

Each type of laboratory has its own special problems connected with the installation of equipment, and the best opportunity of overcoming them occurs when the laboratory is designed. Difficulties invariably present themselves in later years, however, when items are purchased to provide for new techniques and the natural expansion of the department. One of the major problems which arise is that of space.

When new equipment is positioned the adverse effects it may have on equipment already in existence, and conversely any which the existing equipment may have on the new, are important considerations. In many cases special provision must be made for supporting either delicate or weighty items. The effects of local conditions such as humidity, temperature, and the effects of sunlight, dust, draughts, noise, and vibration, must also be given attention. Similarly, any hazards which may be involved must not be overlooked. The placement of the equipment, too, must be related to the position of the laboratory service outlets.

Although the limits of this chapter allow only a few selected examples of the installation of laboratory equipment to be given, I hope they will suggest ways and means of providing for other similar equipment.

VIBRATION

The mounting of most precision equipment is greatly influenced by the effects of vibration and a number of methods of overcoming this problem, by the introduction of various types of vibration supports, have been advocated. According to Strong[1] the horizontal components of vibration are much more serious than the vertical components, which are comparatively harmless. He points out that the shielding effect of the support, which can be considered as an oscillating system loosely coupled to the walls, ceilings, or floor of a room, is determined by the resonance between the support and the wall. To eliminate the vibration it is necessary to interpose a system with a natural frequency lower than that of the vibration. To damp the natural oscillations of the support itself, piers which have a separate foundation from the building are best and on ground floors these should be sunk into the ground. On upper floors they may have to be taken into the solid concrete floor slab below the floor boards. The extent to which anti-vibration measures need be employed naturally depends upon the amount of vibration encountered in the building and this is normally associated with its location. Vibrations may be due to many causes such as machinery, passing traffic, underground trains, and in some cases the movement of persons in the neighbourhood of the apparatus. It is sound policy to investigate and if possible remove or damp the source of vibration, before taking expensive steps to remedy its effects.

BALANCES

The positioning and installation of balances cannot be considered without giving due attention to the balance room itself. It is obvious, too, that more rigorous control conditions are necessary for rooms in which precision balances are housed than for those in which student-quality balances are used. In considering, therefore, the ideal conditions for the housing and mounting of balances, the reader is asked to use his discretion in modifying the following conditions to suit his own particular requirements.

BALANCE ROOM

Ideally the balance room should be situated in the proximity of the people who will use it. Where it is necessary to share a balance room between several laboratories this necessitates a central position. The room should not be situated on an outside wall, nor should the direct rays of the sun enter it. It should be well away from possible sources of vibration such as moving machinery or places where there is a great deal of personnel traffic. Balance room and other neighbouring doors should be fitted with springs to avoid vibrations when they are slammed.

Draughts should be avoided, and for this reason it is preferable to have only one door opening into the room and to ventilate it by means other than the opening of windows. Air conditioning is the best means of ventilating and maintaining the constant humidity and temperature which are necessary for very accurate balances. A constant temperature is particularly important and tubular heaters fitted with a thermostat have been used for this purpose. These are not particularly suitable because they create undesirable air disturbances due to convection. The recommended humidity for balance rooms is 50% and a suitable temperature is 25 °C. The exclusion of dust is a further problem and although it cannot be entirely prevented, it can be minimised by making the balance room easy to clean. The floor should not be swept but should be washed. All corners between the floor and the wall should be coved and other steps taken to prevent the accumulation of dust.

BALANCE SUPPORTS

The major difficulty to overcome in balance rooms is that of vibration. Because vibration makes it impossible to read a balance accurately, and shortens its life by excessive wear on the knife edges, this problem merits full attention.

It is generally agreed that the best way to prevent vibration is to interpose a number of dissimilar materials between the balance supports and the source of vibration. The materials used must themselves be of a stable nature.

Haslam[2] has recommended a method whereby holes are made in the floors of the balance room which penetrate the foundations. The holes are filled with concrete but interposing paper-like material between the concrete and the floor of the building creates a 13–25-mm space. When the concrete has set the material is removed and the space filled with a bituminous preparation. Above the concrete foundations brick pillars are erected, on the top of which is a further 13 mm of concrete surmounted by a 38-mm layer of cork. Finally a heavy slab of polished slate is laid on the cork.

Boos[3] suggests that if the amount of vibration is small balances should be placed on rubber stoppers and supported on wooden tables with 'alberene' stone tops. For floors above ground level he recommends a method whereby individual stone-topped tables are used and the legs of these are supported on hardwood blocks which rest on 13-mm lead plates. Between the stone table top and the balance a smaller stone slab is supported by corks, and on this the balance rests on feet which have hard rubber bases.

In some laboratories, individual concrete bench units faced with terrazzo have been used in preference to fixed slabs supported in the wall. The units are heavy enough to reduce vibration but are free-standing and can be moved to new positions.

Oertling Ltd[4] have made a number of suggestions for mounting balances. They recommend, for conditions of vibration on a ground floor, that the balance table be mounted on a concrete block set 0.9–1.2 m into the ground and isolated from the building foundations. The top of the block is left flush with the floor with a space of approximately 150 mm clear around it. Concrete and brick piers are built up from the block and support a slate slab about 38 mm thick fixed to the top of the piers with 13-mm Cooper felt and No. 65 Croid glue.

In instances where such measures are not possible or where in spite of solid mountings excessive vibrations persist, an eliminator should be placed below the feet of the balance. Such eliminators may be cork and lead sandwiches, Equi-flex spring units or rubber pads, commercial antivibration tables, or a 13-mm layer of filter papers.

Also recommended are slate tops resting on concrete pads, which are in turn supported on dry sand contained in hollow piers made of brick, sheet metal, or concrete. The slate is supported by the sand and should not come into contact with the container.

In my own laboratories it was necessary to arrange solid balance benches on a ground floor in a building where the amount of vibration was limited. For this purpose, it was found sufficient to take brick piers down on to their own foundation, and to interpose a layer of cork 13 mm thick between two pieces of lead sheet, on the top of the brick piers. The piers supported a slate top 63 mm thick. A small space was left between the back of the balance bench top and the wall to avoid transmitted vibration (*Figure 2.1*).

The problem of vibration has received considerable attention in recent years and better methods are now employed. Many kinds of vibration isolators are being commercially produced to cope with vibration problems surrounding sensitive equipment, or to isolate moving machinery which may be responsible for setting up vibrations.

In the case of balances of the less-sensitive type, for instance, non-skid insulators made from cork, Neoprene, rubber, or springs can be used in place of cork and lead pieces to isolate slabs from supporting brick or concrete piers.

For more sensitive balances, tables are available which are

98 INSTALLATION OF LABORATORY EQUIPMENT

supported on a tubular steel column. The balance sits on a plate, which is itself isolated from the table top by means of rubber isolators.

For microanalytical balances, galvanometers, or other sensitive equipment, Christie and Grey Ltd, 1 Finsbury Square, London EC2, have produced a spring-isolated table top which is supported at three points by helical steel springs on levelling screws. A heavy

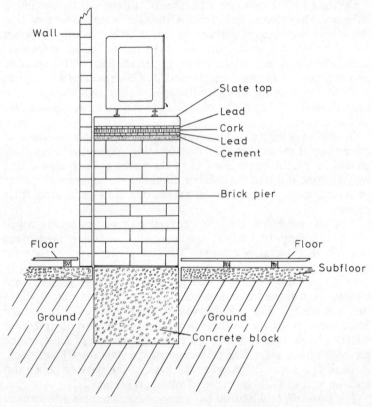

Figure 2.1. A method of supporting balances to reduce vibration

steel bar bolted to the underside of the table top is immersed in a viscous damping fluid contained in a tubular column. Methods for isolating balances from vibration have appeared in a wealth of diverse literature, but for those wishing to pursue antivibration methods still further the findings of Macinante and Waldersee[5], who have made an intensive study of the problem, are well worth attention.

For the use and protection of delicate instruments, a very sophisticated piece of equipment is the vibration-isolated table made by the Lansing Research Corporation, Ithaca, N.Y., U.S.A. It consists of a specially designed table top which automatically levels itself for various loads and is isolated by means of air pistons from vertical and horizontal vibrations.

MERCURY BAROMETER

It is well worth considering, in addition to the details of the procedure for the installation of barometers, the methods of transporting and moving these instruments[6].

The adjusting screw at the base of the mercury cistern should be adjusted so that the glass tube is full of mercury and the barometer is carried with the cistern uppermost.

If the barometer is to be dispatched by rail it should be packed in a strong wooden case, the top of which should be secured by wood screws. The backboard of the barometer should be screwed to the inside of the case. The person to whom the instrument is consigned should be advised of the date of dispatch of the barometer so that the train may be met and the instrument taken care of. Barometers should be sent by passenger train and placed in the parcels van, where they should be secured against the wall in a vertical position.

FORTIN BAROMETERS

When the Fortin barometer is unpacked the cistern must be kept in a raised position while the end of the box is being unscrewed. The barometer board should be unscrewed from the box and the instrument lifted out with the board attached. The barometer is unscrewed from the board and placed on a bench. Throughout these operations the cistern is maintained in a raised position.

The barometer should be hung in a position which is free from vibration and where there is little traffic. The base of the instrument should be 0.75 m from the floor. The room selected for housing the barometer must be well lit and direct sunlight should be avoided. Its temperature should remain fairly constant and there must be no corrosive fumes. Local sources of heat may upset the working of the instrument and it should be kept well away from radiators and hot water pipes.

The backboard should be firmly secured to the wall and its position verified by means of a plumbline to ensure that it is vertical. The barometer is now gently and slowly raised to a vertical position with the reading scale uppermost. A small space may be visible between the meniscus of the mercury column and the top of the tube. If there is no space the adjusting screw below the cistern is given a few turns to lower the mercury level. The instrument should now be turned from the vertical position until the mercury contacts the top of the glass tube. If a sharp click is heard this indicates that air has not entered the tube during transit and, in that respect, the barometer is satisfactory.

KEW BAROMETERS

Kew-type barometers have no backboard and are packed for transit in rubber shock absorbers in an inner case. A bracket is provided with the barometer which must be firmly secured vertically on the wall and at right angles to the floor. The general method of unpacking and installing is the same as for the Fortin-type barometer.

A supporting arm, supplied with the barometer, is attached to the gimbal ring on the instrument and its other end inserted in the bracket slot. The gimbal ring allows the barometer to hang vertically provided the bracket itself is reasonably vertical on the wall.

GLASSBLOWING EQUIPMENT

The setting up of glassblowing equipment requires careful consideration, and since the room it will be housed in largely determines the positioning of the equipment a suitable one is now described.

To assist ventilation, and to allow the heat, water vapour, and gases given off by burners to escape, glassblowing rooms should have high ceilings with ventilators placed high up. The room should be free from draughts, especially at working level.

The acoustic properties should be such that the noise from burners is reduced to a minimum and it may therefore be necessary to use acoustic tiles on the walls and ceiling. The relative merits of various burners in terms of their noisiness should be investigated before a purchase is made.

Adequate daylighting should be provided but direct sunlight should be avoided. The windows should be so arranged that the glassblower does not face them directly; rather, the light should enter either from behind him or at right angles to him. Artificial lighting is also required and fluorescent lighting is best. The enclosed type of lighting fittings should be used and since the carbon from burner flames blackens the exterior of the fittings, resulting in a loss of light efficiency, they should be regularly cleaned on the outside. The overall intensity of illumination in the room should not be too high or it will mask the blowpipe flame. Local lighting over machines should also be provided.

The walls should be painted green or some other suitable colour which enables the burner flame to be easily seen.

A fume cupboard designed to permit the use of hydrofluoric acid and a wash-up sink with draining board are also essential features. Metal rubbish boxes raised off the floor by short feet are necessary to allow hot pieces of glass to be safely discarded without the possibility of fire or damage to the floor.

Wood blocks are suitable for the floor since they are hard enough to prevent pieces of glass being trodden in and do not burn easily.

Adequate room should be allowed for the horizontal storage of glass tubing and rod. The storage rack, which should be fitted with doors to exclude all dust, should be made up of divisions of suitable size to enable the various types of glasses to be segregated. The divisions should not be too deep and great depths of glass tubing should not be kept in them. This avoids the possibility of glass tubes being scratched when neighbouring pieces are withdrawn from the rack.

To serve the various power-driven appliances electrical sockets should be provided. They should be sufficient in number to cope both with the machinery such as drills, cutters, and grinders initially installed and with any other equipment which may be purchased later.

An ample supply of fuel gas is essential and the number of cubic feet per hour likely to be used should be carefully calculated. The gas installation should be so arranged that the supply will be unaffected by any draw-off due to neighbouring laboratories or workshops.

A supply of oxygen from individual cylinders or from a piped supply is required. An adequate volume of compressed air supplied at a pressure of 0.7–1.0 bar is also necessary. If an air compressor is used it requires a reservoir tank to smooth out the air supply to burners.

A number of well-placed low-vacuum points are also very useful in the glassblowing workshop, one of them situated on the glassblowing bench itself.

Oxygen cylinders are best situated outside the room and from them the supply is fed to the glassblowing bench and to the equipment. In larger glassblowing rooms a number of burners may be involved. In this case, since the required volume of oxygen should be supplied at a constant pressure of 1.0–1.4 bar, more than one oxygen cylinder may be required. This arrangement, in terms of the amount of gas used, is also more economical. On the cylinder's manifold a reducing valve is required. Reducing valves are also necessary on the particular appliance inside the room to obtain the correct working pressure, but in some cases one valve may serve several burners. In addition, needle valves should be provided on each appliance for precise control.

In the main fuel gas line, at some point before the gas is distributed to the various appliances, a non-return valve should be fitted to prevent oxygen from entering the line and forming an explosive mixture.

BENCHES

The design of the glassblowing bench is important. The bench itself should be 1 m high, and not less than 1.8 m long and 0.8 m wide. The top of the bench should be covered with hard black cement asbestos composition and should have a raised ledge around its edge to prevent glass rolling off. A clear working space should be left, preferably on three sides of the bench, for the manipulation of long lengths of glass tubing.

The controls to various services, such as fuel gas, oxygen, compressed air, and vacuum, may be placed on a rail under the front edge of the bench and in this position are easily available to the operator.

BURNERS

The gas burners should be suitable for the type of fuel gas to be used. Butane and similar gases, for instance, require burners with special jets and this matter should be thoroughly investigated before burners are purchased.

LATHES

Since glassworking lathes rotate at relatively low speeds, they do not vibrate and require no special rigid fixing. They should, however, be set level and true and are best placed on felt which is glued to the bottom of the feet and to the floor. Shimstock is used for levelling purposes. The gas supply pipe to the lathe must be adequate in size to feed the burners. The bed and the swing of the lathe must be large enough to accommodate the work envisaged and future possible work requirements should be investigated. The universal type of lathe is the most suitable to cope with the diverse nature of the work for research purposes, since permanently horizontal lathes restrict the work to straight seals and similar horizontal operations. The lathe should be well protected from draughts and should not be near doors or windows. It should have all-round accessibility and plenty of space should be left at the ends so that long tubes may project through the mandrel. A vacuum point close to the lathe may be necessary for vacuum chucks and for operations involving the shrinking of glass.

OVENS

Ovens for glass annealing should be level and firm so that hot glass apparatus will not be distorted by gravity effects. The weight of the oven and its support may be considerable, in which case it should stand on feet of large area on a firm floor. Since ovens suitable for larger pieces of glass may have a load of 20 kW or higher, a three-phase electrical supply is usually required.

The oven should be insulated to prevent heat losses. The base should be at normal bench height to allow large items of glassware to be easily placed in it. For the same reason the doors should open the full length and height of the longest side of the oven. The oven elements should be grouped in banks and each bank should be individually controlled to allow the rate of heating and cooling to be carefully adjusted and the temperature over the oven as a whole should be thermostatically controlled. For convenience, the oven should be positioned as conveniently near to the glassblowing bench as possible.

HEAVY EQUIPMENT

In many laboratories the problem of positioning large heavy items of equipment presents itself. It is desirable, therefore, when laboratories are designed, to leave sufficient space for the installation of equipment which may be purchased in later years.

Because of vibration, noise, and other annoying features, heavy moving equipment may need to be accommodated in rooms separate from the laboratories, and in some cases should be situated outside the laboratory areas altogether. If possible it should be housed on a ground floor.

LIQUID AIR PLANT

As an example of the type of heavy equipment likely to be encountered in the laboratory we may consider the liquid air plant, which is becoming increasingly popular in universities and research institutions for high-vacuum and low-temperature work.

The size of machine most suitable for university work can produce up to six litres of liquid air per hour and production begins within a few minutes of starting the machine. The rate of production is sufficient to serve a number of laboratories, and in many cases several departments. Since it is not normally essential that the machine be strictly adjacent to any particular laboratory it may, without causing undue inconvenience, be situated in an outside building.

The site chosen for the outbuilding, presuming a choice is possible, should be central and from an economic point of view should be as close as possible to an existing electrical and water supply. The machine should be placed in the care of one person and others should be prohibited from interfering with it in any way. No-smoking and other safety rules should apply. The room must be well ventilated and an ambient temperature of 20 °C is normally recommended. The room should also have a solid cement floor and should be large enough to allow all-round access to the machine. which may be approximately 2.1 m long by 1 m wide. A floor space of 3.7 m by 2.4 m wide would, therefore, be satisfactory. The building should be of fireproof construction with double doors of sufficient width to admit the machine and to permit the easy passage of liquid air containers.

To serve the machine a three-phase electrical supply and a water supply with an uninterrupted pressure of about 3.5 bar are normally required. The machine is commonly fitted with a pressure contact switch to shut it off in the event of a water supply failure. A drainage system for the water effluent is also required. The electrical and water supply lines can be brought to the machine in floor chases.

The machine is mounted on a concrete block to which it is fastened by means of its baseplate. Bolts are grouted into the top of the block and are positioned by a template when the block is cast. The thickness of the block varies with the weight of the machine installed, and for the size of machine quoted as an example is about 300 mm thick and is mounted on rubber blocks of suitable size and hardness. As well as providing a solid foundation, the block raises the level of the machine and facilitates the introduction of Dewar vessels below the liquid air outlet.

LABORATORY STILLS

Some items of equipment have to be fixed to the laboratory walls and modern water stills fall within this category. The Manesty pattern stills, which are commonly used, are heavy and need to be well supported. Although large screws are normally supplied with the still for fixing purposes, it is safer to use Rawlbolts, or similar fixing devices, to secure the supporting bracket vertically on the wall.

The still should be mounted in a room where the damp atmosphere created will have no harmful effects on other apparatus, and should be positioned so that the top of the still is about 2.1 m above the floor level. Mounting the still in this way allows the distilled water container to be supported on a wooden stand directly beneath it and provided top ventilation is available the room is kept reasonably free from vapours. Assuming the capacity of the distilled water container to be of the order of 100–150 litres, the stand should be stoutly constructed to take the weight and should be about 560 mm high. If it is raised up in this way a reasonable clearance is allowed beneath the container outlet cock when tall vessels are filled. A lead-lined box or cement recess connected to a drain outlet should be formed in the floor below the front edge of the container support and covered with a flat metal grille set flush in the floor. This drain serves to catch the drips from the container outlet tap and permits on-the-spot flushing of vessels. It may also accommodate an overflow pipe led from the side of the container and, should the container overfill, prevents flooding.

The effluent from the still itself should be discharged into a small copper funnel soldered into the top of a copper tube about 13 mm in diameter. Sufficient space is left between the top of the funnel and the still overflow so that the rate of flow of water discharging from the still can be easily seen when adjustments have to be made. The copper tube is led down the wall and under the container support so that it discharges into the drain.

To avoid wastage of water and electricity the still can be made fully automatic by simply introducing a micro switch and a glass float. The method of doing this has been described by Joliffe[7].

The electrical supply socket (or the gas outlet in the case of gas-heated stills) should also be set high on the wall level with the body of the still. This way, the electrical leads to the still elements are kept as short as possible. Similarly, the water supply tap should be situated on a level with the fine control inlet valve so that the connecting hose to the still is also kept short.

To indicate that the still is switched on, two red signal lamps should be provided.

106 INSTALLATION OF LABORATORY EQUIPMENT

The distillate should be led from the still via a glass or polythene tube which should be taken through a small hole formed in the lid of the distilled water container.

SPECTROGRAPHIC EQUIPMENT

In most laboratories there is a need for a room which can be completely or partially darkened and in which optical instruments may

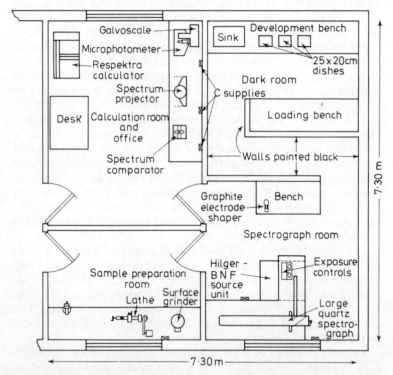

Figure 2.2. Spectrographic laboratory. This laboratory, somewhat more elaborate than the one described, represents a suggested plan for a fully equipped photographic unit. The laboratory has a separate room for the preparation of samples and the spectrograph is in an adjacent room. Plates are taken straight to the darkroom and, when processed, to the calculation room (Courtesy Hilger & Watts Ltd)

be used (*Figure 2.2*). Modern spectrographs may themselves be operated in rooms which are not completely darkened, but some means of reducing light from the windows is often necessary.

The quality of the furniture and various interior finishes in the spectrographic laboratory should be comparable with the expensive equipment housed. This is mainly to provide the necessary conditions for the preservation of the equipment, but also helps to induce a careful attitude in persons working in the room. The finish of the ceilings, walls, and floor should be such that they are easy to clean and there should be no ledges or projections on which dust can accumulate. The room should be dry, free from fumes, and able to be ventilated without causing draughts. Substances likely to give rise to fumes should not be kept in the laboratory.

To serve the equipment, sufficient electrical outlets, both a.c. and d.c., may be required and should be available in appropriate positions. The laboratory should be electrically safe, a point stressed in Chapter 6. If water is to be used in the laboratory, the outlets and feed pipes should be kept well away from electrical sockets.

The actual layout of the benches depends on the amount of equipment installed, but since the shape of spectrographic equipment may in turn determine the actual shape and size of some of the bench tops, the measurements and shape of the equipment should be investigated before the benches are manufactured.

Because spectrographs and some other items of spectrographic equipment are heavy, the laboratory benches should be strongly constructed. Similarly, because it is undesirable, and in any case difficult, to move these instruments once they are in position, thought should be given to a bench layout which allows all-round accessibility to them. In teaching laboratories this may be a very desirable arrangement.

Other laboratory furniture should include a desk for calculations and other written work, a filing cabinet for recorded data and instructions, and some bookshelves. Movable cupboard units may be provided under bench tops and kneeholes should be left in suitable places. A number of drawers, including some small ones for electrodes, photographic plates, and other small parts, and a set of wall shelves for samples and reagents are also necessary.

In conjunction with experimental work a photographic darkroom, adjacent to and directly accessible from the laboratory, is required for the development of plates. If the circumstances are such that the room has also to be used for other photographic work of a general nature it should have a separate entrance. This prevents unnecessary traffic through the laboratory.

Other provisions should if possible also be made for the installation of microphotometers, spectrum projectors, plate reading boxes, and other instruments connected with plate reading. For this purpose, a separate room which can be completely darkened is desirable.

For cleaning electrodes a grinding wheel, a vice, and a number of files are also required. A small workbench for this purpose is best positioned just outside the laboratory.

REFERENCES

1 Strong, J., *Modern Physical Laboratory Practice*, Blackie, London (1939)
2 Haslam, J., 'Analytical Laboratories', in Report of a Symposium on Laboratory Layout and Construction, *R. Inst. Chem. Report*, No. 6, p. 14 (1949)
3 Boos, R. N., *Laboratory Design*, Reinhold, New York (1951)
4 Oertling, *The Elimination of Vibration on Balance Tables*, pamphlet issued by L. Oertling Ltd, St Mary Cray, Kent (n.d.)
5 Macinante, J. A., and Waldersee, J., 'The Vibration Isolation of Knife-edge Balances', *J. scient. Instrum.*, **41**, 1 (1964)
6 Casella, *Instructions for Handling and Erecting Fortin, Kew and Test Barometers*, leaflet No. 3041/RC, C. F. Casella and Co. Ltd, London (n.d.)
7 Joliffe, G. O., 'A Fully Automatic Laboratory Still', *Lab. Pract.*, **12**, 446 (1963)

3
Stores management

TYPES OF STORE

Stores must be in keeping with the nature and size of the establishment they are designed to serve and may accordingly be classified into three types: (a) central stores, (b) main stores, (c) dispensing stores.

CENTRAL STORES

Central stores are to be found in very large research or teaching establishments and in industry, and hold large stocks to meet the varied requirements of such establishments. The individual requirements of the departments concerned necessitate that the stocks from central stores be distributed to subsidiary stores. Central storage has the advantage that bulk purchasing arrangements may be made for items which are common to a number of departments. There are, however, certain disadvantages. It is desirable, for instance, that the storeman should have a very intimate knowledge of all the apparatus held in the stores, and to a great extent the use to which it is put. When a large volume and variety of stock is held on behalf of a number of departments this knowledge is often lacking. Also, the various departments which draw supplies from the store may not be satisfied with the type of material which has been purchased and with which they are issued. Storage difficulties may also be encountered due to the mixed nature of the materials.

MAIN STORES

The second type of store is one which serves a particular department or division and which within the limits of the division may be called the main store. The materials housed are usually applicable to one technological subject although the range of items stocked may be quite comprehensive.

Such stores are to be found, for instance, in universities and serve, from a central position, many laboratories within the same department.

Petty issues to individuals are not permitted from the main store, and similarly only major issues, such as complete containers and case lots, are made to dispensing stores. In some instances a case lot may be divided up between several dispensing stores. Chemical containers which have been opened are not normally kept in the main store.

DISPENSING STORES

Within a particular laboratory area a third type of store is required for dispensing small quantities of materials for local needs. The storeman in charge is responsible for maintaining his stocks by indenting on the main store. In educational establishments students go to the dispensing store for requirements additional to the initial set of apparatus with which they are issued in their locker. The student is required to sign for apparatus at the store and these items, plus the initial locker issue, are debited to him, and may be charged for if not returned in good condition at the end of the final term. Students are allowed to use the particular store in their area but no other.

DESIGN OF THE STORES

MAIN STORES

In designing the main store the location is of great importance. Ideally, it should be situated at the rear of the building and on the ground floor. This facilitates the delivery of goods. A goods lift is desirable when the laboratories serviced by the store are dispersed on floors above.

The store must be large enough to accommodate all the necessary stock and yet leave ample room for the movement of boxes, steps, and ladders and to provide a freeway for trolleys so that no dangerous hazards are presented.

The floor of the store must be strongly constructed so that sagging does not occur under the concentrated weight of the stored materials.

Unpacking facilities are essential and should be provided close to the main doors at which the goods are delivered. It is advantageous if facilities are provided to permit delivery vans to back close up to the door of the store so that goods may be offloaded near to the checking point.

For checking purposes counter tops with shelves below are suitable, or alternatively, long tables. Whatever provision is made, ample room for unpacking and checking the goods is required. As the boxes are unpacked, the contents are placed on the counter top. After checking these may be stored on the shelves below until they are subsequently recorded and stowed away.

To comply with the fire regulations, inflammable substances should be kept in an outside store. Acids, ammonia, and gas cylinders should also be in specially designed rooms. Chemicals should be stored separately from apparatus. Partitions or walls in the store should be fireproof so as to confine fire outbreaks.

Shelves for chemicals

Wooden shelves are best for the storage of chemicals. These should not be too wide or more than 2.1 m high. A slight lip to the front of the shelves is desirable to prevent bottles being crowded off.

A very easily erected type of continuous shelving now marketed consists of slotted steel 'U' channel uprights into which detachable cantilever brackets may be fitted. Metal shelves may then be clipped into the brackets, or alternatively wooden shelves may be used and held in position with screw bushes and strip endplates. Rearrangement of this type of shelving may be effected very quickly indeed.

The uprights which support the shelving are normally fixed to a wall, although freestanding ones are available. The freestanding type is supported in floor and ceiling sockets, which thus allow shelving to be mounted on both sides of the upright.

Also used in the construction of lightweight laboratory shelving and storage units are the now familiar metal angle sections which may be purchased in various section sizes and gauges. Steel shelves are made in a number of different sizes to suit the metal sections, and by means of shelf-adjuster and shelf-divider clips the adjustment of both shelves and dividers is an easy matter. This 'handy angle' material with its wide range of accessories, plus a polychromatic

112 STORES MANAGEMENT

satin bronze finish which is extremely durable, is a valuable aid in the laboratory.

Shelving for apparatus

It is recommended that apparatus be stored on metal shelving. The type sold in units is admirable for this purpose, and provided the units installed are all the same size, general and local rearrangements may easily be carried out. The units have bolt holes at 25-mm centres and accommodate adjustable shelves between angle posts.

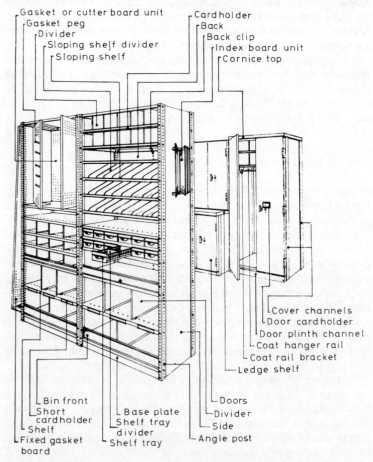

Figure 3.1. Closed metal shelving (Courtesy Constructors Ltd)

The units may be open or can be fitted with sides and backs. Bin dividers, doors, and numerous other accessories are available (*Figure 3.1*).

If the units are to be accessible from one side only, or are backed on to a wall, they can be erected as single-sided stacks. If placed back to back as double-faced stacks they can be used for forming storage bays in the centre of a room or peninsular bays from the walls. Units which back on to each other have a common back. Counter-high units provide storage accommodation as well as acting as table tops for receiving and dispensing apparatus and for clerical work.

If materials exceed the storage space it is possible, by reinforcing the corner posts of the lower stacks and introducing multi-tier shelving in a mezzanine style, to double the available storage space.

Doors with a lock should be provided on some units. This is a security measure to allow the storeman to accept more cheerfully the responsibility for expensive and dangerous items.

Lighting

Good lighting is essential in the store. The fittings must be placed high enough above the stacks to give a good distribution of light

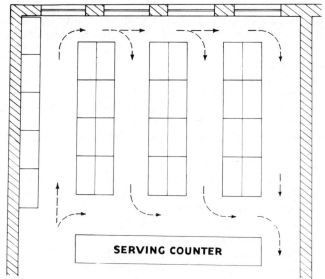

Figure 3.2. Typical arrangement for storage units. The stacks are arranged so that full use is made of the natural lighting (Courtesy Constructors Ltd)

in the bays and should where possible be spaced between stacks so as to illuminate the gangway areas. In the same way the stacks should be so arranged as to take full advantage of natural light (*Figure 3.2*).

Proper arrangements should be made for the clerical duties of the storeman, which should include a place for files, books, and a desk.

Other requirements

A proper serving hatch is among the other stores requirements to enable the storeman to set out his material methodically, to count the items, and to accept the signature of the recipient.

All necessary safety precautions must be adopted such as suitable exits, firefighting appliances, protective clothing, and means for neutralising acid spillages. Safety steps or ladders are essential. These may be of the portable or the travelling variety. The latter are attached to a guide rail near to the top of the stack and have wheels on the feet.

FUNCTION OF THE STORE AND STOREKEEPER

The functioning of the store depends on the ability of the storekeeper. He should maintain a good relationship with the laboratory personnel and, perhaps more important, the laboratory personnel should get to know the storekeeper.

The storekeeper must have a prodigious memory and may often be responsible for fulfilling the requirements of an urgent and immediate project by remembering the whereabouts of an important piece of apparatus when others have long forgotten it.

The storekeeper often has an unenviable task in satisfying the needs of many (and often impatient) people, and he must therefore himself be a patient person.

To be efficient he must have a technical understanding of the way the apparatus is used in the laboratory, since he is often asked to recommend the right item for a particular job. In addition, then, to his other attributes, he must be a veritable mine of information in order to assist laboratory workers.

The storeman accepts the responsibility for all the materials in the stores, and by keeping abreast of the general work and acquainting himself with the individual research projects he is able to meet current demands and to anticipate future requirements.

ACCOUNTING FOR STORES

The day-to-day task of the storeman is to keep an accurate account of the stock in hand, to order materials, and to check the goods which are delivered.

For the head of the section to predict budget requirements accurately, the stock consumed must be fully known. In many departments the laboratory expenses incurred by a particular group must also be known, and through the accounting system the storeman should be able to provide the necessary figures. With this type of information at his disposal, the head of department can issue a timely word of warning to any group who may be near their limit of spending.

Accounting is further complicated by grants of money which may be given in the form of research grants by outside bodies. In some cases this means that issues to individuals, or to certain groups, must be recorded in order that the expense for any particular project may be computed. For these reasons, special groups and individuals should be required to present written orders at the store.

ECONOMY

By co-operating with heads of sections, the storeman can help to keep requirements down. This is the case when several research groups are demanding the same apparatus, which, by means of a suitable work schedule, might easily be shared.

REPAIRS TO APPARATUS

A great deal of the apparatus which finds its way back to the store is in need of repair and should be referred to the workshops or to the glassblower. In many cases apparatus must be sent away for repair and knowing the best place to send it and the method of dispatch are all part of the storeman's duties.

LOANS AND TRANSFERS OF APPARATUS

The emergency loan of materials to other departments is a common occurrence in scientific work. This is not to be deplored since, generally speaking, the economy of the institution as a whole is assisted. Reciprocal loans may in turn assist the storeman himself and his own department. The loaning of apparatus should be done only on a signature basis and for this purpose a loan book should be kept.

To facilitate the work of the institution as a whole, expensive items of equipment may be transferred permanently to other departments. Such transfers should be recorded in the stockbook as a permanent record.

DEPARTMENTAL STOCKBOOK

In educational establishments a general stockbook must be kept. The object of this is to maintain a record of all items of permanent value, and to enable their value to be assessed for the purpose of insurance. It is customary to record all items above a certain value. The general apparatus held in bulk in the main stores may also be included on the stockbook, but once such items leave the storeroom they are no longer an inventory item.

The difficulty found with this system is in deciding exactly what constitutes a permanent item. For instance, does a large Pyrex aspirator, which costs a considerable sum of money, fall within this category? There is no strict line of demarcation and the items listed must be based on the local understanding of the storekeeper.

The stockbook is made up for inspection annually and the storekeeper must locate the various items which may have been dispersed to the far corners of the department. The task is onerous and difficult. To overcome this difficulty, and to track the movements of portable furniture and other equipment, each item is given an identification number which should be marked on it. The number is noted in the stockbook against the item, and to further assist identification the instrument serial number should also be noted together with its full description.

The outside bodies who from time to time donate sums of money for the purchase of apparatus, to the department or to individuals working in the department, often regard such apparatus as being on permanent loan. A list of such equipment may be periodically requested by the donating body. In such cases their initials are painted on the apparatus and the details of the donation are recorded against the item in the stockbook.

Since items which give satisfactory service may need to be reordered, the stockbook can be of advantage to the department if the order number and date of purchase are entered against each item. The particulars recorded in the book thus enable the original order to be found quickly. The stockbook information helps also to locate equipment, already existing in the department, which may be required for new projects.

As equipment wears out it must be struck off the stockbook. Some items never wear out but become so old as to become redundant; these too should be excluded.

The stockbook itself may be lost or destroyed by fire and duplicate records should be kept and stored in a place safe from fire or water.

SIZE AND LAYOUT OF STORES

The size of stores is governed by the following requirements: (a) space available, (b) efficient layout and space utilisation, (c) bulkiness of items stored, (d) amount of stock to be held. Whatever size is decided upon it must allow for the further expansion of the department.

When laboratories are in the planning stage the storage requirements are sometimes glossed over or forgotten. Alternatively, an arbitrary area is set aside with no thought as to its actual storage capacity. Stores must be planned in the same way as laboratories and the space allocated must be realistic in relation to the needs.

The amount of stock to be held affects the size of the store. This depends upon the local availability of equipment and on the period of time which elapses between major purchases. To be able to purchase goods in bulk, at say yearly intervals, enables much cheaper purchases to be made but the storage accommodation must be larger on this account.

An efficient store layout must be arrived at by first planning the arrangement on paper so as properly to utilise the space available. Thought should be given to aisle sizes, heights and layout of stacks, and other similar matters. There is, however, a limit as to how much space should be utilised and a point is reached where the difficulty of stowing material exceeds any advantage in space gained. The ease with which material may be stowed away depends on the height and depth of the racks and bins and the layout of the aisles.

For complete safety and for ease of removal of material, certain general rules apply. Heavy items must be stored low down, and smaller items which may often be required should be within arm's reach. Materials which are not often used, and bulky items of light weight, may be kept on top of the racks. Similarly, if separate storage is not available and bulk package storage and bin storage are used in the same store, the bulk packages may be kept on top of the stacks. Racks may be made higher if the bin fronts are reinforced so that they may be used as a ladder.

The type of material stored will quite obviously affect the stores layout because of the need for different shelf and bin arrangements. The material kept in the physics department store, for example, necessitates quite different bin design as compared with a chemistry department store.

PURCHASE OF STORES MATERIALS

AMOUNT OF MATERIALS TO BE PURCHASED

To replenish his stocks, the storeman must prepare an order. In preparation for this, the stocks are checked at, say, yearly intervals to see they agree with the stock cards. After due allowance has been made for outstanding orders, the difference between the stock in hand and the maximum stock requirements is ordered. The storeman should, however, first consult senior staff members as to their special needs for the coming year. Responsible members of staff should indicate any changes they propose to introduce in their courses which might affect the stocks to be held. Since bulk quantities of material may be purchased under the yearly system of reordering it is undoubtedly the best.

In many institutions, where owing to poor planning or other reasons storage facilities are poor, it is not possible to hold stocks of any magnitude. In these cases materials may have to be ordered more or less as required. If this method of ordering is necessary, a convenient way to keep the limited stocks at a working level, other than by the continual checking of numerous stock cards, is to enter the goods at the time of issue on a reorder list. Alternatively, the various written indents which have been placed on the stores may be referred to. The items entered on the reorder list are consolidated and orders for goods are placed at frequent intervals as necessary. This method is particularly applicable to the reordering of chemicals, where unprecedented demands may easily deplete limited stocks. It has the additional merit of ensuring fresh stocks are always available. Its disadvantages are the loss of discounts and the extra time spent on the requisitioning and invoicing of goods. A card index system suitable for reordering chemicals in small laboratories has been described by McEwen[1].

PURCHASING ARRANGEMENTS

How to purchase

The various aids to purchasing, such as the *British Instruments Directory and Buyers' Guide*[2], are of great assistance but these must be supplemented, for discriminating purchase, by personal experience gained over the years. Popular brands of glassware and other standard apparatus may be brought from any large supply house, but there are many items which do not appear in their catalogues and which may be required in the laboratory and workshop for special work. To meet this difficulty the storeman must help himself by keeping a notebook or card index in which the particulars of the specialist suppliers, including their speed of delivery and prices, are kept. It is useful, too, to record the name of the representative with whom one actually deals, for reference on future occasions. These and other particulars eventually add up to an invaluable and irreplaceable list.

What to purchase

For deciding what to purchase, up-to-date catalogues and various pamphlets should be collected and consulted by the storekeeper. When specialised equipment is to be purchased such literature is also of assistance to the person who works with the equipment and, as is often the case, he will approach the storekeeper for information and advice. The knowledge of the research worker, in terms of his requirements, combined with the experience of the storekeeper in respect of delivery, quality, and cost, ensure that the right apparatus for the work is made available at the right time.

Factors affecting purchase

In deciding where to buy apparatus the storeman considers: (a) cost, (b) quality, (c) delivery.

Cost

The cost of the goods must be considered in relation to the amount of money available.

Quality

The disciminating buyer compares the quality of the goods with the price. The goods must stand up to the use for which they are intended. For example, the cost of crucible tongs varies considerably according to the material of which they are made. In terms of length of life and their suitability for gravimetric work, however, nickel tongs are infinitely better than the cheaper iron variety. It is grossly uneconomical to purchase cheap glassware for use in laboratories. Those who have had the experience of heating soda glass tubes, for example, will appreciate the ultimate saving which can be made when hard glass ones are used.

Delivery

However cheap and whatever the quality, the goods are of no use if they do not arrive when required. For this reason the delivery time is an important consideration.

It is necessary to evaluate the three above-mentioned factors when purchasing goods, but if storage arrangements are adequate, delivery time in most cases may be ignored.

Conditions of sale

The conditions of sale as laid down by the supplier may affect the purchase and should be considered. Most firms are accommodating when wrong goods are delivered or items are lost or damaged in transit. Nevertheless, the customer should be aware of the conditions such as whether or not the cost of carriage is borne by the vendor, the method of signing for goods, and whether the prices as quoted in the suppliers' catalogues are subject to alteration without notice.

SPECIAL PURCHASES

Alcohol and poisons are special purchases and are subject to control. It is absolutely essential that the regulations and procedure appertaining to these be clearly understood.

Purchase of alcohol

Alcohol may be purchased duty free by teaching, research, and experimental establishments, by permission of the Commissioner of Customs and Excise. The types of alcohol issued under licence are ethyl alcohol and industrial methylated spirits. Ethyl alcohol is defined by the Customs as 'pure spirits for use in art or manufacture where Industrial methylated spirits are unsuitable'.

Applications to receive duty-free alcohol are made by the institution concerned on the appropriate form. The purpose for which it is to be used must be stated and other details given. These include the address of the premises on which the alcohol will be stored and the estimated annual requirements. The application may be approved by the Customs and Excise, subject to certain conditions. These conditions provide that the alcohol is kept under lock, the key to be held by a responsible person, and that all issues are made by him. The recovery or redistillation of alcohol is allowed only by the sanction of the Commissioners of Customs and Excise, and duty-free spirits must be kept separate from duty-paid spirits, and from methylated spirits.

Requisition books are issued to the approved establishment, and when alcohol is ordered a requisition covering the amount should accompany the order sent to the supplier. When alcohol is received it is accompanied by an official permit which must be sent or handed to the local Excise Officer. In the case of ethyl alcohol the package should not be opened until the Excise Officer has been notified that the spirits have been received intact. Should the spirits not be intact they must be held in the condition in which they were received until inspected by an Officer who will visit the premises.

For duty-free spirits a special stockbook must be kept where the spirits are stored, and in it must be recorded all receipts and issues. The book should be available at any time for inspection by the Excise Officer, who may also take account of the amount of spirit in stock. A return of all alcohol used must be made annually on the official form.

Purchase of poisons

Certain items in the chemical suppliers' catalogues are marked to define the schedule of rules (made under Section 35 of the Pharmacy and Poisons Act of 1933) under which they are controlled. Substances included in the first schedule of the rules can only be supplied against an order signed by the purchaser stating

Table 3.1 WEIGHT PER METRE OF PLATINUM WIRE

diameter (mm)	weight (g)	diameter (mm)	weight (g)	diameter (mm)	weight (g)	diameter (mm)	weight (g)
0.0250	0.0108	0.650	7.2738	2.40	99.164	4.50	348.62
0.0275	0.0130	0.675	7.8440	2.45	103.34	4.55	356.41
0.0300	0.0155	0.700	8.4358	2.50	107.60	4.60	364.29
0.0325	0.0182	0.725	9.0492	2.55	111.95	4.65	372.25
0.0350	0.0211	0.750	9.6840	2.60	116.38	4.70	380.30
0.0375	0.0242	0.775	10.3404	2.65	120.90	4.75	388.44
0.0400	0.0275	0.800	11.0182	2.70	125.50	4.80	396.66
0.0425	0.0311	0.825	11.7176	2.75	130.20	4.85	404.96
0.0450	0.0349	0.850	12.4386	2.80	134.97	4.90	413.36
0.0475	0.0388	0.875	13.1810	2.85	139.84	4.95	421.84
0.0500	0.0430	0.900	13.7900	2.90	144.79	5.00	430.40
0.0550	0.0521	0.925	14.7304	2.95	149.82	5.05	439.05
0.0600	0.0620	0.950	15.5374	3.00	154.94	5.10	447.79
0.0650	0.0727	0.975	16.3660	3.05	160.15	5.15	456.61
0.0700	0.0844	1.00	17.216	3.10	165.45	5.20	465.52
0.0750	0.0968	1.05	18.981	3.15	170.83	5.25	474.52
0.0800	0.1102	1.10	20.831	3.20	176.29	5.30	483.60
0.0850	0.1244	1.15	22.768	3.25	181.84	5.35	492.76
0.0900	0.1394	1.20	24.791	3.30	187.48	5.40	502.02
0.0950	0.1554	1.25	26.900	3.35	193.21	5.45	511.36
0.100	0.1722	1.30	29.095	3.40	199.02	5.50	520.78
0.125	0.2690	1.35	31.376	3.45	204.91	5.55	530.30
0.150	0.3874	1.40	33.743	3.50	210.90	5.60	539.89

his name and address in full, trade, business, or profession, and the purpose for which the article is required. Substances controlled by the Dangerous Drugs Act are also marked and these may only be supplied to a person holding a licence or who is otherwise authorised to purchase dangerous drugs.

Purchase of platinum

Although the price of platinum varies from day to day, the current price of the metal or of standard manufactured platinum ware may be readily obtained. It is, however, often necessary to calculate

Table 3.2 IMPERIAL STANDARD WIRE GAUGE IN MILLIMETRES

S.W.G.	mm	S.W.G.	mm
1	7.620	26	0.457
2	7.01	27	0.417
3	6.401	28	0.376
4	5.893	29	0.345
5	5.385	30	0.315
6	4.877	31	0.295
7	4.47	32	0.274
8	4.064	33	0.254
9	3.658	34	0.238
10	3.251	35	0.213
11	2.946	36	0.193
12	2.642	37	0.173
13	2.337	38	0.152
14	2.032	39	0.132
15	1.829	40	0.122
16	1.626	41	0.112
17	1.422	42	0.102
18	1.219	43	0.091
19	1.016	44	0.081
20	0.914	45	0.071
21	0.813	46	0.061
22	0.711	47	0.051
23	0.610	48	0.041
24	0.559	49	0.030
25	0.580	50	0.025

the weight, and consequently the price, of equipment to be made in the laboratory. Since such items are normally constructed from wire or foil, the information given in *Tables 3.1* to *3.3* will be useful.

Table 3.3 WEIGHT PER SQUARE CENTIMETRE OF PLATINUM SHEET

thickness (mm)	weight (g)	thickness (mm)	weight (g)	thickness (mm)	weight (g)	thickness (mm)	weight (g)
0.0250	0.0550	0.650	1.4300	2.40	5.280	4.50	9.900
0.0275	0.0605	0.675	1.4850	2.45	5.390	4.55	10.010
0.0300	0.0660	0.700	1.5400	2.50	5.500	4.60	10.120
0.0325	0.0715	0.725	1.5950	2.55	5.610	4.65	10.230
0.0350	0.0770	0.750	1.6500	2.60	5.720	4.70	10.340
0.0375	0.0825	0.775	1.7050	2.65	5.830	4.75	10.450
0.0400	0.0880	0.800	1.7600	2.70	5.940	4.80	10.560
0.0425	0.0935	0.825	1.8150	2.75	6.050	4.85	10.670
0.0450	0.0990	0.850	1.8700	2.80	6.160	4.90	10.780
0.0475	0.1045	0.875	1.9250	2.85	6.270	4.95	10.890
0.0500	0.1100	0.900	1.9800	2.90	6.380	5.00	11.000
0.0550	0.1210	0.925	2.0350	2.95	6.490	5.05	11.110
0.0600	0.1320	0.950	2.0900	3.00	6.600	5.10	11.220
0.0650	0.1430	0.975	2.1450	3.05	6.710	5.15	11.330
0.0700	0.1540	1.00	2.200	3.10	6.820	5.20	11.440
0.0750	0.1650	1.05	2.310	3.15	6.930	5.25	11.550
0.0800	0.1760	1.10	2.420	3.20	7.040	5.30	11.660

Table 3.3 WEIGHT PER SQUARE CENTIMETRE OF PLATINUM SHEET—*continued*

thickness (mm)	weight (g)	thickness (mm)	weight (g)	thickness (mm)	weight (g)	thickness (mm)	weight (g)
0.0850	0.1870	1.15	2.530	3.25	7.150	5.35	11.770
0.0900	0.1980	1.20	2.640	3.30	7.260	5.40	11.880
0.0950	0.2090	1.25	2.750	3.35	7.370	5.45	11.990
0.100	0.2200	1.30	2.860	3.40	7.480	5.50	12.100
0.125	0.2750	1.35	2.970	3.45	7.590	5.55	12.210
0.150	0.3300	1.40	3.080	3.50	7.700	5.60	12.320
0.175	0.3850	1.45	3.190	3.55	7.810	5.65	12.430
0.200	0.4400	1.50	3.300	3.60	7.920	5.70	12.540
0.225	0.4950	1.55	3.410	3.65	8.030	5.75	12.650
0.250	0.5500	1.60	3.520	3.70	8.140	5.80	12.760
0.275	0.6050	1.65	3.630	3.75	8.250	5.85	12.870
0.300	0.6600	1.70	3.740	3.80	8.360	5.90	12.980
0.325	0.7150	1.75	3.850	3.85	8.470	5.95	13.090
0.350	0.7700	1.80	3.960	3.90	8.580	6.00	13.200
0.375	0.8250	1.85	4.070	3.95	8.690	6.05	13.310
0.400	0.8800	1.90	4.180	4.00	8.800	6.10	13.420
0.425	0.9350	1.95	4.290	4.05	8.910	6.15	13.530
0.450	0.9900	2.00	4.400	4.10	9.020	6.20	13.640
0.475	1.0450	2.05	4.510	4.15	9.130	6.25	13.750
0.500	1.1000	2.10	4.620	4.20	9.240	6.30	13.860
0.525	1.1550	2.15	4.730	4.25	9.350	6.35	13.970
0.550	1.2100	2.20	4.840	4.30	9.460	6.40	14.080
0.575	1.2650	2.25	4.950	4.35	9.570	6.45	14.190
0.600	1.3200	2.30	5.060	4.40	9.680	6.50	14.300
0.625	1.3750	2.35	5.170	4.45	9.790		

Bulk purchase

Special quantity prices for single deliveries are allowed on acids, solvents, and some chemicals. A greater saving may be effected for some goods, such as acids, if these are purchased by yearly contract. Contract and quality prices apply also to acids bought in Winchesters. Winchester packing is more expensive but is generally worth the extra cost in teaching establishments since the transfer of acids from carboys is a difficult and hazardous procedure.

Discounts are allowed on glassware purchased in case lots. For reference purposes the quantity constituting a case lot should be marked on the stock record card, but in any event the information is available in the catalogues. It is advantageous if glassware can be purchased in case lots and a considerable sum of money may be saved in this way. Another important consideration is the time and trouble saved in receiving, entering, and accounting.

The size of containers of chemicals to be purchased depends upon the stability of the chemical, its price, and the quantity used. It may be cheaper in the long run to purchase a number of small containers rather than one large one, since large quantities in single containers may deteriorate or become contaminated by frequent handling. A large container usually suggests that the material is cheap and this may deceive people into using it wastefully. Large quantities may be broken down into smaller containers, but this is troublesome and not particularly economical.

Past consumption figures should be consulted to determine how large an order should be placed in order to last for a specific period. Materials known to deteriorate rapidly in storage should be ordered only when the stock position demands it.

CUSTOMS AND EXCISE

OVERSEAS PURCHASES

Owing to the special requirements of the research work in scientific establishments it may be necessary for the superintendent of laboratories or other responsible person to arrange the importation of scientific apparatus from overseas sources. Although this may be accomplished by placing the order with a laboratory supplier or by enlisting the aid of an agent, it is cheaper and sometimes more convenient for the user to make his own arrangements.

The importation of goods from overseas is restricted under the terms of the Import Duties Act 1958, the importation procedure being detailed in various publications such as Notices Nos. 340 and 342 issued by the Board of Trade and the Commissioner of Customs and Excise. Further detailed information may also be found in the complete Tariff and in *Customs Regulations and Procedure in the United Kingdom of Great Britain and Northern Ireland*[3]. Further assistance will be given by the Customs and Excise officials, who will be found considerate and helpful with any problem which may arise.

Scientific apparatus originating and consigned from anywhere other than the Sino-Soviet Bloc is admissible under Open General Licence (which means that no licence is required), but any radioactive materials for use with the apparatus are subject to import control whatever their sources. Under the Import Duties Act all imported goods are subject to import duty. Scientific establishments may, however, apply for relief from duty by means of a Treasury direction.

It is sometimes necessary to obtain also an import certificate for the use of the overseas exporter and this too must be applied for by the importer. The import certificate amounts to an undertaking by the importer that the goods are not intended for re-export. The import certificate may also be required for use by the overseas exporter to enable him to obtain an export licence in his own country. Applications for import certificates must be made on form IC/A.

Import licence

Applications to import scientific apparatus from the Sino-Soviet Bloc must be made on form ILB/A to the Import Licensing Branch, Board of Trade, at Hillgate House, 35 Old Bailey, London E.C.4. The completed form should be accompanied by a quota certificate issued by the appropriate foreign state trading organisation. Import licences are normally valid for a period of six months.

Treasury Direction

It is necessary to make an application to the Board of Trade for duty-free importation of the goods. If the application is successful, relief from payment of duty is granted by a Treasury Direction. In order to obtain the Direction it is essential to show that the articles are required specifically for use in scientific research and are not for sale by the user. If articles similar to those being ordered and which fill the purpose for which the imported goods are to be used are available from United Kingdom sources, then the importer must give satisfactory reasons for requiring the foreign goods.

An application for a Treasury Direction must be made by the intended user of the goods on form DFA.3 which is obtainable from the Board of Trade. The applicant must also obtain four sets of detailed specifications or descriptions of the goods including photographs or illustrations. Three sets of the specification must be attached to the application form and the fourth set is retained by the applicant for reference.

If the application is successful one white copy and one green copy of the Direction are issued. The green copy is retained by the applicant and the white copy is for presentation to the Customs. The Direction states that the payment of duty on the goods is not required.

It is possible to import goods without applying for a Treasury Direction. In this case the amount of duty on the goods has to be paid. If relief from duty is desired it is advisable to obtain the Treasury Direction before the order for the goods is placed. This avoids any complications for the importer and enables him to send the exporter the reference numbers of the documents held.

Customs procedure with goods under a Treasury Direction

Goods imports other than by post have to be formally 'entered' to Customs on the forms appropriate to the goods and this is usually done by clearing agents. If the Treasury Direction is available, the white copy and supporting documents should be attached to the Customs entry, and notice given on the entry, *before* the goods are released from Customs control, that exemption is claimed under the Treasury Direction.

If the Treasury Direction is not received before the goods arrive, the Customs entry should be endorsed 'Treasury Direction has been/will be applied for.' Should the goods be released from Customs control without this being done the Direction, when subsequently received and submitted to the Customs, may not be recognised as valid by them. In such cases it is possible only in exceptional circumstances to obtain a refund of the duty paid. On the other hand, if the Customs entry has been correctly endorsed, there is no difficulty in reclaiming the duty paid. When the Treasury Direction is received the duty is reclaimed simply by sending the receipt for payment, the white copy of the Direction, and all documents attached thereto to the Customs office at the port of importation with a letter claiming repayment. The number and date of the entry for the goods should also be quoted.

In the case of goods imported by post, it is important that overseas suppliers endorse the packages and accompanying declarations with the words 'Treasury Direction applied for.' This ensures that a Notice of Arrival (form C.160 or form C.170) is sent to the addressee by the Customs, enabling him to claim relief from duty and/or purchase tax before the goods are released from Customs control. The addressee should return the completed form to the address given by the postal depot claiming exemption under Treasury Direction if the latter is available and can be attached. If the Treasury Direction has not yet been received, the form should be endorsed 'Treasury Direction has been/will be applied for.' Any duty etc. payable will have to be paid before the package can be delivered. Refund can be claimed from the Customs when the Treasury Direction becomes available for submission; the Customs reference number of the form C.160 or form C.170 should be quoted.

RECEIPT OF GOODS

SIGNING FOR GOODS RECEIVED

When goods are received they should be signed for as 'unexamined'. This is a safeguard, since it is not always possible to examine the goods at the time of delivery. When goods are visibly damaged they should be signed for accordingly and the damage should be reported to the carrier and to the supplier. The time limit for reporting damage to carriers is usually three days. If deliveries are made by the suppliers' own transport, damaged packages should be signed for accordingly. The nature of the damage should be reported by telephone to the firm concerned and confirmed in writing.

CHECKING OF GOODS RECEIVED

To receive them efficiently the storeman must be familiar with the incoming goods. This point is emphasised because the complexity of modern scientific apparatus is such that considerable technical knowledge is required to check the components.

All packages must be carefully opened using the right tools. A blunt knife and a bent screwdriver are not the right tools and efficient nail extractors and other aids are available. Most containers are returnable, and if handled carefully when being opened much trouble is saved when they are closed for return.

The contents of the packages are laid out on the receiving counter and grouped according to type for ease and accuracy of counting. They should be checked against the delivery note enclosed in the package. Delivery notes, however, do not always contain full information and it is advisable to use the order form as an additional check. The items found are ticked off on the delivery note, which is then filed for clerical treatment as described later. Any items found to be missing are notified to the supplier by telephone and later the discrepancies are confirmed in writing. The package is then closed, and if returnable is labelled for return to the sender.

The unpacking of goods should be done carefully as an element of danger is involved. Sawdust or other packing should not be delved into too vigorously for fear of broken glass or leaking bottles. The lid of the box should not be closed until the contents have been checked. This avoids reopening the box, in the event that small items, hidden in the packing, have been overlooked.

RETURNED APPARATUS

When apparatus is old and redundant it is usually suggested that it be sent back to the store. This is, of course, the correct procedure, but if every antique was religiously kept by the storeman his store would certainly need to increase in size. The storeman must use his discretion and the majority of these items should be cannibalised and any parts of value saved. The store must not be allowed to develop into a junk heap.

DISPATCH OF GOODS

RETURN OF EMPTY CONTAINERS

Empties returned by carrier

Empty containers are usually charged to ensure that they are returned to suppliers, hence the careful storage of these in dry conditions is important. Sufficient space should be allocated so that they may be stored and time and trouble is saved in dispatching them in quantity. Before they are put into storage the containers should be labelled ready for dispatch by the carrier.

The charges for empty containers and goods usually appear on the same invoice. If the container is returnable these amounts may be refunded by means of a credit note issued by the firm. A strict record of returned empties should therefore be kept in the following way. The particulars of the empty container are recorded including its number, the amount charged, and the date returned to the supplier. At the time of collection a receipt is obtained from the carman, and on the same day a card (or a copy of the consignment note) should be sent to the firm giving full details of the container, date, and method of return. In due course an acknowledgement of receipt should be received from the firm and this is filed. When the credit note for the empties is received this is noted against the appropriate dispatch particulars which have been entered in the records.

Empties returned by firm's own transport

Containers for collection by the firm's own transport should be grouped together under the name of the supplier concerned; bottles should be further sorted according to the price charged on them. These should then be put into boxes and a complete description of the contents affixed to the box.

Bottles and all other intermediate containers are charged for with the exception of sodium hydroxide containers, plastics bottles, tins, and sealed tubes. Some suppliers make no charge for bottles and other small containers but request that larger empty glass containers be returned to their drivers. Boxes are not charged for if returned within a specified time.

When sufficient containers have been gathered together to merit collection the usual procedure is to telephone the dispatch department of the firm concerned and to advise them of the quantity of 'empties' awaiting their attention.

If extra boxes are required for packing the intermediate containers which are ready for return, firms will supply these at no charge, but in some cases will only send them on the day of collection.

When 'empties' are taken by the firm's vanman a receipt should be obtained which should be filed until the credit note for empties is received.

GOODS DISPATCHED BY ROAD

Examples of the conditions affecting the dispatch of goods by road are given in the conditions of carriage of British Road

Services[4]. These indicate the need for a joint effort on the part of the sender and carrier for the safe delivery of a consignment of goods.

Responsibility of the sender

All packages must be clearly addressed in compliance with the carrier's requirements. The goods must be securely packed and it is good policy to enclose further information inside the package in case the label on the package is defaced or lost. The carriers are not liable for losses of packages from an unpacked consignment.

The consignment must be accompanied by a consignment note which should give full details, including the name and address of the sender and consignee, destination and postal district, nature and weight of the article including packing, and the number of packages. If goods are 'to be called for', at a depot for instance, the consignment note must be so marked.

No responsibility is accepted by the carrier for badly packed goods or those which deteriorate in transit, or for goods damaged or lost through some omission on the part of the sender.

Responsibility of carrier

Provided that the consignor has complied with the conditions, the carrier is liable for loss, misdelivery, or damage to the consignment. By signing a document, which must be prepared and presented by the consignor, the carrier will acknowledge receipt of the goods consigned. This does not bind the carrier in any way with regard to the correctness, condition, quality, or weight of the goods accepted.

The carrier is also responsible for advising the consignee of goods which await his collection at a depot or station, and should the consignee be 'not known' or refuse delivery, the carrier is responsible for returning the goods to the sender.

Goods dispatched to a destination and 'to be called for' must be of a type for which the carrier provides accommodation at the place of destination, and the consignment must be claimed and removed within a reasonable space of time or the carrier may sell the goods to cover his charges. The same conditions apply should the name and address of the consignee be 'not known' and the sender fail to give instructions for disposal of the goods.

Claims

The carrier is not liable for damage, deviation, misdelivery, or delay unless notified in writing within three days. Claims must be made in writing within seven days after termination of transit. In cases of non-delivery, the carrier must be advised within fourteen days and the claim must be made in writing within twenty-eight days after the consignment was handed to the carrier by the sender.

Dangerous goods

The road carriers accept dangerous goods subject to special conditions. These are briefly as follows: previous arrangements must be made with the carrier for the conveyance of the goods and a written declaration is required stating their nature. They must be packed safely and in accordance with any statutory regulations for road transport in force at that date. The packing regulations are the same as those for rail transport which are given in the Dangerous Goods'Classification issued by the British Transport Commission. The sender must also indemnify the carriers for any damage to persons through whose hands the goods must pass.

GOODS DISPATCHED BY RAIL

Goods sent by rail must conform with the various regulations of the British Railways Board and with the Statutory Rules and Orders for the terms and conditions of railway carriage as settled by the Railway Rates Tribunal.

Statutory Rules and Orders

The Statutory Rules and Orders apply to a wide range of merchandise but may be interpreted as affecting goods dispatched from laboratories in the following way. All consignments should be accompanied by a consignment note on which must appear the name and address of the sender and the consignee, the number of packages, and their weight. It should also be indicated who will pay the charges—the sender or the consignee—and if the goods are 'to wait order' at a particular station this should also be stated. The railway acknowledges the receipt at the time of handing over of goods, but this is not evidence as to the correctness of the goods at the time of giving the receipt. The form for the acknowledgement of the receipt of goods should be provided by the sender.

After receiving the goods, the railway company is liable for loss, misdelivery, or damage to goods during transit if this is due to no fault of the sender or his agent.

If the value of any article exceeds £25 this should be declared. An increased charge is made in respect of such articles.

Notification of loss or damage to goods

The railway is not liable for damage to goods sent by rail unless they are advised within three days and a claim is submitted within seven days after the goods are delivered. If the goods are not delivered a complaint should be made within twenty-eight days and a claim made in writing within forty-two days from the time when the goods were handed over to the railway company. (This is a temporary rule made in 1957. The permanent rule is fourteen days and twenty-eight days.)

The arrival of the goods consigned to a station should be notified by the railway to the consignee or, if the address of the consignee is 'not known', or the consignee refuses to accept delivery, they should notify the sender. When a consignee refuses to pay charges these must be paid by the sender. Goods sent 'to be called for', unless collected within a reasonable time, may be sold by the railway to pay their charges.

Live animals (other than wild animals)

These may be sent at company's or at owner's risk rates. They must be accompanied by any necessary orders or regulations of any government department or other authority.

If animals are sent at company's risk the number and description should be given on the consignment note. The railway company place their own valuation on the various kinds of animals and these values are published. If a sender considers his animals to be of a higher value then such value should be declared and an increased charge should be paid.

If animals are sent at owner's risk this should be stated on the consignment note.

Goods not properly packed

If goods are sent improperly packed the railway company will not accept liability for them unless any damage caused is due to the wilful misconduct of a railway employee.

Perishable merchandise sent by passenger train

Perishable merchandise 'to be called for' must be removed within twelve hours of its arrival at a station. The railway company must notify the consignee of the arrival of the goods. If perishable merchandise is sent at owner's risk this should be stated on the consignment note.

Animals sent by passenger train

The general conditions for animals sent by merchandise trains apply. If the livestock is to be charged for by weight, the description and weight of the animals, which includes the weight of crates and boxes, should be stated.

Regulations for the addressing of merchandise carried by merchandise trains

Packages should bear the full name and address of the consignee or should have a distinguishing mark with a label (letter card type). On the outside of the label should appear the name of the station or the destination, and on the inside, the name and address of the consignee.

If a package is sent 'wait order' a distinguishing mark, together with the name of the station or destination, and the full name and address of sender should appear on the package. Some items, such as metal bars, are difficult to label and these should have durable tallies fastened to them by wire, or alternatively the material itself may be stencilled. Old labels should be removed from the package.

Regulations for addressing merchandise carried by passenger train

The regulations for labelling packages for passenger trains are generally the same as those for merchandise trains. If tag or tie-on labels are used, according to the regulations these must be metal, wood, linen, parchment, or strong manilla paper. All labels, including adhesive ones, should be clearly marked.

Packing regulations for goods (other than dangerous goods) carried by the British Railway Board services

The packing regulations state that all receptacles and packing material should afford efficient and suitable protection to goods.

Receptacles should be constructed to (a) British Standards Institution BS 1133 (Packaging Code) or BS 1262 (Tins), (b) Fibreboard Packing Case Manufacturers Home Trade Packaging Standards*.

In addition: (a) Receptacles which contain contents liable to set up internal pressure should be effectively vented and bear an indication THIS SIDE UP. (b) Fibreboard drums containing liquids must have a waterproof lining. (c) Lever-type lids of cylindrical tins of half-gallon capacity or above must be secured by either spot welding, not less than two clips, an overall capsule or other equally effective means of securing.

Special requirements must also be complied with in respect of certain goods listed in the packing regulations.

Dangerous goods

Dangerous goods sent by freight train, passenger train or similar service are subject to the conditions of acceptance of the British Railways Board. They are classified according to the nature of the hazard in the list of dangerous goods and conditions of acceptance issued by the British Railways Board. This publication also gives the various methods of packing suitable for dangerous material†. The regulations for the conveyance of dangerous goods by passenger train are more stringent than freight train regulations in respect of standards of packing and quantity limitations.

---

* Obtainable from Fibreboard Packing Case Manufacturers Association, 14 Chantrey House, Eccleston Street, London SW1.
† The list has not yet been metricated but this will be done when it is next revised.

General regulations for packing and labelling dangerous goods

Packing

The regulations assume that any package contains one substance only, but if more than one substance is sent in the same package there must be no possibility of them interacting dangerously. The receptacles must withstand handling risks, and where the goods are thought by the railway to be very fragile these may have to be conveyed at owner's risk.

Receptacles actually containing a substance and having no vent must be unaffected by the contents, able to withstand any pressure set up, and must not leak. Vented receptacles should allow no surging or splashing. A free space must be left for expansion in liquid containers.

Labelling

Special labels must be affixed to packages containing dangerous materials in accordance with the rules of the section in which the material is classified. The labels are distinctively coloured and marked and have printed instructions on them indicating how the goods should be loaded and transported in relation to other goods. All packages must be labelled in accordance with the instructions given on the back of the consignment note concerning the various types of dangerous goods. If a particular substance has a hazard extra to those normally appertaining to its class, additional labelling appropriate to the hazard is required.

Consignment notes

For dangerous goods special consignment notes must be used and printed on the backs of these are the general conditions for the goods they cover.

The consignee is notified of the arrival of dangerous goods at the station of destination and is asked to remove them within a specified time.

GOODS DISPATCHED BY POST

All goods sent by post must conform with the Post Office regulations[5]. Although the packing regulations for inland and overseas post are similar, slight differences, due to the regulations of the country of destination, do exist. At the risk of repetition, therefore, both inland and overseas packing requirements for certain articles are given. The general rule for all packages is that they must be packed in such a way as not to cause damage to other packets with which they may come into contact, nor should they be dangerous to officers of the Post Office. In many countries special regulations exist whereby the use of hay or straw for packing is prohibited.

Certain items are prohibited in the post, such as explosives, corrosive and noxious materials, dangerous drugs, certain radioactive materials and substances with flashpoints less than 32°C. A leaflet describing recommended packing material for goods sent by post is obtainable at Head Post Offices. Fragile items must be packed in strong containers and surrounded with soft material. The package should bear the words FRAGILE WITH CARE.

Packing methods for special substances

Butter, cream, and semiliquids

These must be packed in such a way that the package does not taint other packets. A tin, for example, should be wrapped in greaseproof or corrugated paper and securely tied with string which should cross the lid in two directions. If sent by letter post the substance should be closed in greaseproof paper or similar material in addition to an outer covering of wood or metal.

Celluloid

Seasoned celluloid may be sent by post if packed in cardboard boxes, corrugated cardboard cartons, wooden boxes, or completely enclosed in corrugated cardboard. Raw celluloid and liquid celluloid are not permissible.

Films, cinematographic, and photographic

(a) Inflammable film. Inflammable film should be packed in a tin case and enclosed in a wooden box with dovetailed sides. The bottom of the box should be screwed to the sides. Alternatively, strong vulcanised fibre containers should be used. The inner container must be surrounded with soft packing material and held firmly in position. Outside the package a white label must be attached with FILMS—INFLAMMABLE printed in plain black letters upon it.

(b) Nonflammable or slow-burning film. The special packing conditions as for inflammable films do not apply but the package should bear a white label with the words SAFETY FILMS or FILMS—NONFLAMMABLE in plain black letters upon it.

Glass

Each item of glass should be wrapped separately and sent in rigid boxes of metal, wood, or stout fibreboard with soft packing to prevent movement.

Liquids

Liquids must be sent in sealed tins or bottles. Tins which contain one pint or more should be enclosed in fibreboard or wooden boxes or in wicker cases. Bottles should be separately wrapped and packed in soft material in rigid boxes of fibreboard, wood, or metal, or, if in small packets to be sent by parcel post, should be wrapped in strong corrugated cardboard. If liquids are sent by sample or letter post, they should be in a sealed container around which some absorbent substance must be placed which will contain all the liquid in event of breakage. The whole should be packed in a box which opens at one end and has a tightly fitting lid.

Live bees, leeches, silkworms, and parasites

These should be safely enclosed in boxes.

Maps and drawings

Maps, drawings, and similar items should be enclosed in strong cardboard tubes with a rigid support for the roll.

Nuts, bolts, and small machine parts

Items such as nuts and bolts must be wrapped in hessian or similar material with soft packing to prevent movement and tied with string.

Paint, varnish, and kindred substances

Provided varnishes and paints have a flashpoint of 66 °C or above, they may be sent by letter, sample, or parcel post. The packing conditions are as for liquids. Those with flashpoints between 32 °C and 66 °C may also be sent by letter, sample, or parcel post provided not more than one quart is enclosed in a single postal packet. These substances must be in a container hermetically sealed by the Farwig or similar approved method. Lever-top tins must have the lid fastened to the body by solder. An air space of not less than $7\frac{1}{2}\%$ of the container's total cubic content must be left in each tin. The tin must then be packed in a stout metal or wooden box and a space left between the tin and the sides of the outer container. The space should contain sufficient absorbent material to prevent movement of the inner container and absorb all the liquid contents in the event of breakage.

Small packets not exceeding 8 oz gross weight may be sent by parcel, letter, or sample post under packing conditions as for liquids, but the inner receptacle must be a tin with a rolled-on cap, sealed by the Farwig or similar approved method, or a tin with its lid secured by soldering, or a hermetically sealed bottle.

Articles of thin section moulded from plastics material: wireless apparatus, etc.

Plastics materials, such as Bakelite, must be packed in soft packing material and sent in rigid boxes made of wood, fibreboard, or strong corrugated cardboard. Heavy attached component parts, in the case of electrical or wireless goods, need additional support. Valves should be taken out from wireless apparatus and packed separately in the same way as plastics materials.

Powders and fine grain

These must be in a securely closed inner covering. A strong outer covering of metal, wood, fibreboard, or cardboard is also necessary. For quantities above 3 lb weight cardboard boxes are not suitable.

Sharp instruments

All sharp instruments must have their edges and points covered.

Samples

A sample must be a specimen of goods for sale and not sent in execution of an order. Living creatures and pathological specimens cannot be sent by sample post.

Inland post

Samples may be sent by parcel or letter post but the maximum weight must not exceed 8 oz and the limits of size are 2 ft long, 18 in wide, and 18 in deep. For articles sent in the form of a roll, the limits of size are 3 ft 3 in for the length and twice the diameter combined and 2 ft 8 in for the greater dimension.

Packing and address. Samples must be able to be easily examined and the package must be clearly marked FREE SAMPLE: NOT FOR SALE. When samples are being returned they may be sent at sample rate if the name and address of the original sender is put on the outside of the packet and the wrapper bears the words RETURNED SAMPLE.

Overseas post

For Commonwealth countries the maximum weight is 5 lb and for other countries 1 lb.

Weight and size. For Commonwealth countries the limits of size are 2 ft long, 1 ft wide, and 1 ft deep. For other countries limits of size are 3 ft in length, width, and depth combined. The greatest dimension must not exceed 2 ft and the minimum size for all destinations is 4 in long and $2\frac{3}{4}$ in wide. If samples are packed in a roll form, the length and twice the diameter combined must not exceed 3 ft 3 in and the greatest dimension must not exceed 2 ft 8 in.

Definition. Sample post is generally restricted to trade samples but certain other items may be sent. These include dried and preserved animals or plants, geological specimens, tubes of serum and vaccine, urgently required medicine, and pathological objects which are rendered innocuous by the mode of preparation and packing.

Packing and address. The sample itself should be indelibly marked FREE SAMPLE: NOT FOR SALE or defaced in some other way so as to make the article unsaleable in the ordinary way of trading. The packet must be easy to examine and the top left-hand corner of the address side should be marked FREE SAMPLE.

Special packing regulations for certain articles

Regulations for special packing include:

Dry colouring powders

These must be packed in a strong tin box which should be placed inside a wooden box with sawdust between the covers.

Fatty substances which do not easily liquefy

These should be enclosed in a box or bag of cloth or parchment or similar material and the whole placed in a second box made of wood, metal, or stout thick leather.

Glass or fragile material

These must be securely packed in a box made of metal, wood, or strong corrugated cardboard.

Liquids: oils or semiliquids

These substances must be packed in a hermetically sealed container which should in turn be placed in a box constructed of metal, strong wood, or corrugated cardboard. The outer box should contain sufficient sawdust, or spongy material other than cotton, to absorb the liquid should the container break.

Live bees, leeches, parasites

These must be in a box so constructed that all danger is avoided and the contents can be ascertained.

Paints, varnishes, enamels, and kindred substances

These should be packed as stated in the packing methods for special substances, but the amount in each packet should not exceed 1 quart and the total weight should not exceed the sample post weight limit which is applicable to the country to which the package is being sent.

Small consignments less than 8 oz gross weight may be packed as 'liquids' but a hermetically sealed bottle must be used in place of a metal container or tin.

Letter post

Articles for medical examination or analysis

Perishable biological specimens for medical examination or analysis may only be sent by letter post between officially recognised laboratories. They must be enclosed in a receptacle which is hermetically sealed or otherwise securely closed. The receptacle itself must be placed in a strong wooden, metal, or leather case so as not to be able to move about. It must have a sufficient quantity of some absorbent material packed about it to prevent any leakage and must be conspicuously marked FRAGILE WITH CARE and bear the words PATHOLOGICAL SPECIMEN.

Parcels

Inland

Weight and size. The maximum weight allowed is 22 lb for inland parcels. The limits of size are maximum length 3 ft 6 in, length and girth combined 6 ft.

Addressing. The address should be written on the parcel itself and not merely written on a label since this might become detached. It is also desirable that the name of the sender appear inside or on the cover of the packet.

Overseas

The maximum weight is generally 22 lb but to some countries it is lower. Overseas parcels may need a customs declaration form on which must be stated the nature and value of the contents.

STORES MANAGEMENT

DOCUMENTATION FOR STORES

Various documents are encountered in connection with the ordering, receipt, and payment for goods.

ORDER FORM

The material is ordered on an official order form and this is sent to the supplier. In addition to the list of goods required it usually indicates to whom, and at what address, the goods should be invoiced. A carbon copy of the order is kept for reference in the order book.

ACKNOWLEDGEMENT

In due course the supplier acknowledges the order form. The acknowledgement states that the order has been received and may tell the customer when the goods will be dispatched. The acknowledgment form should be filed and the duplicate order form marked to record the fact that the order has been received by the supplier.

ADVICE OF DISPATCH

When the supplier dispatches the goods he may send the customer an advice of dispatch. This is not always done and does not usually apply when goods are dispatched by the firm's own transport.

DELIVERY NOTE

A delivery note which accompanies the goods assists the storeman when he checks the goods received. The delivery note bears the customer's order number so that he can readily refer to his order in his own order book. Delivery notes should be filed until the invoice for the goods is received.

INVOICE

The supplier submits to the purchaser an invoice which is a copy of the delivery note but which also indicates the amount of money to be paid for the goods. It bears the customer's order number, which allows reference to be made to the customer's duplicate of his original order.

The invoice is checked against the delivery note and, if correct, the invoice is certified and sent to the accounts section for payment. The number, date of the invoice, the sum charged, and the fact that the invoice has been certified for payment, may be recorded in the order book. If departmental accounts are kept, further entries will be made in the appropriate ledgers.

The accounts section sends the invoice with a cheque for payment and the receipted invoice is returned to them for their records.

STATEMENT

From time to time a firm may submit a statement indicating the outstanding invoices which remain to be paid. The accounts section checks the invoices held against the statement to see if they agree.

CREDIT NOTE

A firm may inadvertently overcharge for goods delivered, or in some cases goods may be returned by the customer as unsuitable. In such cases a credit note is issued to the customer which bears the order number to which it refers. Credit notes may also be issued in respect of empty containers returned by the customer. These usually refer to the receipt note issued by the firm's carman when the empties were collected.

DEBIT NOTE

If a customer is undercharged for goods the matter is rectified by a debit note issued by the suppliers. This indicates the amount payable which is the difference between the amount shown on their invoice and the actual value of the goods delivered.

RECORDING STOCK

RECORDS

The keeping of accurate records in the main store is essential, but this does not necessarily mean an involved system is necessary. Paper work should be kept to a minimum. For stock purposes an alphabetical card index system will suffice. The details on the stock card should be complete, however, and those shown on the card

Maximum Stock	Nomenclature						Minimum Stock	
	Location :							
Date	In Stock	Date	Received	Price	Supplied by	Total	Consumed	Initials

Figure 3.3. A typical stores stock card for consumable items

(*Figure 3.3*) could be further enlarged to embrace details of outstanding orders and other particulars, but generally speaking, the order book, if well kept, will furnish these details.

PRESERVATION AND STORAGE OF MATERIALS

The simple methods of preservation and storage of apparatus, as given in the following notes, may apply to the dispensing store as well as to the main store.

CHEMICALS

Mention has been made of the necessity for the segregation of chemicals which may be dangerous when stored together. Other chemicals tend to deteriorate with age or because of their hygroscopic qualities and for other reasons. A constant check at regular intervals on the condition of stock is, therefore, necessary. The old stock should be used up before the new, and a methodical turnover should be effected. The easiest way to do this is to stamp the date on each bottle as it is received and to place the new stock at the back of the shelves.

In teaching establishments it is also advisable to label dangerous chemicals, such as some chlorates, in a distinctive way when they are received so that they do not find their way into the laboratories without control. Periodically bottles should be inspected for drooping or faded labels and these should be replaced.

Classes of chemicals

The chemicals may be grouped into two main categories: inorganic and organic. If space permits further subdivision into grades of purity such as A.R., M.A.R., and O.A.S. grades is desirable as this permits greater speed, efficiency, and accuracy when issues are made.

In school laboratories the quantity of A.R. grade chemicals is usually limited and provision for separate storage may be somewhat easier. If the various grades are kept together in a common store the practice of separating A.R. grades from others of lesser purity will not only prevent accidental contamination but will also ensure that costly chemicals are not used for exercises where chemicals of lower purity are suitable. Within their particular classes the bottles must be grouped in such a way that they may be quickly located for issue and it is easier to do this for inorganic chemicals than for organic ones.

Inorganic chemicals

These should be arranged under the name of the metal. The shelves should be labelled accordingly. Double compounds such as ammonium nickel sulphate or ferrous ammonium sulphate tend to create some confusion but this can be overcome if the normal method of labelling, as adopted by the usual supplier of chemicals to the store, is followed, and the chemicals are placed on the shelves accordingly. In any case such matters are put right by the adoption of a procedure suitable to the storeman and by the cross-reference and location marks on his index cards.

The prefixes to the names of chemicals, such as di-, tri-, ortho-, and meta-, are ignored for storage purposes and such chemicals are stored in the usual way under the name of the metal. Tri-ammonium ortho-phosphate, for example, would be stored with the ammonium compounds. Ferrous and ferric, cuprous and cupric salts are stored under iron and copper respectively.

Organic chemicals

Organic substances present much more difficulty. The storage of these chemicals in classes such as alcohols, ethers, and acids may be convenient for the selection of substances for certain class exercises, but the system is not convenient for storage purposes. Organic chemicals should be kept in alphabetical order. In employing this system the prefixes such as o-, m-, and p- are disregarded. Other prefixes, however, such as di- and tri- should be taken into account for purposes of alphabetical location. This is also generally done in suppliers' catalogues so that if the normal system of the chemical supplier is adopted it simplifies the stores arrangement. This is also helpful when chemicals are being reordered and the suppliers' catalogue may be easily related to the items of stock held.

For organic chemicals the shelves should be lettered A to Z and the containers dispositioned accordingly. The stock cards are marked with the letter which corresponds to that on the shelf under which the substance is located.

Substances with dual titles such as Guaiacol (o-methoxy-phenol) should also be located under the title as normally used by the supplier. Two cards are kept in the index which are cross-referenced and marked with the letter of location.

A further complication may arise with substances such as sodium benzoate. These may be kept with the inorganic or organic group, whichever suits the storeman best, provided they are all treated in the same way.

Dangerous poisons should be kept in a strong unglazed cupboard under lock. The key should be held by the storeman and the poisons issued on signature only. The complete signature of the recipient, who should be a responsible person, must be recorded in a poisons book kept soley for this purpose.

APPARATUS

The numerous small items kept in the store necessitate an orderly system. One system employed is to give each stack an identifying letter and each row of bins within the stack a number. If necessary, each bin may also be given a letter so that B.1.B would indicate that the goods were in stack B, top row, second bin. Other similar systems may be used for locating stock in glass racks, and in this case, the pigeon holes in the racks would be numbered or lettered appropriately. The stock card in the alphabetical index is given a similar marking to enable the article to be found.

The various methods of storing all the articles of equipment is beyond the scope of this book but the following general methods of storing items of various compositions may be helpful.

Glassware

The size of the items of glass apparatus stored should be standardised as far as possible. These items need careful storage and several rules apply. They should not be placed too high or mixed with heavy apparatus or metal articles. Tall glass apparatus should be stored at the back of shelves and smaller pieces in the front. Special pieces of glassware, for example Kipps apparatus, are stored as far as possible in their original packing and paper containers, such as are supplied on Emil glassware, should not be removed. This is particularly important if the glass is stacked in bins. Case lots of glassware kept in main stores may be retained in the original boxes or cartons as delivered. Simple expedients, such as the adequate lining of drawers for the storage of glassware and devices to prevent materials toppling off shelves, do much to save on glassware expenses.

Glassware is stored according to its type and size. All flasks, for instance, should be stored in neighbouring bins but separated according to size. Flat-bottom vessels may stand upright but round-bottom vessels should be stored in a bin with a high front. Small glassware, such as clock glasses, specimen tubes, Petri dishes, and microscope slides, are best kept in shelf trays. Burettes require a long padded drawer. The glass taps in separating funnels should be prevented from sticking by means of a fragment of tissue paper placed between the tap and the barrel or by greasing. The taps should be secured by a rubber band or they may fall out and become mixed. Clock glasses, if stacked, need a piece of paper between the interfaces. All expensive glassware, such as special pieces with groundglass joints, should be separately packed in soft wadding or similar material. Thermometers should be kept in their cardboard cases and stored according to type and range.

Standardised glassware, if certificated, should be given specially careful treatment and the certificate should be attached to each piece. The certificates should be issued with the apparatus and the name of the person to whom it is issued should be recorded.

When glassware is returned to the dispensing store it should not be accepted unless clean and dry.

Groundglass joints

Conical joints (cones and sockets) should be stored according to size. Such joints are nowadays internationally interchangeable. *Tables 3.4* and *3.5* give the sizes of Quickfit joints, which agree

Table 3.4 CONICAL JOINTS (CONES AND SOCKETS)*

Size designation	Nominal diameter of wide end (mm)	Nominal diameter of narrow end (mm)	Nominal length of engagement (mm)
B5 and B5/13	5.0	3.7	13
B7 and B7/16	7.5	5.9	16
B10 and B10/19	10.0	8.1	19
B12 and B12/21	12.5	10.4	21
B14 and B14/23	14.5	12.2	23
B19 and B19/26	18.8	16.2	26
B24 and B24/29	24.0	21.1	29
B29 and B29/32	29.2	26.0	32
B34 and B34/35	34.5	31.0	35
B40 and B40/38	40.0	36.2	38
B45 and B45/40	45.0	41.0	40
B50 and B50/42	50.0	45.8	42
B55 and B55/44	55.0	50.6	44
C10 and C10/13	10.0	8.7	13
C14 and C14/15	14.5	13.0	15
C19 and C19/17	18.8	17.1	17
C24 and C24/20	24.0	22.0	20
C55 and C55/29	55.0	52.1	29
D24 and D24/10	24.0	23.0	10
D40 and D40/13	40.0	38.7	13
D50 and D50/14	50.0	48.6	14

Table 3.5 SPHERICAL JOINTS (BALLS AND CUPS)*

Size designation	Nominal diameter (mm)	Minimum diameter of wide end (mm)	Maximum diameter of narrow end (mm)
S13	12.700	12.5	7.0
S19	19.050	18.7	12.5
S29	28.575	28.0	19.0
S35	34.925	34.3	27.5
S41	41.275	40.5	30.0
S51	50.800	50.0	36.0

* Courtesy Quickfit & Quartz Ltd. Stone, Staffs.

with the essential recommendations of the International Standards Organisation and of the various national standardising authorities.

Glass tubing and rod

Horizontal storage for glass tubing or glass rod is undoubtedly the best. The tubing must be well supported along its length to prevent sagging, which is often noticeable in glass tubing which has been stored vertically. Large-diameter tubing should be plugged at the ends to keep out dust. Soda glass and hard glass should be kept as far apart as possible.

Should soda and Pyrex glass tubing become accidentally mixed it is extremely difficult to separate them by visual examination. They may be identified, however, by applying a little aqueous solution of phenolphthalein to a piece of porous plate. The plate quickly dries out, and if the end of the glass tubing is wetted and drawn across it soda glass produces a red line. If the tubing is Pyrex the indicator shows no change.

Another means of identification is to immerse part of the unknown tubing in a glass-walled vessel containing trichlorethylene. When looked at through the wall of the vessel, Pyrex tubing will be almost invisible whereas soda glass tubing is easily seen.

Glass tubing should be stored by 'weight' (i.e. light, medium, and heavy wall) and each weight of tubing should be kept separately. Within these groups the tubing is stored according to its outside diameter. It is convenient if a table of sizes, including wall thicknesses and allowed tolerances, is hung close to the glass tubing storage racks. The tubing is normally delivered in 1.5-m lengths and keeps much cleaner if stored in its original packing.

Rubber

Rubber tubing

Many recommendations have been made for the storage of rubber such as the inflation of black rubber tubing and corking of the ends, dusting with talc, and so on. This particular commodity, however, leaves the stores at such a rate that the preserving treatment given it by the manufacturer is usually sufficient to keep it for the limited periods involved.

When space is available, it is advisable to hang rubber tubing in coils on a pegboard and to separate the coils according to type and size; alternatively, it may be stored in bins. A manufacturer's sample chart should be kept for reordering purposes and for display to persons who find difficulty in describing the size and type of rubber tubing they require.

Rubber teats

Rubber teats and policemen should be kept in a shelf tray and dusted with talc powder occasionally.

Rubber bungs

Rubber bungs should be stored in bin trays and segregated according to numbers. Each size of bung should be given a number which corresponds to its particular measurements and in this way they are more easily referred to. The measurements and the corresponding number are entered on the appropriate index card. The bungs are further segregated into the one-hole and the two-hole varieties.

Corks

Corks are stored the same way as rubber bungs. Large shives or bungs, which may be seldom called for and which take up considerable space, can be kept in boxes. If one of the corks or bungs is affixed to the outside of the box, it is easy to see the size of those contained in the box.

At the dispensing store a board, with a sample of each cork screwed to it, should be set up. The corks are screwed on so that the narrow end is pointing away from the board. Persons requiring corks may test the neck of the apparatus they wish to fit on the corks affixed to the board, and thus identify the size they want by a number painted on the board next to the cork.

Plastics

Plastics apparatus should be stored away from heat sources. If it is brittle, for example Bakelite, it should not be stored with heavy apparatus.

Paper

The general rules for storage of paper are that it should be kept clean, dry, and away from fumes.

Filter paper

Filter paper should be stored according to its grade and size and kept in sealed boxes. If opened, the boxes should always be closed again after use. The expensive grades of paper should be carefully issued to students and to others who may not appreciate the cost of special-quality filter paper.

The papers should not be exposed to laboratory fumes, damp air, or strong sunlight. Chromatographic papers should be stored flat and handled as little as possible.

It is extremely helpful if the storeman has an intimate knowledge of the filter papers in his care. This enables him to assist users in their selection and when necessary to recommend the grade most suitable for their requirements. The small information booklets[6] issued by various manufacturers should also be made available to students.

Unfortunately, some laboratory workers are under the impression that good results are obtainable only by using expensive filter papers. This is far from the truth, of course, and the most expensive filter paper may be the worst for the work in hand and may ruin the experiment altogether.

It is therefore profitable to examine briefly the various types of filter papers available and to consider their chemical and physical capabilities. But before we make these comparisons, we should first make certain we understand what is meant by filtration.

Filtration

Generally speaking, the term filtration means the separation of a precipitate or an insoluble residue from a liquid medium. The action of a filter paper is simply that of a porous membrane which, according to its texture, can retain coarse or fine precipitates and yet allow the liquid phase to permeate through it.

Although the retention of the precipitate or insoluble residue is of prime importance, the scientist is also concerned with the speed of filtration. In many cases he must accept some sort of compromise between good retentivity and a fast speed of filtration.

At first sight it seems easy to select the correct filter paper simply by using one which has the maximum retentivity and the minimum filtration speed. Unfortunately, these two characteristics are not the only ones to be considered, and many other factors concerned with both the nature of the liquid medium and that of the insoluble matter must also be taken into account.

In qualitative analysis a recovered precipitate is normally redissolved so that papers with a low ash content are unnecessary. Such papers are known as the qualitative grades.

For quantitative analysis, however, the ash content of the paper is of great importance because the precipitate is usually ignited. In this case the greater the amount of ash left behind after ignition the greater the possibility of error in subsequent weighing operations. For gravimetric work, therefore, the ideal paper is one which when ignited leaves the minimum amount of residual ash.

Ashless papers

Papers suitable for quantitative work are known as ashless papers and in the course of their manufacture all mineral constituents present in them are removed. The removal of the minerals is effected by washing with acid and this gives rise to the term acid-washed paper. The papers may be single-acid-washed or double-acid-washed. Single-acid-washed grades are used when the lowest ash weights are not required and double-acid-washed grades whenever the lowest ash weights are necessary. Double-acid-washed papers are made by first washing with hydrochloric acid and then washing again with hydrofluoric acid to remove any silica which may be present. A good-quality double-acid-washed paper of 11 cm diameter may have an ash weight as low as 0.000 06 g.

Hardened papers

For many quantitative operations the use of the bench filter pump and a Buchner funnel is essential. The first requirement of papers used in gravimetric analysis is therefore that they be capable of withstanding the suction of the pump. Hardened papers, consisting of almost pure cellulose and toughened by an acid-washing process, are used for this purpose. In some cases hardened papers which have been double-acid-washed are used for they are even stronger than the single-acid-washed grades.

Hardened papers not only possess great wet strength but are

also resistant to acid or alkali action and can thus be used for the filtration of strong acid or alkali solutions. Furthermore, their hard, smooth surface permits a precipitate to be easily washed or scraped off and their extra wet strength also allows both the paper and precipitate to be removed together from the funnel when necessary. These papers are especially suitable for accurate gravimetric work.

Folded filter papers

For some operations pleated or fluted filter papers may be used. Their main advantage is that they increase the rate of filtration and for some operations, such as the filtration of hot organic substances for subsequent crystallisation, they are extremely useful.

Table 3.6 shows the main characteristics of various grades of filter paper of different manufacture.

Table 3.6

Green's	Whatman	Schleicher and Schull	Characteristics
			Qualitative grades
795 (or 401)	1	595	thin general-purpose papers; medium retention and filtration speed; used for general qualitative work; unsuitable for fine precipitates
797 (or 402)	2	597	thick papers with better retentive qualities than the 401, 595, or No. 1 grades, but with slower filtering speed; used for qualitative analysis
798	3	598	strong thick papers with high retentive qualities, medium filtering speed; capable of retaining fine precipitates; normally used as large circles for large volumes of liquids, e.g. oils, syrups; also used for the analysis of nonferrous alloys
704 (or 904)	4	604	softer papers of loose texture, suitable only for coarse or gelatinous precipitates, e.g. ferric or alumina hydroxides; very fast filtration speeds; not recommended for use with suction pump
702 (or 902)	5	602	very retentive, suitable for the finest precipitates, e.g. barium sulphate; slow filtering but can be used with a filter pump

(continued overleaf)

Table 3.6 *continued*

Green's	Whatman	Schleicher and Schull	Characteristics
$488\frac{1}{2}$	12	588	suitable for general qualitative work; similar texture to Nos. 401, 1, and 595; folded papers with medium-fast filtration, unsuitable for fine precipitates
$788\frac{1}{2}$	13	550	similar to $488\frac{1}{2}$, 12, and 588 (see above), but with a higher speed and more retentive
904	15 (flat)	520a	used mainly for filtration of agar-agar and sticky liquids and juices; fast filtering
$904\frac{1}{2}$	15	$520a\frac{1}{2}$	same characteristics as 904, 15 (flat), and 520a (see above), but folded
960	—	520b	thicker and stronger than $904\frac{1}{2}$ and 52; very fast filtering; also used for agar-agar filtration

Low-ash papers—Single-acid-washed

Green's	Whatman	Schleicher and Schull	Characteristics
81F	30	589/2*	used for general qualitative work; good retention but not recommended for barium sulphate; medium filtration speed, may be used for alumina or iron hydroxide and gelatinous silica; all have a low ash
82F	31	589/1*	not very retentive, suitable for coarse or gelatinous precipitates; loose texture, therefore filter rapidly; low ash, not recommended for use with suction pumps
807 (or 902)	32	589/3*	used for very fine precipitates such as lead or barium sulphate; good retention but slow filtering speed unless used with a suction pump; low ash; 803 and 807 are similar papers, 803 having the lower ash

Double-acid-washed filter papers for accurate gravimetric work—ashless

Green's	Whatman	Schleicher and Schull	Characteristics
801	40	589/2	suitable for quantitative work or as general-purpose papers for gravimetric work; medium retention and filtering speeds
802	41	589/1	open textures, therefore filter very fast; not recommended for the finer precipitates, but suitable for coarse flocculent ones such as iron and aluminium hydroxide and many of the metal sulphides
803 (or 808)	42	589/3	very dense papers, highly retentive; slow filtering but suitable for the finest precipitates such as barium sulphate; can be used with filter pumps; 808 is more closely textured than 803

Table 3.6 *continued*

Green's	Whatman	Schleicher and Schull	Characteristics
—	43	—	medium retention, medium-fast filtering; intermediate to Nos. 40 and 41
800	44	590	thin papers, extremely low ash content; suitable for accurate work; retentive, slow-medium filtering; retain fine precipitates, although 800 is not recommended for barium sulphate in cold dilute solutions
Single-acid-washed hardened papers			
975 (or 902)	50	—	close-textured, slow filtering, and highly retentive; suitable for precipitates such as barium sulphate and lead sulphate; can be used with a filter pump
81F	52	—	good general papers for quantitative work; close-textured, medium retention and filtration speed; not recommended for very fine precipitates but suitable for alumina, iron hydroxide and gelatinous silica; low ash
82F	54	—	open-textured; very rapid filtering but low retention; suitable only for coarse and gelatinous precipitates and metal sulphides; low ash
Double-acid-washed hardened filter papers—ashless			
802 (or 904)	541	589/1	fast filtering, open texture; suitable only for gelatinous or large-particle precipitates; resistant to alkalis, may be used for hydroxides of iron and aluminium
801 (or 995)	540	589/2	ash practically negligible; medium filtration rate; very suitable for quantitative work; unsuitable for finer precipitates such as barium sulphate
803 (or 902)	542	589/3	very highly retentive, suitable for the finest precipitates; slow filtration; suitable for accurate work, can be used with a filter pump
808 (or 975)	542	589/3	very close texture indeed, very slow rate of filtration; specially recommended for the finest precipitates such as barium sulphate; may be used with filter pump
800	544	590	lowest possible amount of ash; thin papers with medium filtration rate, immense wet strength; slow filtering but not suitable for the very finest precipitates

* Double-acid-washed papers.

Indicator papers

Litmus and other indicator papers should be kept in their original wrappings until required for use as these soon deteriorate in a chemical atmosphere.

Metalware

All items made of metal should be stored together. Those kept in dispensing stores, and which may have been in use in the laboratory, should be repainted occasionally. Clamps, bossheads, and the collars of Bunsen burners need oiling periodically. The individual pieces which comprise a set of water bath rings should be wired together when being stored, as should cork borer sets.

Electrical parts

Electrical equipment and components should be stored as a group, away from fumes and chemicals. Valves and other delicate items should be wrapped in cotton wool.

Expensive items

Expensive items such as platinum-ware should be kept under lock and should be stored with their formers in place. Weight boxes and other expensive items should also be under lock. Fractional weights and riders should be kept in small pill boxes.

High-vacuum greases and waxes

High-vacuum greases and waxes should be stored at an even temperature in a cool place. Because of the large variety used today for high-vacuum work, the storeman should have an intimate knowledge of them and be in a position to give advice when necessary on the properties of the materials in his care. Some of the general uses for these materials are given in *Table 3.7*.

MANAGEMENT OF PETTY CASH ACCOUNTS

To meet the cost of carriage for packages dispatched by road or rail and the charges which occasionally have to be paid on goods received, a petty cash fund should be established in the department. The fund is also convenient for the payment of other minor expenses such as the local purchase of miscellaneous items of small value. By making cash payments for items of this nature, the necessity for making out orders and the clerical work involved in the execution, receipt, and payment for such goods is avoided.

PETTY CASH BOOK

The petty cash should be kept by a responsible person who should be in sole charge of the fund. All payments should be made by him and recorded in a petty cash book. For complete security the money should be held in a cash box and kept in a locked cupboard or drawer.

PETTY CASH VOUCHERS

Documentary evidence must be obtained by the person in charge of the petty cash for all payments made by him, and persons to whom the payments are made should be required to obtain, and submit, a receipt from the establishment from which the goods were purchased. This procedure should be the general rule but may not always be practicable and the petty cashier should exercise his judgement when demands for payment are made. In addition, persons receiving money should be required to complete a petty cash voucher which is retained by the petty cashier as a receipt for the amount paid out.

The petty cash voucher should bear the complete details of the purchase to enable the auditor to see that the payment was made for purposes connected with the running of the department. They should be dated, consecutively numbered, and signed by the payee. The amount of the cash paid out should be clearly indicated and the numbers of the vouchers should be entered against the applicable items in the petty cash book. All current vouchers should be kept in the book and if an envelope is stuck inside the front cover of the book it will be found useful for this purpose. The vouchers are retained until the petty cash book is submitted to the chief cashier for checking purposes, when further petty cash grants

Table 3.7 GREASES, SEALING COMPOUNDS, AND WAXES FOR HIGH-VACUUM WORK

	Use
Shell Chemical Co.	
Apiezon L	greasing temporary ground-glass joints and taps on high-vacuum apparatus; m.p. 47 °C; max. working temperature 30 °C
Apiezon M	sealing ground joints; similar to L but has somewhat higher vapour pressure and is less expensive; m.p. 44 °C; max. working temperature 30 °C
Apiezon N	rubber grease for the sealing and lubrication of glass conical taps; m.p. 43 °C; max. working temperature 30 °C
Apiezon T	where operating temperatures necessitate a high-melting-point grease; mobile, and can be applied at low temperatures; m.p. 125 °C; max. working temperature 110 °C
Q (sealing compound)	sealing unground joints; easily moulded into position, yet remains firm up to 30 °C; useful for temporarily blanking off; m.p. 45 °C; max. working temperature 30 °C
W (wax)	W, W100, and W40 grades of wax may all be used for permanent high-vacuum joints; W grade is a hard wax; has the highest softening point and is useful on parts where a temperature rise may occur; m.p. 85 °C; max. working temperature 80 °C
W100	softer than W and can withstand vibration; m.p. 55 °C; max. working temperature 50 °C
W40	softest, flows easily when warm; can be made to flow easily in and around joints but should not be used on joints where the temperature is likely to exceed 30 °C; m.p. 45 °C; safe max. working temperature 30 °C
Edwards high-vacuum	
vacuum grease (soft)	for lubricating stopcocks; flows freely at 40 °C
vacuum grease (hard)	suitable for flat or conical joints of glass or metal; flows freely at 50 °C
top-seal sealing compound	suitable for metal stopcocks
Picein (wax)	black wax which adheres to glass and metal and is therefore suitable for glass-to-metal joints; unsuitable for ground-glass joints; joints can be separated by heating; dropping point 90 °C; max. working temperature 50 °C
W.E. Wax 6	softening point about 80°C; all-purpose cement for industrial applications
silicone high-vacuum grease (Midland silicones)	used for ground joints and taps; suitable for pressures less than 1.33×10^{-8} pascal

Table 3.7 GREASES, SEALING COMPOUNDS, AND WAXES FOR HIGH-VACUUM WORK—*con.*

	Use
Leybold	
P	greasing temporary ground-glass joints and taps on high-vacuum apparatus; dropping point 65 °C; max. working temperature 25 °C
R	used for sealing ground joints and taps; more viscous than P; constituents of high vapour pressure have been removed to an even greater extent than P; dropping point 65 °C; max. working temperature 30 °C
Lithelen	undergoes little change in viscosity with temperature variation; can be used from below 0 °C up to 150 °C; dropping point 210 °C; max. working temperature 150 °C
joint grease DD	for rotary transmission valves
Ramsey grease (soft)	for lubricating stopcocks; dropping point 56 °C; max. working temperature 25 °C
Ramsey grease (viscous)	lubrication of ground joints
Pizein (wax)	black wax which adheres to glass and metals and is therefore suitable for glass-to-metal joints; unsuitable for ground-glass joints; joints can be separated by heating; dropping point 90 °C; max. working temperature 50 °C
V (wax)	seals unground joints only; does not require heat for working; Pizein is best for high-vacuum work
white sealing wax	adheres well to glass, wood, paper, and metal; similar to Pizein and suitable for glass-to-metal joints; unsuitable for ground-glass joints; will not resist organic solvents or water; dropping point 108 °C; max. working temperature 50 °C
Helmitin vacuum sealing compound	polyester used to make vacuum-tight joints by cementing together metal parts; must be mixed with a hardener before use; max. working temperature 100 °C
silicone high-vacuum grease	suitable for ground joints and taps; very stable even with temperature fluctuations but does not maintain its lubricating properties as well as Lithelen at high and low temperatures; max. working temperature 150 °C; polymerises above 200 °C

1970		£	p	1970		£	p
9th Oct	To cash	10	00	11th Oct	Ink		55
				16th "	Fares		63
				20th "	Postage	1	00
				22nd "	Cartage	1	00
				23rd "	Sellotape		50
				28th "	Pliers		50
				1st Nov	Washing Soda		55
				3rd "	Postage		16½
				8th "	Ice		29½
				15th "	Glue		57½
				16th "	Sellotape		50
16th Nov	Balance c/d	10	00	16th "	Balance c/d	6	26½

1970		£	p	1970		£	p
16th Nov	Balance b/d	10	00	16th Nov	Balance b/d	6	26½
				20th "	Glue		57½
				21st "	Railage	1	50
				23rd "	Fares	1	41
					By balance c/d		25
		10	00			10	00
24th Nov	To balance b/d		50				
	To cash	10	00				

(a)

Figure 3.4. Shows the pages of a memorandum book in which a simple petty cash account has been kept. In this instance the fund has been replenished when the original amount of petty cash has been expended. Example (a) shows the original amount of cash is underspent by 50p. The balance of the cash in hand on 24 November is therefore £10·50. Example (b) (opposite) shows the original amount of cash is overspent by 25p. The balance of cash in hand on 24 November is therefore £9·75.

are made by him. The chief cashier should initial the vouchers which should then be filed for a reasonable length of time by the petty cashier.

PETTY CASH SYSTEMS

Two systems for petty cash accounting are used. The first involves the replenishment of the petty cash fund when the previous grant has been expended. An example of this system is given in *Figures 3.4*

1970		£	p	1970		£	p
	Balance b/d	10	00	16th Nov	Balance b/d	6	26½
				20th Nov	Glue	1	7½
				21st Nov	Railage	1	50
				23rd Nov	Fares	1	41
						10	25
					By balance		25
	To balance		25				
		10	25				
24th Nov	To cash	10	00				

(b)

(*a*) and (*b*). In smaller establishments where petty cash payments are few, the account is often kept in a rough memorandum book and not in a dissected petty cash book. In the first illustration this simple system of book-keeping has therefore been shown. The consequent irregularity with which the petty cash fund is replenished, and the necessity for special checks by the main cashier to ascertain the balance of cash in hand, make this system generally unsuitable.

It is desirable that departmental expenditure be brought to the notice of the chief cashier at regular intervals, and, similarly, this information may be required for departmental purposes. This is particularly the case where petty cash expenditures are heavy in

Dr	Receipts					Payments			Cr.
Date Rec'd 1970	Amount received	Voucher No	Day	Details	Total	Postage & telegrams	Fares	Cartage	Sundries
	£.p				£.p	£.p	£.p	£.p	£.p
1st Jan	20.00	15	3	Zeta College. Stamps	3.00	3.00			
		16	3	Fee. Research students	.20				.20
		17	7	Fares. Plant material	.50		.50		
		18	9	Ink. Office use.	.20				.20
		19	12	Telegram. H. Smith	.90	.90			
		20	13	Railage. British Rail. Southern Region.	3.75			3.75	
		21	18	Washing soda. 1st year experiment.	.50				.50
		22	20	Screwdriver. Workshop use.	.60				.60
		23	20	Fares. Visit to scientific suppliers.	1.00		1.00		
		24	21	Railage. British Rail. Southern Region.	1.25			1.25	
		25	23	Sellotape. Departmental office.	.95				.95
		26	25	Rubber bands. Burette.	.30				.30
		27	27	Railage. British Rail.	2.00			2.00	
					15.15	3.90	1.50	7.00	2.75
				Balance c/d.	4.85				
	20.00				20.00				
1st Feb	4.85								
1st Feb	15.15								

Figure 3.5. Imprest system. A page from a dissected type of petty cash book is shown in the example

relation to the departmental budget as a whole. For these and other reasons it is generally desirable that a dissected petty cash book be maintained and that the imprest system be used.

Imprest system

When the imprest system is used a sum of money, sufficient to meet the needs of the department for a certain period of time, is held by the petty cashier. The accounting period may vary in accordance with the internal procedure of various establishments, but is normally one month. At the end of the period the petty cash book, the receipt vouchers, and the balance of cash in hand are submitted for scrutiny by the chief cashier. The cash balance is checked to see that it agrees with that shown in the petty cash book and the analysis columns and extension totals are verified. The total of the payments made, when added to the balance of cash in hand, should add up to the amount of cash originally received by the petty cashier. When the account has been checked a further cash payment is made to the petty cashier to reinstate the imprest amount to the original allowance. This system is illustrated in *Figure 3.5*. Special petty cash audits may be made at any time. The petty cash book entries and the petty cash vouchers should therefore be completed at the time when payments are made so that the account is ready for inspection at all times.

REFERENCES

1 McEwen, J. E., 'A Card Index System for Laboratory Chemicals', *Lab. Pract.*, **4**, 428 (1955)
2 *British Instruments Directory and Buyers' Guide*, United Science Press Ltd in association with the Scientific Manufacturers' Association of Great Britain, London (1968)
3 *Customs Regulations and Procedure in the United Kingdom of Great Britain and Northern Ireland*, obtainable from H.M.S.O., London
4 British Transport Commission, *Conditions of Carriage, British Road Services*, London (1955)
5 *Post Office Guide*, London
6 Barcham Green, *Signposts to Better Filtration*, pamphlet issued by J. Barcham Green Ltd, Maidstone (1969)

4

Preparation and storage of reagents

PREPARATION ROOM

LOCATION

In departments with a number of laboratories, it is advisable to prepare all the reagents in a central room under the supervision of a senior laboratory technician. There are several advantages in this, chiefly that uniformity in the strength of the solutions used throughout the department is ensured.

Since all preparative work demands a reasonable degree of personal concentration, entrance to the preparation room should be restricted to those persons whose presence is absolutely necessary.

SERVICES

All the main services including gas, water, and electricity should be available. At least one large sink should be provided. A good supply of distilled water in the room is very necessary.

FURNITURE

The furniture for the preparation room should be of such a nature that ample cupboards, drawers, and shelf space are provided. Some of the shelving, because of the weight involved, should be well supported and should be designed to accommodate stocks of solutions stored in Winchesters or aspirators. The shelf provided for the storage of common bulk solutions in aspirators should be wide and situated at low level at the back of a bench. Directly below the front edge of the shelf, a Vulcathene channel should be let into the bench to catch any drips. The channel should be covered with a plastics grille to allow full use of the bench top and should be drained by a sink waste. A fume cupboard, writing desk, and bookshelf are also required.

APPARATUS

In addition to the normal apparatus, other larger items are required in the preparation room. This is particularly true of the volumetric ware, flasks of five-litre capacity being especially useful items.

To store the prepared bulk solutions, a number of containers for the dilute acids and other solutions in constant demand are required. Polythene aspirators are the most suitable for this purpose and may be purchased with a tubular and a fitted stopcock. These stopcocks do not tend to stick as easily as glass ones owing to the action of alkaline solutions, and they are certainly preferable to the rubber tube and pinch cock method for emptying aspirators. Winchester bottles may be used for storing smaller amounts of stock solutions.

Most solutions keep for long periods if correctly prepared and stored. This is true even for standardised solutions provided pure chemicals and distilled water are used. All solutions must be well corked.

An interesting table showing the keeping properties of various volumetric solutions giving their stability over long periods of storage time has been prepared by Durham[1].

It is advantageous if all containers for solutions are graduated to show the volume of liquid which they contain. This may be done by sticking a gummed paper strip vertically on the container and marking it at the various levels with marking ink. The paper is afterwards waxed over. Similar durable methods may also be used. Using the same method the strength of the solution should also be marked on the container since this knowledge often saves a considerable amount of time when solutions are prepared. In some cases the actual method of preparation of the solution should also be written on a label affixed to the container. If this method of marking the strengths of solutions is extended to the laboratory reagent bottles, the information will be of great assistance to persons working in the laboratories.

Reeves[2] recommends the use of an ungraduated measuring

cylinder which has been marked with a file. The marks correspond to the volumes of various solutions which when added to a Winchester of distilled water provide bench reagent of the strength normally used in the laboratories. Alongside the marks are written the names of the solutions to which they refer.

The apparatus in the preparation room should also include balances. A 'rough balance', or scales of sufficient capacity for weighing large quantities of materials, and a balance of the analytical type for more accurate work are essential. A large stirrer for mixing bulk solutions is another useful item of equipment.

STANDARD SOLUTIONS

In teaching laboratories standard solutions are normally required in large quantities. The solid is dissolved in distilled water, introduced into a five-litre volumetric flask and the volume made up to the mark. Larger quantities are made by pouring further fillings of the five-litre flask into an aspirator containing the stock solution. The flask is allowed to drain each time.

A convenient method of ensuring a thorough final mixing is to lay the aspirator on its side on the bench top. One hand is placed on the top closure while the other grips the tubulure and prevents damage to the stopcock. The aspirator is then rocked backwards and forwards.

The method of preparation of various common bulk standard solutions for examination purposes has been described in full detail by Mitchell[3].

CLEANLINESS

The cleanliness and orderliness of the preparation room will be reflected in the quality of the reagents, which may become contaminated if dirty conditions prevail. The purity of the chemicals used should never be in doubt; substances which may have been subjected to contamination or which may have absorbed moisture should not be used.

DISTRIBUTION OF REAGENTS

For laboratory stock purposes most solutions will be issued in Winchester quart quantities ($2\frac{1}{2}$ litres). It is, therefore, convenient if the laboratory stock bottles are duplicated and fresh solutions are obtained from the preparation room in exchange for an empty container. The stock bottle should be labelled in accordance with the laboratory to which it belongs. This system not only ensures that solutions are readily available, but also allows the dispenser ample time to prepare further solutions.

RECORD BOOK

A record book, in which directions for the preparation of reagents are noted, should be kept. Such a book will prove to be of great assistance to the person in charge of preparation, and more particularly to those who may succeed him. Whenever changes are made in the method of preparation, owing to new laboratory procedures or for other reasons, the fact should be noted. Notes concerning the particular requirements of persons in charge of sections may also be included. For preparation rooms in which several persons may be employed a card filing system may be used instead of a record book. The face of the cards should be protected by a plastics material and a strict procedure followed indicating by whom a particular card has been removed.

STRENGTH OF SOLUTIONS

The strength of the bench solutions depends upon the general policy of the department concerned. A number of systems are used for the preparation of dilute acids and alkalis and for a number of other common reagents. The one most commonly used now is the molar system, which is replacing the twice normal (N) and the five times normal (5N) systems which were formerly widely used.

In many cases no agreement is to be found for the preparation of a reagent. The strength, method of preparation, and, in some cases, even the constituents of the reagent differ according to the source of information consulted. It appears that much confusion would be avoided if standard methods for the preparation of these reagents were adopted.

Although it is not possible within the limits of this book to list the innumerable reagents in use today, an effort has been made to give the method of preparation of the more common ones. Others may be found by consulting the appropriate literature mentioned in the bibliography.

For convenience, the reagents are grouped under two headings, namely chemical reagents and biological reagents. It is obvious, however, that no strict line of demarcation can be drawn.

Many reagents require no special method of preparation and are simply prepared by dissolving an appropriate amount of the solid in water. For these solutions, the weight of the substance contained in a litre of a 'molar' solution is given and from this figure other strengths may be calculated.

Because some chemicals are expensive, or may be sparingly soluble, special strengths of such solutions are generally used. These are therefore included in the list of reagents requiring special preparation. Distilled water should be used for all aqueous solutions unless otherwise stated. The term alcohol should be interpreted as meaning 93.5% alcohol by volume (64 O.P.) unless a particular strength is given.

CHEMICAL REAGENTS

Reagent	Weight (g) contained in 1000 ml normal solution	Weight (one mole) contained in 1000 ml molar solution
Ammonium acetate	77.08	77.08
Ammonium chloride	53.49	53.49
Ammonium dichromate	42.01	252.07
Ammonium nitrate	80.05	80.05
Ammonium persulphate	114.10	228.20
Ammonium thiocyanate	76.12	76.12
Barium chloride $BaCl_2.2H_2O$	122.14	244.28
Calcium chloride $CaCl_2.6H_2O$	109.54	219.08
Calcium nitrate $Ca(NO_3)_2.6H_2O$	59.04	236.15
Lead nitrate	165.60	331.20
Magnesium nitrate $Mg(NO_3)_2.6H_2O$	128.21	256.41
Manganous chloride	98.96	197.91
Oxalic acid $(COOH)_2.2H_2O$	63.04	126.07
Potassium chloride	74.56	74.56
Potassium cyanide	65.11	65.11
Potassium dichromate	49.03	294.19
Potassium hydroxide	56.11	56.11
Potassium iodide	166.01	166.01
Potassium permanganate	31.61	158.04
Potassium thiocyanate	97.18	97.18
Silver nitrate	169.88	169.88

Reagent	Weight (g) contained in 1000 ml normal solution	Weight (one mole) contained in 1000 ml molar solution
Sodium acetate	136.08	136.08
Sodium carbonate Na_2CO_3	53.00	105.99
Sodium carbonate $Na_2CO_3.10H_2O$	143.07	286.14
Sodium chloride	58.44	58.44
Sodium hydroxide	40.00	40.00
Sodium nitrate	84.99	84.99
Sodium sulphide $Na_2S.9H_2O$	120.09	240.18
Sodium thiosulphate	248.18	248.18
Zinc nitrate $Zn(NO_3)_2.6H_2O$	148.74	297.47

Acetic acid (2M = 2N)
Dilute 116 ml glacial acetic acid to 1000 ml with water (glacial acetic acid contains 96.6% CH_3COOH).

Alcoholic potassium hydroxide
10 g potassium hydroxide pellets is refluxed for 30 min with 100 ml alcohol. Filter through glass wool.

Alizarin S
Dissolve 1 g in 1000 ml water.

Aluminon (ammonium salt of aurine tricarboxylic acid)
Dissolve 1 g in 1000 ml water.

Ammonium carbonate (2M = approx. 4N)
Dissolve 160 g in a mixture of 140 ml of 0.880 ammonia solution and 860 ml water.

Ammonium hydroxide dilute (2M = 2N)
Dilute 135 ml ammonia solution (0.880) with water to make 1000 ml.

Ammonium mercuric-thiocyanate
Dissolve 8 g mercuric chloride and 9 g ammonium thiocyanate in 100 ml water.

Ammonium molybdate
Dissolve 40 g molybdenum trioxide in a mixture of 70 ml ammonia solution (0.880) and 140 ml water. Add slowly with constant stirring to a mixture of 250 ml concentrated nitric acid and 500 ml water. Dilute with water to 1000 ml. Allow to stand 48 hours. Decant the clear solution.

Ammonium oxalate (0.5N)
Dissolve 35.5 g of the crystalline salt in 1000 ml water.

Ammonium sulphate (saturated solution)
Dissolve 750 g in 1000 ml water.

Aqua regia
Mix 3 parts concentrated hydrochloric acid with 1 part concentrated nitric acid.

Barium nitrate (0.25M = 0.5N)
Dissolve 65 g in 1000 ml water.

Barfeod's reagent
Dissolve 13 g neutral crystalline copper acetate in 200 ml 1% acetic acid.

Baryta water
Dissolve 32 g barium hydroxide in 1000 ml water. Siphon off the clear liquid. Store in a bottle fitted with a soda lime guard tube.

Benedict's solution
Dissolve 86.5 g sodium citrate crystals and 50 g anhydrous sodium carbonate in about 350 ml water. Filter if necessary. To the solution add 8.65 g copper sulphate crystals in 50 ml water with constant stirring. Dilute to 500 ml and filter if necessary.

Benzidine reagent
Dissolve 0.05 g benzidine or benzidine hydrochloride in 10 ml glacial acetic acid. Dilute to 100 ml with water. Filter.

Bromine in carbon tetrachloride
Mix 5 g liquid bromine with 100 ml carbon tetrachloride.

Bromine water
Add bromine a few drops at a time to water and shake after each addition. Continue until no more will dissolve.

Brucine
Dissolve 0.2 g in 100 ml concentrated sulphuric acid.

Calcium sulphate (0.016M = 0.03N)
A saturated solution is prepared. Shake 3 g $CaSO_4.2H_2O$ with 1000 ml water. Filter or decant the solution after several hours.

Chlorine water
Pass chlorine into water until the water is saturated with the gas. Store the solution in a dark-coloured bottle. The chlorine gas may be prepared by dropping hydrochloric acid on potassium permanganate crystals.

Chloroplatinic acid
Dissolve 2.653 g $H_2PtCl_6.6H_2O$ in 10 ml water.

Cobalt nitrate (0.25M = 0.5N)
Dissolve 73 g $Co(NO_3)_2.6H_2O$ in 1000 ml water.

Cobalticyanide papers (Rinmann Green test)
Soak filter paper in a solution containing 4 g potassium cobalticyanide and 1 g potassium chlorate in 100 ml water. Dry at 100 °C or at room temperature.

Copper ammonium chloride
Dissolve 110 g NH_4Cl and 170 g $CuCl_2.2H_2O$ in hot water. Make up to 100 ml.

Copper sulphate $CuSO_4.5H_2O$ (0.5M = 1N)
Dissolve 125 g $CuSO_4.5H_2O$ in water to which 5 ml concentrated sulphuric acid has been added. Dilute to 1000 ml.

Cupferron reagent (qualitative test reagent)
2% aqueous solution.

Cupferron reagent (Baudisch's reagent for iron analysis)
6% aqueous solution.

Cupron reagent
Dissolve 5 g benzoin oxime in 100 ml alcohol.

Denige's reagent
Dissolve 5 g mercuric oxide in 100 ml water to which 20 ml concentrated sulphuric acid has been added. Cool and filter.

Dimethylglyoxime
Dissolve 10 g in 1000 ml alcohol.

Diphenylamine (nitrate test)
Dissolve 0.5 g in 100 ml concentrated sulphuric acid diluted with 20 ml water. Care! Add the acid to the water.

2,2'-Dipyridyl reagent
Dissolve 1 g in 10 ml dilute hydrochloric acid. Dilute to 100 ml with water.

2,4-Dinitrophenylhydrazine (alcoholic)
Dissolve 2 g of the solid in 15 ml concentrated sulphuric acid with constant stirring. Add 150 ml alcohol and dilute to 500 ml with water. Allow to cool. Filter.

Diphenylcarbazide
Dissolve 1 g in 50 ml glacial acetic acid and dilute to 500 ml with absolute alcohol.

EDTA (ethylenediaminetetra-acetic acid disodium salt) (0.1M)
Dissolve 37.21 g in 1000 ml water.

Fehling's solution
Solution No. 1: Dissolve 34.66 g of pure hydrated cupric sulphate in water to which 0.5 ml concentrated sulphuric acid has been added. Dilute with water to 500 ml.

Solution No. 2: Dissolve 70 g sodium hydroxide and 175 g potassium sodium tartrate (Rochelle salt) in water and dilute to 500 ml. Store in bottle fitted with a rubber stopper.

Equal volumes of solutions No. 1 and No. 2 are mixed immediately before use.

Ferric chloride (0.5M = 1.5N)
Dissolve 135.2 g $FeCl_3.6H_2O$ in water containing 20 ml concentrated HCl. Dilute to 1000 ml with water. Filter if necessary.

Ferron reagent (7-iodo-8-hydroxyquinoline-5-sulphonic acid)
0.2% aqueous solution.

Ferrous ammonium sulphate (0.5M = 0.5N)
Dissolve 196 g $Fe(NH_4SO_4)_2.6H_2O$ in water containing 10 ml concentrated sulphuric acid. Dilute to 1000 ml.

Ferrous sulphate (0.5M = 1N)
Dissolve 139 g in 1000 ml water to which 7 ml concentrated sulphuric acid has been added.

Fröhde's reagent
Dissolve 1 g ammonium molybdate in 100 ml concentrated sulphuric acid.

Fusion mixture
Mix together 10 parts by weight of sodium carbonate and 13 parts by weight of anhydrous potassium carbonate.

Hydrochloric acid dilute (2M = 2N)
Dilute 172 ml concentrated acid s.g. 1.18 to 1000 ml with water; *or* Dilute 200 ml concentrated acid s.g. 1.16 to 1000 ml with water.

8-Hydroxyquinoline reagent (oxine)
Dissolve 5 g in 100 ml alcohol.

Indigo
Grind 4 g indigo with 20 ml concentrated sulphuric acid. Leave for 24 hours and pour into 1000 ml water. Filter.

Iodine solution (0.05M = 0.5N)
Dissolve 12.7 g iodine in solution of 20 g potassium iodide in 30 ml water. Dilute to 1000 ml with water.

Lanthanum nitrate
5% aqueous solution.

Lead acetate (0.25M = 0.5N)
Dissolve 95 g $Pb(C_2H_3O_2)_2.3H_2O$ in 1000 ml water. Add sufficient dilute acetic acid to clear the solution.

Lime water
Boil water vigorously for one hour. Cool in a flask fitted with a soda lime guard tube. Add 2 g calcium hydroxide to one litre water

and shake well. After several hours siphon or filter off the clear liquid and store in a bottle fitted with a guard tube.

Lithium hydroxide reagent (0.4M = 0.4N)
Dissolve 1.0 g solid lithium hydroxide (or 1.75 g $LiOH.H_2O$) and 5.0 g potassium nitrate in 100 ml water. Fit the storage bottle with a soda lime guard tube.

Lucas reagent
Dissolve 136 g anhydrous zinc chloride in 105 ml concentrated hydrochloric acid.

Magnesia mixture
Dissolve 55 g $MgCl_2.2H_2O$ and 135 g NH_4Cl in water. Add 350 ml 0.880 ammonia solution and dilute to 1000 ml.

Magnesium sulphate (0.25M = 0.5N)
Dissolve 62 g $MgSO_4.7H_2O$ in 1000 ml water.

Magneson 1 (*p*-nitrobenzeneazoresorcinol)
Dissolve 0.25 g in 1000 ml 1N sodium hydroxide solution. Store in a rubber-stoppered bottle.

Magnesium uranyl acetate
Dissolve 100 g uranyl acetate in 60 ml glacial acetic acid and dilute to 500 ml. Dissolve 330 g magnesium acetate in 60 ml glacial acetic acid and dilute to 200 ml. Heat the solution to boiling point until clear. Pour the magnesium solution into the uranyl solution, cool, and dilute to 100 ml. Allow to stand overnight and filter if necessary.

Marme's reagent
Dissolve 33 g potassium iodide and 16 g cadmium iodide in 50 ml water. Add saturated potassium iodide solution to make up to 100 ml.

Marquis's reagent
Add 10 ml formaldehyde to 50 ml concentrated sulphuric acid.

Mayer's reagent
Dissolve 13.58 g mercuric chloride in 600 ml water. Pour into a solution of 50 g potassium iodide dissolved in 100 ml water. Make up to 1000 ml with water.

Molisch's reagent
Dissolve 15 g α-naphthol in 100 ml alcohol or chloroform.

Mercuric chloride (0.1M = 0.2N)
Dissolve 27 g of the solid in water and dilute to 1000 ml. A saturated solution may be prepared by adding 68 g to 1000 ml water.

Mercurous nitrate (approx. 0.1M = 0.1N)
Add 25 g $HgNO_3.H_2O$ to 800 ml water and carefully add nitric acid

drop by drop with stirring until all the solid just dissolves. Make up to 1000 ml with water.

Millon's reagent
Dissolve one part by weight of mercury in two parts of concentrated nitric acid. Dilute the solution with twice its volume of water. Decant the supernatant liquid after several hours.

α-Naphthol (20%)
Dissolve 20 g α-naphthol in 100 ml alcohol.

Nessler's solution
Dissolve 50 g potassium iodide in 50 ml ammonia-free water. Add a saturated solution of mercuric chloride (22 g in 350 ml water) until an excess is indicated by the formation of a precipitate. Add 200 ml 5N sodium hydroxide and dilute to 1000 ml. Shake occasionally for several days, allow to settle, and decant the clear liquid.

Nickel uranyl acetate
Dissolve 175 g nickel uranyl acetate in 1000 ml 2N acetic acid by heating. Leave 24 hours, decant or filter.

Nitric acid dilute (2M = 2N)
Dilute 128 ml of the concentrated acid (s.g. 1.42) with water to 1000 ml.

Nitron (diphenyl-endo-anilo-dihydrotriazole)
Dissolve 10 g in 100 ml 5% acetic acid.

α-Nitroso-β-naphthol
Dissolve 1 g in 50 ml glacial acetic acid and dilute to 100 ml with water.

Nylander's reagent
Dissolve 4 g Rochelle salt and 2 g bismuth nitrate in 100 ml 8% sodium hydroxide solution.

Obermeyer's reagent
Dissolve 4 g ferric chloride in 1000 ml concentrated hydrochloric acid.

Phosphoric acid–sulphuric acid mixture
Mix together 150 ml concentrated sulphuric acid and 150 ml phosphoric acid (s.g. 1.75) and pour slowly with stirring into 600 ml water. Cool and make up to 1000 ml with water.

Picric acid (Hager's reagent)
1% aqueous solution.

Potassium antimonate
Add 11 g potassium antimonate to 500 ml boiling water. Boil until nearly all the salt is dissolved. Cool rapidly adding 18 ml of 10% potassium hydroxide solution. Allow to stand overnight and filter.

Potassium chromate (0.25M = 0.5N)
Dissolve 49 g of solid in 1000 ml water.

Potassium cyanide (0.5M = 0.5N)
Dissolve 32.5 g in 1000 ml water.

Potassium ferricyanide (0.167M = 0.5N)
Dissolve 55 g in 1000 ml water.

Potassium ferrocyanide (0.125M = 0.5N)
Dissolve 53 g $K_4Fe(CN)_6.3H_2O$ in 1000 ml water.

Potassium permanganate (for Baeyer's test)
1% aqueous solution.

Potassium thiocyanate (0.5M = 0.5N)
Dissolve 49 g in 1000 ml water.

Quinaldinic acid reagent
Neutralise 1 g quinaldinic acid with sodium hydroxide solution and dilute to 100 ml.

Quinalizarin
Dissolve 0.02 g in 100 ml alcohol.

Rhodamine B
Dissolve 0.01 g in 100 ml water.

Rubeanic acid (dithio-oxamide) reagent
0.5% in alcohol.

Schiff's reagent
Dissolve 0.2 g rosaniline in 40 ml of a freshly prepared aqueous solution of sulphur dioxide. Allow to stand in a closed vessel for some hours until the pink colour disappears. Shake the solution with 0.2 g decolorising charcoal. Filter and dilute with water to 200 ml. Keep in well-stoppered dark-coloured bottle.

Schweitzer's reagent
To a concentrated solution of copper sulphate add 0.880 ammonia solution with stirring until the precipitate first formed redissolves.

Sodium bisulphite (aldehydes and ketones)
Just cover sodium carbonate crystals with water. Pass in sulphur dioxide until the crystals dissolve and the cloudiness disappears, leaving an apple-green solution.

Sodium cobaltinitrite (reagent for potassium)
Dissolve 17 g A.R. $Na_3[Co(NO_2)_6]$ in 250 ml water, or mix a solution of 12 g $Co(NO_3)_2.6H_2O$ in 30 ml water with a solution of 20 g sodium nitrite in 30 ml water, stir vigorously. Then add 5 ml glacial acetic acid, shake well, dilute to 250 ml, and filter after a few days.

Sodium hypobromite
Chill to less than 5 °C in ice 1000 ml 20% sodium hydroxide. Add slowly 50 ml bromine with constant stirring.

Sodium nitroprusside (test for sulphur)
Prepare freshly a 1% aqueous solution.

Sodium perchlorate
Dissolve 20 g in 50 ml water. Add 50 ml alcohol.

Sodium phosphate ($Na_2HPO_4.12H_2O$), (0.167M = 0.5N)
Dissolve 60 g in 1000 ml water.

Stannous chloride
Dissolve 56 g $SnCl_2.2H_2O$ in 100 ml concentrated hydrochloric acid and dilute to 1000 ml with water. A piece of tin kept in the bottle preserves the solution.

Starch solution
Make a paste with 1 g soluble starch and 10 ml cold water. Add drop by drop with stirring to 90 ml boiling water. Continue boiling for 5 minutes. Add a few drops of chloroform. The solution can be preserved for a long time under a layer of toluene in a stoppered bottle.

Sulphanilic acid reagent
1% solution in 30% acetic acid.

Sulphuric acid dilute (M = 2N)
Dissolve 55 ml concentrated acid by pouring into 800 ml water. Cool and make up to 1000 ml with water.

Sulphurous acid
Pass sulphur dioxide slowly into water until a saturated solution is obtained.

Thiourea
Dissolve 10 g in 100 ml water.

Titan yellow
0.1% aqueous solution.

Tollen's reagent
Prepare fresh. Mix equal volumes 10% sodium hydroxide solution and 10% silver nitrate solution. Add concentrated ammonia until precipitate just redissolves.

Zinc uranyl acetate
(a) Dissolve 20 g crystalline uranyl acetate in 100 ml water and 4 ml glacial acetic acid. Warm if necessary.
(b) Dissolve 60 g crystalline zinc acetate in 2 ml glacial acetic acid and 100 ml water. Mix the solutions in equal proportions.

Zirconium nitrate solution
Dissolve 200 g zirconium nitrate in 2000 ml N nitric acid by agitation. Filter or leave for 24 hours and if clear decant from sediment.

Zirconyl chloride
Dissolve 17.7 g zirconyl chloride in 100 ml water.

INDICATORS

Alizarin
Dissolve 0.25 g in 250 ml water.

Barium diphenylamine sulphonate
Dissolve 0.3 g in water. Dilute with water to 100 ml.

Brilliant orange
Dissolve 0.1 g in 50 ml alcohol.

Bromocresol green
Dissolve 0.1 g in 2.9 ml N/20 sodium hydroxide by warming. Dilute to 250 ml with water.

Bromocresol purple
Dissolve 0.1 g in 2.7 ml N/20 sodium hydroxide by warming. Dilute to 250 ml with water; *or*
Dissolve 0.1 g in 50 ml alcohol and make up to 250 ml with water.

Bromophenol purple
Dissolve 0.1 g in 3.8 ml N/20 sodium hydroxide by warming. Dilute to 250 ml with water.

Bromophenol blue
Dissolve 0.1 g in 3.0 ml N/20 sodium hydroxide by warming. Dilute to 250 ml with water; *or*
Dissolve 0.1 g in 50 ml alcohol and make up to 250 ml with water.

Bromopyrogallol red
Dissolve 0.05 g in 50 ml alcohol and make up to 100 ml with water.

Bromothymol blue
Dissolve 0.1 g in 3.0 ml N/20 sodium hydroxide by warming. Dilute to 250 ml with water; *or*
Dissolve 0.1 g in 50 ml alcohol and make up to 250 ml with water.

Calcon
Dissolve 1.0 g in 100 ml alcohol containing 0.8 g anhydrous sodium carbonate.

Chlorophenol red
Dissolve 0.1 g in 4.7 ml N/20 sodium hydroxide by warming. Dilute to 250 ml with water; *or*
Dissolve 0.1 g in 50 ml alcohol and make up to 250 ml with water.

Chromazurol S
Dissolve 0.1 g in 100 ml water.

Chromotrope F4B
Dissolve 0.1 g in 100 ml water.

Congo red
Dissolve 1 g in 1000 ml water.

Cresol red
Dissolve 0.1 g in 5.3 ml N/20 sodium hydroxide by warming. Dilute to 250 ml with water; *or*
Dissolve 0.1 g in 100 ml alcohol and make up to 500 ml with water.

m-Cresol purple
Dissolve 0.1 g in 5.2 ml N/20 sodium hydroxide by warming. Dilute to 250 ml with water; *or*
Dissolve 0.1 g in 100 ml alcohol and make up to 500 ml with water.

Dichlorofluorescein
Dissolve 0.1 g in 100 ml 70% alcohol.

Di-iododimethyl fluorescein
Dissolve 1.0 g in 100 ml 70% alcohol.

3,3'-Dimethylnaphthidine
Dissolve by warming 0.2 g in 100 ml glacial acetic acid.

Dimethyl yellow
Dissolve 0.1 g in 225 ml alcohol and make up to 250 ml with water.

Diphenylamine
1% solution in concentrated sulphuric acid.

Diphenylcarbazide
1% solution in absolute alcohol.

Diphenylcarbazone
Dissolve 0.1 g in 100 ml alcohol.

Dithizon (diphenylthiocarbazone)
Dissolve 0.025 g in 100 ml absolute alcohol.

Eosin
Dissolve 0.1 g in 100 ml 70% alcohol or 0.1 g of the sodium salt in 100 ml water.

Eriochrome black T
Dissolve 0.2 g eriochrome black T in a mixture of 15 ml ethanolamine and 5 ml absolute alcohol. The solution remains stable for several months. *Alternatively*, a solid mixture consisting of
    eriochrome black T   1 part
    sodium chloride   99 parts
ground together to a fine powder may be used. 0.2 g–0.4 g of the solid is added to the titration mixture when required.

Eriochrome blueblack B
Dissolve 0.2 g in 100 ml methanol.

Erythrosin
Dissolve 0.1 g in 70 ml alcohol and make up to 100 ml with water.

Fluorescein
0.1 g in 100 ml 70% alcohol or dissolve 0.1 g of the sodium salt in 100 ml water.

Ferric alum
Dissolve 10 g ferric ammonium sulphate in 100 ml hot water. Cool and add concentrated nitric acid drop by drop with shaking until the brown colour just disappears.

Methyl orange
Dissolve 1.0 g in 1000 ml water.

Methyl orange–xylene cyanol
Dissolve 0.9 g in 50 ml alcohol and make up to 250 ml with water.

Methyl red
Dissolve 1 g in 600 ml alcohol. Dilute with 400 ml water.

Methyl violet
Dissolve 0.25 g in 500 ml water.

Murexide (Ammonium purpurate)
A saturated aqueous solution of murexide may be used but must be freshly prepared.
A solid mixture consisting of:
   murexide   1 part
   sodium chloride   199 parts
ground to a fine powder keeps indefinitely.
0.2 g–0.4 g of the solid mixture is added to the titration mixture when required.

PAN (1-(2-Pyridyl-azo)-2-naphthol)
Dissolve 0.5 g in 100 ml alcohol.

PAN (4-(2-Pyridyl-azo)-resorcinol monosodium salt)
Dissolve 0.1 g in 100 ml water.

Phenolphthalein
Dissolve 5 g in 500 ml alcohol; add 500 ml water with constant stirring.

Phenol red
Dissolve 0.1 g in 5.7 ml N/20 sodium hydroxide by warming. Dilute to 250 ml with water; *or*
Dissolve 0.1 g in 100 ml alcohol and make up to 500 ml with water.

Phenosafranine
0.25% aqueous solution.

N-Phenylanthranilic acid
Dissolve 1.07 g in 20 ml N Na_2CO_3 and dilute with water to 1000 ml.

Phenylthymolphthalein
Dissolve 0.25 g in 125 ml alcohol and make up to 250 ml with water.

Phthalein purple
Dissolve 0.1 g phthalein purple in 2 ml 0.880 ammonia solution and make up to 100 ml with water.

Potassium chromate
1% aqueous solution.

Pyrocatechol violet
0.1% aqueous solution.

Quinaldine red
Dissolve 0.1 g in 250 ml alcohol.

Rhodamine 6G
Dissolve 0.1 g in 100 ml water.

Screened methyl orange
Dissolve 1 g methyl orange and 1.4 g xylene cyanol FF in 500 ml 50% alcohol.

Solochrome red B
Dissolve 0.2 g in 100 ml water.

Sulphosalicylic acid
Dissolve 2.0 g in 100 ml water.

Tartrazine
0.5% aqueous solution.

Thymol blue
Dissolve 0.1 g in 4.3 ml N/20 sodium hydroxide by warming. Dilute to 250 ml with water; *or*
Dissolve 0.1 g in 50 ml alcohol and make up to 250 ml with water.

Thymolphthalein
Dissolve 0.1 g in 50 ml alcohol and make up to 250 ml with water.

Tiron
Dissolve 2.0 g in 100 ml water.

Tri-ortho-phenanthroline ferrous sulphate
Dissolve 1.485 g tri-ortho-phenanthroline in 100 ml 0.025M ferrous sulphate solution.

Tropaeolin O
Dissolve 0.1 g in 600 ml alcohol and make up to 1000 ml with water.

Variamin blue B
Dissolve 1.0 g in 100 ml water.

Xylenol orange
Dissolve 0.1 g in 50 ml alcohol and make up to 100 ml with water.

BIOLOGICAL REAGENTS

FIXATIVES

Acetic alcohol (Farmer's formula)
Glacial acetic acid	1 part
Absolute alcohol	3 parts

Acid alcohol
Hydrochloric acid	1 ml
Alcohol (70%)	99 ml

Altman's fluid
Potassium dichromate (5% aqueous)	1 vol
Osmic acid (2% aqueous)	1 vol

Bouin fixative
Picric acid-saturated solution	75 ml
Formalin (40% formaldehyde)	25 ml
Glacial acetic acid	5 ml

Bouin's fluid
Picric acid (saturated aqueous solution)	75 ml
Glacial acetic acid	5 ml
Formaldehyde (40%)	25 ml

Carnoy's fluid
Absolute alcohol	60 ml
Chloroform	30 ml
Glacial acetic acid	10 ml

Champy's fluid
Potassium dichromate (3%)	7 ml
Chromic acid (1%)	7 ml
Osmic acid (2%)	4 ml

Chromic acid
1% aqueous solution chromic acid
Glacial acetic acid
A usual strength for this solution is 1 ml acetic acid to 100 ml chromic acid, but as weaker solutions may be used the two are kept separately and mixed as required.

Chromo-acetic fixative (for marine algae)
Chromic acid	1.0 g
Glacial acetic acid	0.4 ml
Sea water	400 ml

Flemming's fluid
Chromic acid (1%)	15 ml
Osmic acid (2%)	4 ml
Glacial acetic acid	1 ml

Add the osmic acid just before use.

Formol alcohol
Formalin (40% formaldehyde)	100 ml
Alcohol (70%)	900 ml

Formol saline
Sodium chloride	0.9 g
Water	90 ml
Formalin (40% formaldehyde)	10 ml

Gilson's fluid
Water	880 ml
Alcohol (60%)	100 ml
Mercuric chloride	20 g
Glacial acetic acid	4 ml
Nitric acid (concentrated)	18 ml

Helly's fluid
Mercuric chloride	5 g
Potassium dichromate	2.5 g
Sodium sulphate	1 g
Water	100 ml

Add 5–10 ml of formalin immediately before use.

Iodine
Saturated solution of iodine in 5% aqueous potassium iodide.

Lugol's iodine
Iodine	1 g
Potassium iodide	2 g
Water	100 ml

Dissolve the potassium iodide in 20 ml water. Add the iodine and dissolve. Make up to 100 ml with water.

Muller's fixative
Potassium dichromate	2.5 g
Sodium sulphate	1 g
Water	100 ml

Osmic acid
1% aqueous solution. Store in amber bottle.

Picric acid
Picric acid	1 g
Water or 70% alcohol	100 ml

Picric alcohol (Masson)
Picric acid-saturated alcoholic solution	60 ml
Alcohol (95%)	30 ml

Schaudin's fluid
Mercuric chloride-saturated aqueous solution	100 ml
Absolute alcohol	50 ml
Glacial acetic acid	1 ml

Susa fluid (Heidenhain)
Distilled water saturated with mercuric chloride	50 ml
Trichloracetic acid	2 g
Formalin (40% formaldehyde)	20 ml
Glacial acetic acid	4 ml
Distilled water	30 ml

Zenker's fluid
Mercuric chloride	5 g
Potassium dichromate	2.5 g
Sodium sulphate	1 g
Water	100 ml

Add 5 ml glacial acetic acid to above solution just before use.

STAINS

Acetocarmine (Schneider's)
Carmine	0.4 g
Glacial acetic acid	45 ml
Water	55 ml

Heat the carmine and water to boiling; then add the acetic acid. Boil again, cool, and filter.

Acid fuchsin (Mallory)
0.5% aqueous solution.

Aniline blue stain
Aniline blue	1 g
Absolute alcohol	100 ml

Aniline chloride
Aniline hydrochloride	10 g
Water	100 ml
Hydrochloric acid	1 ml

Azo carmine (for Heidenhain Azan stain)
Azo carmine B (or G)	0.5 g
Glacial acetic acid	1 ml
Water to	100 ml

Basic fuchsin
Basic fuchsin	1 g
Alcohol (95%)	100 ml
Water	100 ml

Best's carmine stock solution
Carmine	2 g
Potassium carbonate	1 g
Potassium chloride	5 g
Distilled water	60 ml

Boil gently in a beaker with stirring for a few minutes. Cool and add 20 ml 0.880 ammonia solution. Filter. Keep in a stoppered bottle in a refrigerator.

Bismarck brown stain (Vesuvine)
Bismarck brown	2 g
Alcohol (70%)	100 ml

Borax carmine (Grenacher)
Borax	4 g
Carmine	3 g
Distilled water	100 ml
Alcohol (70%)	100 ml

Add the carmine powder to the borax dissolved in the distilled water. Simmer for 30 min. Cool. Add the alcohol. Filter.

Carbol fuchsin (Ziehl Neelsen)
Basic fuchsin	1 g
Phenol	5 g
Alcohol (95%)	10 ml
Water	100 ml

Dissolve the basic fuchsin in the alcohol and dissolve the phenol in the water. Mix the two solutions well and allow to stand overnight. Filter.

Carmalum (Mayer's)
Carminic acid	1 g
Ammonium alum	10 g
Water	200 ml

Boil the solution for an hour. Restore the volume to 200 ml. Allow to cool, filter, and add a crystal of thymol to prevent growth of moulds.

Chlorazol black
Saturated solution in 70% alcohol.

Crystal violet stain
Crystal violet	0.5 g
Water	100 ml

Cotton blue in lactophenol
Cotton blue 1 g
Lactophenol 100 ml
For use 5 ml of this solution should be diluted to 100 ml with lactophenol.

Delafield's haematoxylin
Solution I. Prepare 400 ml of a saturated aqueous solution of ammonium alum (56.5 g).
Solution II. Dissolve 4 g haematoxylin in 25 ml absolute alcohol.
Add II a few drops at a time to I and allow the mixture to stand exposed to the light and air in an unstoppered bottle for 4 days.

Ehrlich's haematoxylin
Haematoxylin 2 g
Absolute alcohol 100 ml
Glacial acetic acid 10 ml
Water 100 ml
Glycerine 100 ml
Potassium alum in excess

After dissolving the haematoxylin in the alcohol add the acid. Add the glycerine and water. Close the flask with a well-vented stopper so that the solution is open to the air. Keep in the light until the stain ripens to a dark red colour.

Eosin
1% aqueous solution. Add a few drops of chloroform as a preservative. Alternatively a 1% solution in 70% alcohol may be used.

Gentian violet
Gentian violet 1 g
Alcohol (95%) 15 ml
Water 80 ml
Aniline 3 ml

Giemsa
Giemsa stain powder (compounded) 3.8 g
Pure methyl alcohol 250 ml
Pure glycerine 250 ml

Warm on a water bath for one hour. Cool and filter. This stain is best purchased ready-made.

Heidenhain's iron haematoxylin
Solution I. Dissolve 4 g iron alum in water (freshly distilled) and dissolve by shaking.
Solution II. Add 5 ml well-ripened haematoxylin (10%) in absolute alcohol in 95 ml water.

Iodine solution (Gram's)

Iodine	1 g
Potassium iodide	2 g
Water	300 ml

Dissolve the potassium iodide in water and add the iodine and dissolve. Make up the volume to 300 ml with water.

Iron acetocarmine (Belling's)
To some Schneider's acetocarmine add a trace of ferric hydrate dissolved in 45% acetic acid. When the liquid becomes just bluish-red dilute the solution with an equal amount of acetocarmine solution. If further dilution is necessary use 45% acetic acid.

Light green in clove oil

Light green	1 g
Clove oil	75 ml
Absolute alcohol	25 ml

Leishman's stain

Leishman's stain powder	0.15 g
Pure methyl alcohol (acetone-free)	100 ml

Dissolve by heating in a flask lightly plugged with cotton wool on a water bath. Cool, filter.

Loeffler's methylene blue

Methylene blue	0.5 g
Potassium hydroxide 1% solution	1 ml
Absolute alcohol	30 ml
Water	100 ml

Heat the water to 50 °C and add, with stirring, the methylene blue. Add the other materials and filter.

Mallory's triple stain

Solution A	0.5% Acid fuchsin	
Solution B	Aniline blue (water-soluble)	0.5 g
	Orange G	2.0 g
	Oxalic acid	2.0 g
	Water	100 ml
Solution C	Phosphomolybdic acid	1%

Mayer's haemalum

Haematoxylin	1 g
Water	1000 ml
Sodium iodate	0.2 g
Ammonium alum (powdered)	50 g
Chloral hydrate	50 g
Citric acid	1 g

The haematoxylin is dissolved in the water with gentle heat. Add

the sodium iodate and the alum. When dissolved add the citric acid and the chloral hydrate. Allow to stand 7 days.
Alternatively, the following solution can be used immediately.

Water	1000 ml
Ammonium alum	50 g
Haematein	1 g
Alcohol (95%)	100 ml

Dissolve the alum in the distilled water and the haematein in 100 ml rectified spirit. Add the haematein slowly to the alum. Add a thymol crystal to prevent growth of moulds.

Methylene blue
For living organisms:

Methylene blue	1 g
Water	110 ml
Add sodium chloride	0.5 g

For dead material:
Add 30 ml 95% ethanol to 100 ml water. Dissolve 0.3 g methylene blue.

Neutral red (Jensen)

Neutral red	0.1 g
1% Acetic acid	0.2 ml
Water	100 ml

Orcein (aqueous)

Orcein	2 g
Glacial acetic acid	2 ml
Water	100 ml

Orcein (alcoholic)

Orcein	1 g
Absolute alcohol	100 ml
Hydrochloric acid	1 ml

Orcein (acetic)

Orcein	2 g
Glacial acetic acid	90 ml
Water	110 ml

Dissolve the orcein in the acetic acid by heating and when cool add the water. Shake and filter.

Safranin

Safranin	1 g
Alcohol (95%)	50 ml
Water	50 ml

A few drops of aniline should be added to the water.

Scarlet R (for fat in animal tissue)
Saturated solution in equal parts 70% alcohol and acetone.

Sudan III
Sudan III	0.5 g
Alcohol 70%	100 ml

Dissolve in a warm water bath; then filter

Van Gieson's stain
Acid fuchsin (1%)	5 ml
Picric acid, saturated aqueous solution	100 ml

Weigert's iron haematoxylin
A stock solution of 10% haematoxylin in absolute alcohol is allowed to ripen for at least five weeks.
Solution I. Add 10 ml of the stock solution to 90 ml alcohol. Filter.
Solution II.
Ferric chloride (hydrated) 30% aqueous solution	4 ml
Hydrochloric acid pure	1 ml
Water	95 ml

Filter.
Solutions I and II are mixed at the time of use. The mixture is allowed to stand a few minutes after mixing.

MOUNTING MEDIA

Berlese's fluid
Dissolve 15 g gum arabic in 20 ml water and add 10 ml glucose syrup. Add 160 g chloral hydrate and finally 2 ml glacial acetic acid.

Canada balsam (neutral)
Dissolve solid Canada balsam in xylol by putting it in an incubator set at 37 °C. When it becomes syrupy add calcium carbonate and mix well by stirring. Allow it to settle and decant off the xylol–balsam.

Farrant's medium
Dissolve 20 g gum arabic in 20 ml water. Add 10 ml glycerine and mix thoroughly. Filter through glass wool. Add a crystal of thymol to prevent growth of moulds.

Glycerine jelly
Gelatine	5 g
Glycerine	35 ml
Water	30 ml

Place the gelatine in water and warm on a water bath for 2 hours.

Add the glycerine and continue to warm for 15 minutes. Filter through paper pulp and add 0.2 g phenol as a preservative.

Gum dammar
Dissolve 100 g in chloroform. Filter through filter paper wetted with chloroform to remove impurities. Remove the chloroform by evaporation. Dissolve the purified gum in 100 ml of xylol. Thin as required with xylol.

Lactophenol
Lactic acid	100 ml
Phenol	100 g
Glycerine	100 ml
Water	100 ml

Dissolve the phenol in water, add the glycerine and lactic acid.

Meyer's albumen fixative
White of egg	50 ml
Glycerine	50 ml
Sodium salicylate	1 g

Separate the white of the egg from the yolk. Add the glycerine and stir to break up the albumen. Add the sodium salicylate dissolved in a little water. Mix the whole well. Filter through coarse filter paper.

AGAR MEDIA

Czapek-Dox agar
Sucrose	30 g
Sodium nitrate	2 g
Potassium phosphate (K_2HPO4)	1 g
Magnesium sulphate	0.50 g
Potassium chloride	0.50 g
Ferrous sulphate	0.01 g
Agar	15 g
Water	1000 ml

Sterilise by heating for 30 minutes at 100 °C on 3 successive days.

Malt agar
Malt extract	20 g
Agar	15 g
Water	1000 ml

The agar is dissolved in water and filtered through paper pulp. Add the malt extract dissolved in a little hot water. Autoclave for 20 minutes at 120 °C.

Oatmeal agar

Fine oatmeal	50 g
Agar powder	20 g
Water	1000 ml

Heat the oats on a water bath at 58 °C in a half of the water. When the oats are sufficiently digested add the remainder of the water and the agar. Filter through muslin and sterilise.

Potato agar

Grate 500 g clean peeled potatoes into 500 ml water. Allow to stand 6–8 hours. Filter off the solution and add 20 g agar dissolved in 500 ml water. Steam for 30 minutes. Filter through paper pulp and autoclave for 20 minutes at 115 °C.

SPECIAL REAGENTS

Antiformin (macerating fluid for preparation of bones)

Sodium carbonate	150 g
Calcium hypochlorite	100 g
Distilled water	1000 ml

Add the sodium carbonate to 250 ml water. Add the calcium hypochlorite to 750 ml water. When almost dissolved mix the two solutions and shake occasionally for 3 hours. Filter.

Add the solution to an equal quantity of 15% sodium hydroxide solution. The stock solution is diluted for use[4].

Cement for tops of glass museum jars

Gutta-percha	4 parts
Stockholm pitch	4 parts
Paraffin wax (54 °C)	1 part

Cement for Perspex jars

1,2-Dichloroethane	100 ml
Glacial acetic acid	10 ml
Perspex	5 g

Collodion for permeable membranes

Add to 40 g shredded pyroxalin 500 ml absolute alcohol and stir until most the pyroxalin is dissolved. Add 800 ml ether gradually and allow to stand before using.

Eau de Javelle

Cream 20 g bleaching powder with water in a mortar and make up to 100 ml with water. Add to this solution about 80 ml of a solution consisting of 25 g potassium carbonate dissolved in 100 ml water. Filter. Add dropwise more potassium carbonate solution until no more calcium carbonate is precipitated. Filter and store in a dark bottle in a cool place.

Freezing mixtures (After Chivers[5])

Ingredients	Parts by weight	°C
Ammonium chloride	5	
Potassium nitrate	5	−12
Water	16	
Ammonium chloride	5	
Potassium nitrate	5	
Sodium sulphate	8	−16
Water	16	
Ammonium nitrate	1	
Sodium carbonate	1	−22
Water	1	
Sodium sulphate	6	
Ammonium nitrate	5	−40
Nitric acid (dilute)	4	

Knop solution for algae
Potassium nitrate	1 g
Magnesium sulphate	1 g
Calcium nitrate	3 g
Potassium phosphate (K_2HPO4)	1 g

Dissolve the potassium nitrate, magnesium sulphate, and potassium phosphate in 1000 ml water. Add the calcium nitrate. A white precipitate is formed. Shake the stock solution before use.

Preservative for algae
Alcohol (50%)	90 ml
Formalin	5 ml
Glycerine	2.5 ml
Glacial acetic acid	2.5 ml
Copper chloride	1.0 g
Uranium acetate	1.5 g

Preservative for green algae
Place the material in a solution of 0.5% or 1% copper acetate in 2% formalin for 24 hours. Preserve in 5% formalin.

Sea water

Sodium chloride	23.427 g
Potassium chloride	0.729 g
Calcium chloride ($CaCl_2.6H_2O$)	2.218 g
Magnesium chloride ($MgCl_2.6H_2O$)	10.702 g
Sodium sulphate ($NaSO_4.10H_2O$)	8.967 g
Sodium bicarbonate	0.210 g
Sodium bromide ($NaBr.2H_2O$)	0.07 g
Water	1000 ml

Schulze's macerating fluid

Potassium chlorate	1 g
Nitric acid (concentrated)	50 ml

Heat the material in this mixture for a time and then wash and mount in glycerine.

Water culture solution
Normal complete solution (Sachs)

Calcium sulphate	0.25 g
Calcium phosphate	0.25 g
Magnesium sulphate	0.25 g
Sodium chloride	0.08 g
Potassium nitrate	0.70 g
Ferric chloride	0.005 g
Water	1000 ml

SPECIAL RECIPES

Ink stain remover
Ballpoint—sponge with alcohol.
Writing ink—cover stain with 2% potassium permanganate solution and leave for a few minutes. Remove permanganate with solution of oxalic acid and wash the material in water.

Marking of glazed porcelain
(a) *Temporary marking.* For gravimetric exercises involving the heating of glazed porcelain crucibles, groups of students may have to share muffle furnaces. This sometimes gives rise to considerable difficulty in identifying individual crucibles when the ignition is complete, because at high temperatures most ordinary marking media, including many so-called china markers, burn off. Since it is usually essential that the identification marks be only temporary, several methods for effective temporary marking of crucibles are given below. In each case, the marks may be easily removed after the porcelain ware has been heated to temperatures up to 850 °C.

1. A small portion of the glazed surface is removed from the crucible by means of an emery wheel. An ordinary lead pencil can then be used for marking on the unglazed portion and may later be removed with a pencil eraser.
2. Mars Omnichrome No. 2424 Staedtler brown pencil may be used (from Blythe Colour Works, Ltd, Cresswell, Stoke-on-Trent, Staffs). It writes directly and legibly on to glazed porcelain and this is the most convenient method for producing small identification marks.

(b) *Permanent marking.* To permanently mark porcelain, chromium oxide is fluxed by the addition of No. 8 flux, which is available from most colour manufacturers. Alternatively, a small quantity of finely ground lead bisilicate and finely fritted borax may be used as flux.

Silvering solution (plane mirrors)
Clean the glass to be silvered thoroughly with chromic acid cleaning mixture. Rinse in tap water and again in distilled water. Polish with a chamois leather.
Solution A. Add 0.880 ammonia solution dropwise to a 5% aqueous solution of silver nitrate until the precipitate first formed is almost entirely dissolved.
Solution B. Dissolve 5.7 g potassium sodium tartrate in 100 ml water.
Mix together 1 part Solution A, 1 part Solution B, 1 part distilled water. Place the glass to be silvered on a level surface and slowly pipette the silvering solution on to the face until it is completely covered. Alternatively, place the solution in a flat dish and suspend the glass so that the face to be silvered is just below the liquid surface. A mirror surface forms in a few hours but to obtain a dense mirror the glass should be left for 12 hours. Wash and leave to dry.

Silvering solution (lecture demonstration)
For lecture demonstration purposes the silvering of glass should be immediate and may be illustrated in the following way. A 1000 ml round-bottom flask is cleaned with chromic acid solution, washed, and dried. 200 ml of a 6% aqueous solution of silver nitrate is placed in the flask and 140 ml of 3% aqueous potassium hydroxide is added. Add 0.880 ammonia drop by drop until the precipitate just dissolves. Add 5 g of glucose shaking continuously. The solution becomes dark and the flask is immersed in a trough of hot water (approximately 70 °C). A silver mirror is formed inside the flask within a few minutes. The silvering solution is discarded and the inside of the flask washed with water.

Soap solution
100 g of dry Castile soap is dissolved in 1000 ml alcohol. The

solution is allowed to stand for several days and then diluted with alcohol until 6.4 ml produces a permanent lather with 20 ml of standard calcium solution. (Standard calcium solution: dissolve 0.2 g calcium carbonate in a little dilute hydrochloric acid. Evaporate to dryness and make up to 1000 ml.)

REFERENCES

1 Durham, B. W., 'Keeping Properties of Certain Volumetric Solutions', *J. chem. Educ.*, **28**, 387 (1951)
2 Reeves, J. E., 'Making up Bench Reagents', *Sch. Sci. Rev.*, **37**, 132 (1955)
3 Mitchell, A. D., 'Standard Solutions and Indicators for Examinations', *Sch. Sci. Rev.*, **33**, 163 (1952)
4 Edwards, J. J., and Edwards, M. J., *Medical Museum Technology*, Oxford University Press, London (1959)
5 Chivers, B., *Sch. Sci. Rev.*, **15**, 238 (1933)

5
Laboratory inspection and maintenance

PERIODIC INSPECTION OF LABORATORIES AND WORKSHOPS

The numerous services which are provided in laboratories and the expensive equipment housed in them require a regular inspection to ensure that a department is working efficiently. The major maintenance work in any busy department should be effected when the laboratories are not in use, and in educational and similar establishments a suitable time for assessing necessary repairs is immediately prior to a holiday period.

INSPECTION

At such times as the departmental budget is being prepared, or at other convenient intervals, it is usual to submit a report concerning any repairs which may be necessary for the upkeep of the building or equipment. Provision for these may then be made in the estimates. The repairs are effected by the maintenance staff or by an outside contractor, but during the course of the inspection useful information is gathered regarding other repair work with which the laboratory staff are required to deal.

The general condition of the laboratory, such as the decoration, will undoubtedly occupy much attention and a regular expenditure in this direction saves much higher costs at a later stage. Deterioration should never be allowed to continue until the paintwork is affected to such an extent that the material it covers is exposed to fumes. This is particularly important in the case of metal windows. It is a common error to make provision for the whole building to be repainted at regular intervals and yet allow no special consideration to the paintwork of the laboratories, which deteriorates at a much faster rate than that of other parts of the building.

Windows

Windows and window frames require special attention during the inspection, particularly the sash cords if fitted. Cords in a dangerous condition are a menace in any circumstances but may have serious consequences in laboratories where urgent maximum ventilation may be required to meet an emergency. Nylon or other fume-resistant cords are recommended for laboratory windows. Damaged window panes are a grave source of danger to people inside and outside the building. Where large numbers of people are involved, the opening and shutting of windows, especially those with metal frames, results in many broken panes. The reason for this is usually found in the form of faulty window fasteners and an early repair of these is worth while to prevent further damage and provide maximum security.

Ceilings

Ceilings also require attention, and if they appear to have deteriorated through dampness the cause must be urgently investigated. Dampness in balance rooms and chemical stores is a very serious matter.

Drainage system

The drainage system is a vital feature of any laboratory and the inspection of the system is of paramount importance. Channels and traps should receive particular attention, for neglect of these may cause untold damage by flooding. Drainage channels soon become silted and this is often caused by the generous use of cleaning powders in the sinks, and a control of the use of such powders may be necessary. The cleaning of chemical sumps must be done as a routine measure. Other blockages are the result of broken or missing waste grilles, and these should be religiously replaced.

Water taps and water filter pumps

Water taps should all be rewashered at the same time, say once a year, and this effects an insurance against the troublesome recurrence of leaking taps. The efficiency of all water filter pumps should be checked. Faulty ones should be dismantled and cleaned, and the jets replaced if necessary.

Gas taps

The gas taps should also be tested for leaks as they tend to loosen with constant use.

Electrical fittings

Light bulbs and discharge tubes should be replaced if faulty or deficient and metal shades repainted if corroded. Artificial lighting troubles when the laboratories are in use are very irksome, since repairs to these often involve the removal of fuses, which consequently interferes with the work of the whole laboratory. The locating of possible causes of trouble in switches, flexes, and starters is an important aspect of the inspection, and sockets and plug tops should also be investigated. Conduits should be checked for signs of corrosion and the electrical leads of the various instruments must be kept in good repair, as short circuits, due to the fraying of the leads, are a common cause of blown fuses. The pieces of equipment mainly affected are water baths, hotplates, stills, etc., in which a rapid deterioration in the condition of the electric leads is caused by heat.

Safety apparatus

Since the safety of the whole building may depend on the efficient upkeep of firefighting and other safety equipment, the conscientious checking of these should leave nothing to chance. Fire extinguishers must be regularly tested, and in some cases, even though they have not been used, should have their contents renewed. The date of the last refilling should be stencilled on the cylinder. The cylinder itself should be checked for signs of corrosion. Carbon dioxide cylinders are checked by weight. Whenever a cylinder is used the fact should be immediately reported and this must be insisted upon as a rigid point of discipline. The safety equipment inventory should be consulted to ensure that all pieces of apparatus, such as respirators, goggles, and gloves, are in good order and in their proper place in the laboratory.

Fume cupboards

Poor fume disposal may cause untold misery and inconvenience in the laboratory. The velocity of air drawn through the fully open front of the cupboard should be checked, and any deterioration in the performance of the fan or the efficiency of the fume ducting should be noted. At regular intervals the bearings of the fan motors should be either oiled or greased, depending on the type of bearings.

Lecture rooms

Lecture room facilities are extremely important and the condition of these should be fully investigated. The inspection should cover lighting, projection equipment, seating, and blackboards. Sliding blackboards may need to have their cords renewed and writing surfaces repainted with special blackboard paint.

Laboratory workshop

A rigorous inspection is necessary in the workshop to make sure that all the required safeguards are being applied. Guards on the machines should comply with the regulations laid down in the Factories Act and machine belts should be tested for wear. Scientific staff do not often, in the course of their normal duties, come into contact with modern workshop machinery and equipment, and it may be for this reason that many laboratory workshops are antiquated. This not only makes the workshop a dangerous place in which to work but also means the up-to-date facilities necessary for research work are absent. It is therefore recommended that during the course of the inspection the age and efficiency of the machinery be investigated and new equipment purchased where necessary.

Stores

Dry storage, particularly where chemical stocks are involved, is essential. Any signs of dampness should therefore be noted and repairs effected as quickly as possible. Of equal importance is the condition of the various safety devices including the efficient functioning of fans and firefighting and other safety equipment. Ladders and steps should be in perfect working order.

Shelf labels should be renewed if necessary and at the same time the stock should be generally rearranged and tidied up, especially with regard to any new stock which is expected to arrive. Any rearrangement, too, should be based on the experience of popular issues made over the past year. The rearranging should also be related to any new work which is to be started in the laboratories and which might entail the storing of new types of materials.

OVERHAUL, MAINTENANCE, AND CLEANING OF FURNITURE, EQUIPMENT, AND APPARATUS

The maintenance, cleanliness, and periodic overhaul of the equipment and apparatus in laboratories and workshops is the key to the efficiency of any department. Since it is not possible to give methods for the upkeep of all the many different items of equipment to be found in various laboratories, it is hoped that the following selected examples will give some indication of the need for care and attention which laboratory equipment requires.

EQUIPMENT

Rotary vacuum pumps

To ensure satisfactory performance, rotary vacuum pumps require checking at regular intervals. A superficial examination of the external parts should be carried out and special attention should be paid to the condition and tension of the belt. The oil level should be adjusted if necessary.

Deterioration of the oil through leaks in the system or contamination by vapours is indicated by a marked fall-off in the pump's performance. Timely renewal of the oil is therefore a wise safeguard and the regularity with which this must be done depends on the amount of work and the type of duty the pump has been called upon to perform. To prolong the life of the pump a log book attached to it is therefore an additional safeguard. Some departments, however, prefer to be on the safe side and renew the pump oil at regular short periods irrespective of the amount of use the pumps may have had.

Table 5.1 GRADES OF OIL FOR ROTARY PUMPS

Pump series number	Oil charge	Remarks
EDWARDS		
RCB2	non-lubricated (has graphite blades)	
RB1	8A	
RBF3	8A	combined rotary compressor
RBF4	8A	and vacuum pump
RB5	17	
RB10	17	
1SP20	18	
1SC450	18	
1SC900	17	
2S20	8A	
2SC20	18	
1SC1500	17	
1SC3000	17	for clean systems number 15
ES35	18	oil should be used in place
ES75	18	of 17 and number 16 oil in
ES150	18	place of 18; this gives the
ES250	18	best possible vacuum
ED35	18	
ED75	18	
ED150	18	
ED250	18	
ED500	18	
ES7500	17	
CENCO		
Portable blower and vacuum unit	Hyvac light-grade	
Air-cooled blower and vacuum pump	Hyvac light-grade	
Hyvac 300	Hyvac light-grade	
Hyvac 225	Hyvac light-grade	
Hyvac 150	Hyvac light-grade	
Hyvac 45	Hyvac light-grade	
Hyvac 28	Hyvac light-grade	two-stage
Hyvac 14	Hyvac light-grade	
Hyvac 7	Hyvac light-grade	
Hyvac 2	Hyvac light-grade	
Hyvac	Hyvac light-grade	
Hyvac 28	Hyvac light-grade	
Hyvac 14	Hyvac light-grade	single-stage
Hyvac 7	Hyvac light-grade	
Pressovac 90515		
Pressovac 90510	Hyvac heavy-grade	
Pressovac 90550		
Megavac		
Hypervac 23	Hyvac light-grade	
Hypervac 25		

Table 5.1 GRADES OF OIL FOR ROTARY PUMPS—*continued*

Pump series number	Oil charge	Remarks
Hypervac 100	Hyvac heavy-grade	
LEYBOLD		
Minni pump	N62	
VP2	N62	
S1	N62	
S2	N62	
S3	N62	
S6	N62	
S12	N62	
S25	N62	
D1	N62	
D2	N62	
D3	N62	in all pumps Protelen should be used instead of N62 when vapours are present
D6	N62	
D12	N62	
D25	N62	
S60	N62	
S100	N62	
S200	N62	
S400	N62	
S800	N62	
S25–S6	N62	
RUTA 60	N62	
RUTA 100	N62	these are tandem pumps and the oil specified is for use in the rotary pump only
RUTA 200	N62	
RUTA 400	N62	
RUTA 800	N62	
E106	N62	
E116	N62	
E126	N62	
E136	N62	
E146	N62	
23	N62	
33	N62	
43	N62	
26	N62	
36	N62	
46	N62	
57	N62	
87	N62	
107	N62	
Z05	Diffalen	
METROVAC		
GDR1		
GDR210		
GS10	Shell rotary pump oil	
GS24		
GS59		

Unfortunately, many laboratory workers disregard the fact that the ultimate performance of their pump depends on the grade of oil used in it and only too often any grade of oil near to hand is used for refilling purposes. To overcome this the correct grade of oil to be used should be clearly marked on the pump body.

Table 5.1 shows the correct grade of oil for pumps of various manufacture.

Electric furnaces

To maintain electric furnaces in good condition they require regular attention. The moving parts, such as the door gear and door switch platforms, require periodic lubrication. Springs and pivot points should also be given a light smear of graphite grease. If the body of the furnace is made of metal the exterior may require a coat of heat-resisting aluminium paint. Hearth trays, which are used to protect the chamber, must be cleaned regularly with emery paper, and all scale removed. The protection of the chamber, too, is of great importance and it should be regularly brushed clean when the furnace is cold.

When the furnace is given a more thorough overhaul, the chamber should if necessary be reglazed and suitable glazing compounds, which are obtainable from furnace manufacturers, may be applied with a brush. After glazing, the heat of the furnace must be slowly raised to a temperature recommended by the manufacturers. It is then allowed to cool, but if the glazing is not smooth the process should be repeated. Thermocouples need occasional checking and for this purpose should be removed. Similarly, the pyrometer itself may need checking. Indicator lamps should be given attention to ensure they are working and should, in any case, be occasionally renewed. The operation of the energy regulator must be checked. If the furnace chamber shows signs of serious wear or damage it is recommended that the furnace be returned to the manufacturer for repairs.

Stills

Laboratory stills must be regularly examined and periodically overhauled and cleaned. The length of time which should be allowed between each descaling depends upon how long and how frequently the still is operated, and the hardness of the water used in it. In electric stills the scale forms rapidly around the heating elements, and regular cleaning not only prevents damage to these and increases the output of distilled water but also saves electricity. Stills in regular use should be descaled at least every three months in districts where the water is hard, and at the same time the fibre washers between the head of the element and the body of the still should also be renewed.

Ovens

Although ovens require careful treatment they are often abused. Materials spilled in them should be removed, or they may corrode the floor of the oven and expose the element. The action and accuracy of the controls should be checked, and if actuated by the making and breaking of points these should be carefully cleaned and at reasonable intervals reset.

Microscopes

To maintain microscopes in perfect condition the following points should be observed. The instrument should be kept, when not in use, in its protective box. If wanted for regular use it should be kept under a plastics cover. The eyepieces should be kept in the draw tube to prevent dust falling into the tube and objectives should be stored in screw-top containers. The moving parts of the microscope require light greasing occasionally but the important part of the microscope is the lens and this requires careful and regular attention.

Microscope lenses

If a microscope lens is cleaned inefficiently more damage will be caused than if the dust had been allowed to accumulate upon it. The correct procedure is first to try to remove the dust by directing a stream of dry air on to the lens with hand bellows or other similar means and never by blowing directly on to it with the mouth. Alternatively, the lens may be gently stroked with a camel hair brush which has been warmed against a hot surface such as an electric light bulb. In this case, the brush tends to pick up the dust by electrostatic attraction. If these efforts fail, the lens must be cleaned by a wet method.

Xylol is used for wet cleaning and a lens tissue paper should be moistened with a minimum quantity of this liquid. The damp tissue is gently wiped across the lens, which is immediately polished with dry tissue paper. In the case of neglected oil-immersion objective lenses on which the oil has been allowed to dry, this method may have to be repeated several times. Nevertheless, heavier applications of xylol must at all costs be avoided, or the solvent action of the liquid may loosen the lens in its mount.

Microscope condensers and mirrors may be cleaned the same way and if lens tissue paper is unobtainable a soft cloth which has been well washed may be used.

Other lenses

Lenses should never be touched by the fingers. They should be kept in a dry, cool, dust-free atmosphere and if possible in their own protective cases. The first rule for cleaning any lens is first to remove all dust particles. Camel hair brushes joined to a rubber bulb which blows air over the lens surface are now available for this purpose. All grease should be removed from lenses by the use of very small quantities of either a proprietary brand of lens cleaning fluid or a pure organic solvent.

Plastics lenses should be cleaned with an appropriate proprietary brand of lens cleaning fluid or with soapy water applied with a lens tissue.

Coated lenses may be cleaned with acetone or xylol but ordinary tapwater, which may contain free chlorine, should not be used.

Razors

The cut-throat razor is still commonly used, especially in botanical laboratories for section cutting. Like all sharp-edged tools it requires careful handling and maintenance.

The genuine botanical razor is hollow-ground on one side and flat on the other with a bevelled edge on the hollow side. To ensure that it is kept in good condition certain rules apply. The most important of these is that the razor should be used only for the job for which it is intended, e.g. section cutting. It is most essential, too, to wipe the blade clean immediately after use and, if the razor is to be stored for any length of time, to smear the blade with Vaseline and wrap the whole in a protective covering. When in use razors should always be closed before they are put down on the bench top.

In spite of precautions some razors are inevitably found to be in bad repair and should be honed on a good-quality natural stone and then stropped. They should not be honed too frequently, however, as this wears the blade. Stropping, on the other hand, should be carried out frequently and if possible each time before the razor is used.

Flexible strops are easily damaged and may become dangerous. The safest strops are the fixed variety made by securing the ends of a leather strip over the ends of a wood block of appropriate length.

Microtome knives

The microtome knife, another sharp-edged tool, also requires careful attention if damage to prepared sections is to be avoided. It should always be cleaned after use with a piece of soft cloth and may be lightly oiled. The oil is cleaned off with xylene prior to further use.

Sharpening

Knives which are slightly blunt require only to be sharpened by stropping. If the edge is very blunt or damaged to the extent that it is ragged, the knife must be honed prior to stropping. A honing stone of good quality should be used and is lubricated with light oil to keep the edge cool and to remove metal particles during the sharpening process. The knife is fitted with its own back for the honing process and, as this becomes worn after several honings, it must be occasionally renewed. A handle is also fitted to the knife before honing. Then place the knife in position with its back resting on the stone and its edge facing away from you. Maintaining a constant angle, push the knife from the near to far end of the stone in a diagonal direction. When the far end of the stone is reached, reverse the blade and draw it in a diagonal direction towards you. Plane–concave knives should be honed only on the concave side. The burred edge which occurs on the plane side when the blade is sharp may be carefully removed by placing the plane side of the knife flat on the stone and rubbing it gently once or twice.

Balances

Since the first half of the nineteenth century rapid strides have been made in the design of balances. This advancement has accelerated in recent years to the extent that many teaching institutions, which use large numbers of balances, face the problem of maintaining many different designs. The problem is further complicated by the requirements of present-day laboratory work, which necessitates the use of balances ranging from student quality with a sensitivity of 2 mg to those used for ultra-micro work with a sensitivity of 10^{-8} g. Intermediate balances of analytical, semi-micro, and micro quality are also used and have sensitivities of the order of 0.1 mg, 0.02 mg, and 1 µg, respectively.

Note that persons other than those specially skilled should not be allowed to 'repair' balances of the higher order of sensitivity, and that generally speaking the overhauling of balances is a matter for a balance mechanic. Arrangements should be made with a reputable firm, and preferably the suppliers of the balances, for regular skilled service.

Cleaning of balances

The occasional cleaning of balances may be necessary and this may be undertaken by persons who have sufficient knowledge of the use and care of the class of balance being dealt with. Assuming the balance is, for example, a free-swinging centre-pivot beam type, the following procedure, which must be carried out in a dust-free atmosphere, is adopted.

At each side of the balance a piece of glass or a sheet of white glazed paper is laid on the bench. White linen gloves or rubber finger stalls are worn by the operator to prevent grease from the fingers being transferred to the balance parts. With the balance in the rest position, and commencing with the pan supports, the parts are removed and placed on the glass or paper on the respective sides of the balance. In this way the parts are not likely to be mixed. Great care should be exercised when the beam is being removed and it should afterwards be supported on the bench in such a way that the pointer is not likely to be damaged.

The balance case should now be thoroughly cleaned. A damp chamois leather is used for cleaning the glass, which is afterwards polished with a well-washed cloth. The interior of the case is brushed out with a camel hair brush and may be finished by seeking out dust from cracks and crevices with a fine vacuum nozzle. It is finally polished with a piece of chamois.

The pillar and other metal parts are also cleaned with chamois. The planes and knife edges need special care and are cleaned with a separate piece of soft chamois kept specially for this purpose. The balance pans should not be cleaned with anything which removes the plating, but if necessary may be wiped with a cloth moistened with alcohol and should be finally polished with chamois. The brass pans on cheap balances may, if badly tarnished, be cleaned with metal polish or Duraglit. Since some metal will be removed by this process, it involves readjusting the balance.

The front of the balance should now be closed and the cleaned parts on the bench covered with glazed paper. After a half-hour has elapsed to allow the dust particles to settle, the balance is reassembled. It is then allowed to 'settle' overnight and the following day is checked and adjusted. A plastics cover should be kept over the balance case at all times when the balance is not in use, as an added precaution against dust.

Note that too-frequent cleaning is deleterious to balances and their best protection lies in exercising the utmost vigilance to keep the balance room free from dust and fumes, and in the strict observance of the rules of weighing by all who enter it.

Accumulators

Acid accumulators are used extensively in most laboratories. Since they become 'tied' to apparatus, however, the regular care and maintenance which they require to prolong their life is apt to be overlooked. To overcome this situation each accumulator, when purchased, should be given a laboratory number and registered in a charging book. A complete record and history of the cells should be maintained. The responsibility for the regular charging, servicing, and occasional overhauling of accumulators should be specifically delegated to a responsible technician. Neglected cells rapidly deteriorate and may be completely ruined. Note that accumulators which are not in use also require a freshening charge at least every two weeks. At all times the cells should be kept topped up with distilled water so that the tops of the plates are kept covered.

Accumulator acid

Sulphuric acid is used in accumulators and must be pure. The concentrated acid is diluted for this purpose until its specific gravity is suitable for the type of accumulator in which it is to be used. The correct specific gravity of the acid usually lies between 1.19 and 1.2, and directions for charging are given with new accumulators. Accumulator acid may be made by slowly adding 222 ml of concentrated sulphuric acid to distilled water and making up to one litre. Heat is evolved and the specific gravity of the diluted acid should not be read until it has reached room temperature.

Charging accumulators

Before accumulators are charged their stoppers should be removed. New accumulators are given a special initial charge in accordance with the instructions which accompany them. Those already in service should be charged at their normal charging rate but in any case never at a rate which is less than half their normal rate of charge. Undercharging and overcharging are injurious to the cells. The cells should never be more than half discharged before they are again brought up to full charge.

State of charge. The fact that the accumulators have reached full charge is indicated by the free gassing of the cells. The voltage of the cells also shows the state of charge, and when fully charged each cell has a voltage of 2.6 V. This is not an entirely satisfactory method of testing, however, since the voltage of accumulators tends to fall off as their age increases.

The proper test is to read the specific gravity of the acid, which increases progressively when the accumulator is being charged. A syringe hydrometer is used for taking the reading. The specific gravity also varies with the temperature of the acid and a true reading can only be obtained if a table, which relates temperature and specific gravity, is consulted. It is convenient if such a table is pasted in the 'charging book'.

Overcharge. It is recommended that accumulators be given an occasional overcharge, which simply means charging at the normal rate for a longer period. When the overcharge should be given depends upon the regularity with which the accumulator receives its normal charge, but it would not be unreasonable to give the overcharge at every sixth charge. When overcharging is being carried out, the specific gravity is closely watched until it reaches a maximum which remains constant over several successive readings taken at 15-minute intervals.

Fast charging. The fast charging of batteries involving special devices and currents of up to 100 A is undoubtedly harmful. In certain emergencies it may prove to be a necessary evil and, provided the battery is healthy, no irreparable damage will result. Nevertheless, it is most important that the charge be terminated at the proper time. Repeated fast charging shortens a battery's life by 75%. Batteries which suffer most serious damage from fast charging are those which are overdischarged; yet those subjected to fast charging are usually in this condition. New batteries undergoing their initial charge should never be charged at an amperage exceeding their normal charge rate.

Cleaning accumulators

The interior of the cells, which accumulate dirt by the ingress of foreign matter and through excess sedimentation, may have to be cleaned to avoid the short circuiting of the plates. The cells should therefore be flushed with water when in a discharged condition and immediately filled with fresh acid. The accumulator is then charged.

Accumulators not in use

If accumulators are not required for use for a considerable period it may be expedient and economical to store them. The accumulator should first be fully charged, and each cell in turn should then be emptied of acid and filled with distilled water, to prevent the drying out of the plates. At this stage the terminals should be bridged with a piece of copper wire, and after 48 hours the water should be poured off. The cell is then ready for storage. When the stored accumulators are restored to service they should be filled with acid of 1.18 specific gravity. The cell should then be treated as a new one and should be given an initial charge.

Alkaline accumulators

Although the acid accumulator has for many years proved to be the most popular, the alkaline accumulator is increasingly more used, especially in schools and educational establishments. In this type of battery nickel hydroxide, nickel hydrate, and nickel flake are employed for the positive plates or tubes used in various types of batteries and iron or cadmium in powder form is used for the negative plate.

Nife accumulators. Probably the most popular alkaline accumulator is the Nife accumulator, so called from the chemical symbols appertaining to the original constituents of the plates. The modern Nife battery is a nickel–cadmium alkaline variety in which a solution of potassium hydroxide is used as the electrolyte. The electrolyte, which is contained in a rust-proof sheet-steel container, is normally supplied ready-mixed in the battery. The battery should be charged in accordance with the working instructions supplied with it. In some instances, e.g. when batteries are exported, the electrolyte may be supplied separately in solid form and is mixed with distilled water before use. The battery is then filled and charged. When charged at the normal (usually 7-hour) rate the battery is fully charged when the charge has been continued for $1\frac{1}{2}$ hours after a steady voltage has been reached. As with the lead–acid accumulator, Nife cells when in use should always be kept topped up with distilled water just sufficient to cover the tops of the plates.

The small portable batteries used for laboratory work of experimental nature have an average discharge voltage of 1.2 V and are housed in a wooden crate containing from one to five batteries. Thus voltages of 1.2, 2.4, 3.6, 4.8, and 6 V are provided.

Care of alkaline accumulators. During the charging process the electrolyte gives off hydrogen and oxygen, so that when batteries are inspected the usual charging precautions concerning naked lights should be observed.

Since the battery containers are made of steel and are live, the crates and the batteries themselves should be kept in a clean condition and metal articles or bare leads should not be left in contact with the steel containers.

Occasionally, the cells may be given a boosting charge to overcome the 'packing' tendencies of the active material in the plates.

Great care should be taken to ensure that traces of acid do not find their way into the batteries and for this reason alkaline batteries should be kept well away from acid accumulators, especially when charging. Also, if contamination is to be avoided the utensils and tools used for servicing and testing acid accumulators should not be used for testing alkaline batteries.

Advantages. Nife batteries have several advantages over lead–acid ones. Not the least of these is the fact that they may be used in conjunction with delicate equipment, since they do not give off corrosive fumes or spray. They may also be left lying idle for long periods in various states of discharge without suffering damage. Lastly, they can withstand heavy discharge currents and can be charged at widely varying rates without damage.

FURNITURE

Blackboards

Blackboards have changed in colour and character so much in recent years that they may now be better described as writing surfaces. Old-fashioned blackboards, however, are still used and the repainting of them is therefore worth describing. To remove grease the board should be first cleaned down with a rag soaked in methylated spirits. One of several proprietary brands of blackboard paint may be used to resurface it, but such paint may be made as follows. Dissolve 50 g shellac in a pint of alcohol and place a mixture of 25 g good-quality lampblack, 25 g emery powder, and 25 g ultramarine blue on a fine strainer. The shellac solution is poured on to the powders and stirred constantly until they gradually pass through the strainer.

Alternatively, the method used by Ansley[1] may also be adopted.

Laboratory benches and tables

Benches and tables constitute the main items of laboratory furniture and it is the tops of these which require the most maintenance. In many laboratories the modern finishes such as Formica, stainless steel, P.V.C., and certain enamels are used. Little can be said about the maintenance of these other than that spilled liquids should be wiped from them to prevent any harmful effect.

In most teaching laboratories, however, where the bench tops are subject to very harsh treatment, wooden tops are still extensively used. The hardwoods used for these contain natural oils, and in physical and similar laboratories need only be treated with raw linseed oil and thereafter with wax furniture polish. Alternatively, the bench may be left unpolished and given light applications of raw linseed oil at regular intervals.

In laboratories where chemicals are extensively used, bench tops require extra protection and the best method is to impregnate them with wax which should be worked well into the grain. A treatment involving the hot application of wax dissolved in xylene is described in the chapter on laboratory design, as is the method of acid-proofing bench tops. The further treatment for bench tops is the repeated application of a good-quality wax polish, which should also be used on the cupboard fronts below the bench tops. With the advent of electric polishers manual polishing of the bench tops is unnecessary and with the aid of these machines the bench top acquires a finish unsurpassed by synthetic polishes. All laboratory workers must be encouraged to use a piece of asbestos beneath heated apparatus, since the waxed surface suffers from heat reflected down on to the bench top. This effect becomes worse if chemicals are spilled on the softened wax surface. Nevertheless, a light rub with steel wool prior to the next application of polish never fails to restore the bench top to normal.

As an alternative to the xylol–wax treatment, high-melting-point paraffin wax may be grated and sprinkled on bench tops. The wax is flamed on with a Bunsen burner and may be ironed into the grain with a heated flatiron.

In spite of the pleasant finish obtained by these methods, the tops of the benches tend to darken over a long period and it is advisable every few years to renovate them. This is done by scraping off the wax and sandpapering the bench top prior to a new application of wax.

APPARATUS

In addition to the regular maintenance required to keep laboratory apparatus in good order, a complete overhaul should be carried out during holiday periods when the laboratories are not in use. This can be in conjunction with an overhaul of the laboratory dispensing stores so that common items of apparatus may be checked and repaired together. All chemicals in the store and laboratories should be inspected for signs of deterioration and any unfit for laboratory use should be discarded.

Clamps

Owing to corrosion, metal clamps often break at the threaded screw. It is perhaps not generally realised that provided the clamps are purchased from a recognised manufacturer the screwed parts are obtainable for replacement. The actual fitting of the new screw is quite a simple matter, but the necessity for these repairs can largely be avoided if clamps are periodically immersed in oil for an hour or two.

To avoid the laborious and unsatisfactory cutting of cork sheet by hand for recorking clamps, spare cork shapes may be purchased in bulk from manufacturers. These should be affixed by a glue which is not affected by heat. After applying the glue to the surfaces and fixing the corks in position, the jaws of the clamp should be tightened on to a bung so that pressure is applied until the glue hardens.

Pestles

Loose heads on pestles should be refixed and an up-to-date adhesive suitable for this purpose and having exceptionally high resistance to impact is the epoxy resin Araldite AW.106. For use the Araldite must be mixed with a hardener:

Araldite AW.106	100 parts by volume
Hardener HV.953U	100 parts by volume

until a uniform colour is obtained throughout. The parts to be bonded should first be cleaned with a grease solvent and the mixture applied with a small brush to both surfaces. Leave to harden for 12 hours at room temperature. Any resin which exudes from the joint should immediately be removed with a little acetone on a rag.

General glassware

All glass apparatus should be checked over. Damaged glassware easily finds its way into the dustbin and yet so much of this can be effectively repaired. A box for damaged glassware should be provided in the laboratory stores where the damaged material should be kept until a quantity sufficient to make repairs worthwhile has accumulated.

Glassware which is cracked, or so badly chipped or scratched that it cannot be repaired, should be thrown away. The item of glassware which has the highest casualty rate is possibly the burette. Burette breakages are largely due to careless washing under water taps, small pieces being broken from the tops of the burettes as they

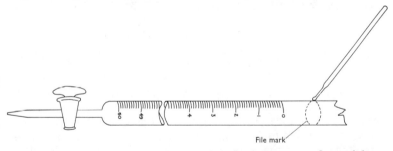

Figure 5.1. The method of squaring off the top of a broken burette prior to flame polishing

are withdrawn. To prevent this, a short piece of P.V.C. tubing should be attached to each of the swanneck outlets in the laboratory. Other damage to burettes is usually restricted to the jets. It is not necessary to have an expert knowledge of glassblowing to be able to effect simple repairs to glassware. If the top of the burette is to be repaired, a mark made right around the stem at a point just below the damaged portion is touched with a fine point of red-hot glass (*Figure 5.1*).

The damaged portion then readily falls off. The unpolished end is rotated in a blowpipe flame and gently reamed out with the tang of a file or with a reamer. Because of its length it may prove difficult for an inexperienced repairer to hold the burette steady in the flame, but a simple adjustable rest, to support the burette at one end, is a great help and can be manufactured easily from scraps of wood and a bolt with a wingnut (*Figure 5.2*).

Burettes with damaged jets are also easily dealt with. If the jet is badly damaged the whole stopcock is cut off and replaced with a new one. The cost of this is less than half the cost of a new burette.

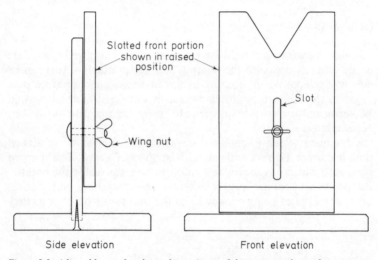

Figure 5.2. Adjustable rest for glass tubing. A rest of this type may be used as a support when unwieldy lengths of glass are worked in a blowpipe flame

Jets which are very slightly chipped may be carefully smoothed and rounded with a carborundum stone. Care should be taken not to grind back the tip too far, as this results in a faster outflow from the burette which makes it unusable. Burette jets may be heated in the flame above the damaged portion and drawn out, but this operation requires skill to get the right thickness of glass and size of jet.

Burettes which have been badly damaged provide a useful source of spare taps or keys. The graduated stem, provided a reasonable portion remains, can serve a number of useful purposes, particularly for experiments in the physics or physical chemistry laboratories. Burettes and other apparatus with interchangeable keys are available, and although their cost is somewhat higher, these items are highly recommended since the nightmare of lost or mixed keys, and the replacement of taps, are avoided.

Seized ground-glass joints

The separation of seized conical joints usually presents considerable difficulty. Most seizures are due to jamming arising from the contraction or expansion of the joint, and this often occurs when glassware is used for vacuum work or when it is subjected to considerable changes in temperature.

Seized joints may also be caused by cementation at the ground faces, since the matt ground surfaces are easily attacked, particularly by strong caustic solutions. It is therefore advisable not to allow caustic solutions to remain in contact with the joint longer than necessary. It is also good practice to separate joints while they are still warm or in any event immediately after use. Lubrication of joints before they are used is often desirable and a light greasing with Vaseline or paraffin–rubber lubricants is often sufficient for this purpose. For work involving high vacuum or high temperatures, special compounds such as Apiezon or silicone greases are necessary. The disadvantage associated with silicone greases is their tendency to spread and the consequent gross contamination of the glassware. This may render the apparatus unsuitable for future operations, especially when volumetric work is involved. If the glassware has been contaminated in this way it should be cleaned as quickly as possible after use by one of the methods described on page 225.

Polytetrafluoroethylene (P.T.F.E.) is sometimes used as a joint sleeve or as a coating on the joint faces instead of lubricants and this facilitates the easy separation of the cone and socket after use.

Releasing seized joints. Generally speaking, the larger the joint, i.e. the larger the surface area of the ground portion, the more difficult it becomes to separate the cone and socket if they seize.

With practice it is possible to determine visually whether or not cementation at the faces has occurred. If the joint appears clear it is unlikely that cementation has taken place but a cloudy or crystalline appearance on the other hand usually indicates that the ground faces of the joint have been affected.

If the cone and socket are cemented together the joint should be soaked for some time in a releasing solution (see page 223) before physical attempts to free it are made. Unfortunately, physical separation is difficult because of the relatively frail nature of the joint and occasionally because of the impossibility of emptying the contents of an attached vessel.

218 LABORATORY INSPECTION AND MAINTENANCE

To separate the components attempts should first be made to 'rock' the joint free. If this fails, try hot water run on to the socket combined with 'rocking'. Should these methods prove unsuccessful the socket should be gently heated in the brush flame of a blowpipe. On withdrawing the joint from the flame the socket flange should be gently tapped with a piece of wood. Alternatively the joint itself may be tapped on a wooden surface. Should this fail to free the joint, it is wise to resort to more soaking before making further attempts to release it.

Filter flasks

Filter flasks are usually damaged at the extreme end of the side arm. In this condition they can be quite dangerous when an attempt

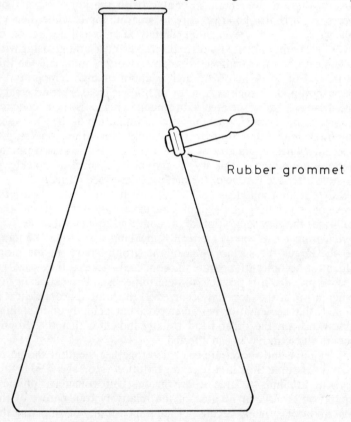

Figure 5.3. Filter flask with replaceable side arm

is made to put pressure tubing on the end of the damaged side arm. Immediately they are seen to be damaged, they should be withdrawn from service. If the end of the arm is merely chipped, the flask can be made serviceable again by smoothing the chipped portion in a blowpipe flame. If the side arm is completely broken off, it is a difficult task to replace it effectively unless one has had considerable glassblowing experience, since it is made of heavy-weight hard glass. An oxy–coal gas flame is necessary to work the glass. Note that flasks are available with loose and replaceable side arms, which are set in the side of the flask through a rubber grommet (*Figure 5.3*).

Measuring cylinders

Measuring cylinders also have a high casualty rate and it is economical to purchase rubber ring guards and to fit these on to the top of the cylinder immediately they are put into use. However, if the tops should suffer damage there are several methods of cutting them off. For smaller sizes up to 25 ml, marking round with a file or glass knife and touching off with a hot glass rod is suitable. Larger sizes present more difficulty because of the thickness of the glass, and in these cases a crack is led around the vessel with a hot glass rod by extending the crack in the direction of the source of heat. Alternatively, hot wire methods described by Ansley[1] may be employed. Another simple method of cracking off large-diameter tubes is first to make a file mark around the tube. As close as possible to the mark, and on each side of it, pieces of blotting paper are wrapped around the tube and are held in position by thin copper wire. The blotting paper is then damped with water and a hot rod or a small pointed flame is applied to the file mark. When the damaged portion has been cracked off the sharp edges of the cylinder are carefully heated and smoothed in the blowpipe flame.

Separating funnels

The stems of separating funnels prove to be their vulnerable point and are constantly being broken off. For this reason it is a sound proposition to pay a little extra and buy hard glass funnels. If the funnel is made of Pyrex or similar hard glass it may with care be repaired even when the stem is broken off close to the stopcock. When fitting a new stem the key must first be removed from the stopcock and the open ends of the keyway plugged with asbestos corks or asbestos wool. During the repair operation the ground-glass barrel is protected from the direct heat of the flame by asbestos string of suitable thickness which is wound around it. Before actually joining on the new stem the area near to the ground-glass barrel should be gently heated before fiercer heat is applied to make the join.

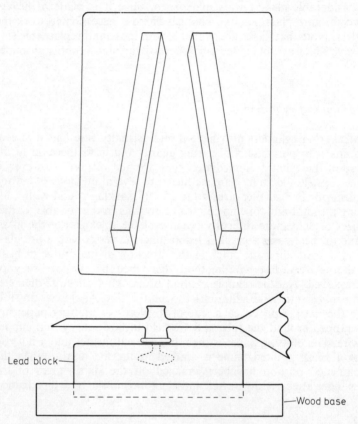

Figure 5.4. Apparatus for removing stuck keys from stopcocks

The majority of separating funnels end their useful life when attempts are made to remove stuck keys. Many methods of doing this have been advocated, yet none have been completely successful and most of them take time. If the simple apparatus illustrated in *Figure 5.4* is used, stuck keys present little difficulty. The apparatus, which has been described by the author[2], consists simply of two diverging blocks of lead set in a wooden base. The barrel of the stopcock rests on the lead shoulders and the key is tapped out with a hardwood drift.

This apparatus is also useful for stopcocks on other pieces of apparatus. Very obstinate stopcocks may be loosened by immersing them in water in a suitable size of vessel and bringing the water to the boil for 10 minutes. Alternatively, if the shape of the apparatus does not allow the immersion of the stopcock, a jet of steam may be directed on it. The key is then removed as previously described.

Thermometers

Thermometer repairs usually entail the joining up of broken mercury threads. This is not a difficult operation and the immersion of the thermometer in a freezing mixture usually suffices. Freezing mixtures, however, may not be readily available, and provided sufficient care is exercised the thermometer may be waved in a yellow Bunsen flame, which drives the mercury into the safety reservoir at the top of the capillary. It requires a little skill to estimate the correct time to withdraw the thermometer completely from the area of the flame, which is just after the mercury enters the reservoir. The mercury in thermometers which have broken stems can be saved by breaking the bulbs against the bottom of a small glass bottle. The mercury is filtered from the glass fragments when sufficient has been gathered to make the operation profitable.

Desiccators

Many samples are left in desiccators with the best of intentions, only to be afterwards forgotten. All samples should, therefore, be labelled. Labelling avoids the unnecessary monopolisation of the desiccator and permits old samples to be regularly discarded.

Old grease on the ground faces should be cleaned off regularly and fresh grease applied. If cleaning is neglected the lid sticks and when efforts are made to open it in the proper way by sliding it sideways it becomes difficult to move. Finally, it moves only too rapidly and an accident often results. The condition of the drying agent, too, is most important and it should be regularly renewed.

Glass tubing

The amount of small pieces of glass tubing which accumulate during the course of the year is usually considerable and the temptation to discard them is very great. Useful items, such as dropping pipettes, and delivery and capillary tubes can be manufactured from this glass, which also provides useful glassblowing material for the making of such items by the junior laboratory staff.

Graduated glassware

Renewal of fillings in graduations

The various media used for filling graduations in glassware may in time deteriorate and the graduations become difficult to read. Several mixtures for refilling the graduations have been recommended, but in each case the preliminary requirements are that the glassware should be thoroughly cleaned. This is done by immersing it for a few hours in sulphuric acid–sodium dichromate cleaning mixture, after which it should be rinsed well and dried.

White graduations. Two methods have been described by Edwards[3] for filling white graduations. A paste of zinc oxide is rubbed into the graduations with the finger and allowed to dry for twelve hours. When it is dry and hard the excess zinc oxide is removed by means of tissue paper.

In the place of the zinc oxide, a mixture of plaster of Paris, white lead, and boiled linseed oil can be used, and the excess removed with a cloth.

Black graduations. The most troublesome items of glassware with regard to black graduations are thermometers, and to renew the filling in these, a method described by Northall-Laurie[4] may be adopted. This consists of making a paste of silicon-ester paint medium and good-quality lampblack. The paste is allowed to harden in the graduations for a few days whereupon it sets hard with the formation of colloidal silica.

Lampblack in shellac mixed to a paste can also be used. The mixture is rubbed into the graduations and the excess removed with a rag.

A third method is to apply boiled linseed oil mixed with manganese dioxide and to clean off the excess with tissue paper. It is well worth while purchasing good-quality glassware with permanent graduations which are fused into the glass during manufacture.

Reagent bottles

When checking the laboratory reagent bottles it is usually found that glass stoppers have in some cases stuck in the necks. To release these the bottle should be grasped in one hand with the thumb pressed firmly on to one side of the stopper. The stopper is then tapped firmly with a piece of wood on the other side. If this is insufficient to free it, the stopper may be soaked out by applying glycerine and allowing it to stand for several hours. A releasing mixture which is successful in obstinate cases consists of 5 parts glycerol, 10 parts chloral hydrate, 5 parts distilled water, and 3 parts hydrochloric acid.

Another releasing solution recommended by Jobling[5] consists of

alcohol	2 parts
glycerine	1 part
sodium chloride	1 part

A further method for releasing stoppers is to heat the neck with hot water. In extreme cases, provided the contents of the bottle are nonflammable, the neck can be heated gently over a yellow Bunsen flame.

Rubber corks are also sometimes found to have become stuck in the necks of bottles or flasks. They may be released by inserting a sharp pointed awl between the stopper and the glass. The awl is then rotated so that it travels around the inside of the neck.

Metalware

In spite of the better finishes used nowadays on retort stands, tripods, and other items of laboratory metalware, they deteriorate very rapidly in chemical laboratories and need repainting at regular intervals. It is essential that all rust be removed before painting. The metalwork should be well scrubbed with a wire brush and afterwards rubbed down with emery paper until a smooth surface is presented. Aluminium paint is best for laboratory use and withstands the effects of fumes and heat better than any other.

CLEANING OF APPARATUS

GLASSWARE

If glassware is cleaned immediately after use the removal of the contents presents little difficulty. Nevertheless, quantities of dirty glassware inevitably accumulate and must be dealt with.

The cleaning of the glassware should be undertaken in a

methodical way and it should first be sorted according to the nature of the 'dirt' with which it is fouled. Generally speaking, it may be divided into two categories: (a) superficially dirty, and (b) containing residues difficult to remove.

Methods of cleaning glassware

General

Glassware which is superficially dirty may be cleaned by the general method of washing it in hot water containing a little detergent. To protect the hands rubber gloves should be worn. If the glassware is washed up in a 'hard' sink, a plastics or rubber mat should be used in the sink. As for any other laboratory operation the correct tools should be used for this work, and an assortment of suitable brushes and cleaning agents should be available. Worn brushes, the wire on which is exposed, scratch the glassware and when in this condition should be discarded.

Obstinate dirt can be removed by a suitable cleaning powder into which the brush is dipped. Tri-sodium phosphate solution together with a little pumice powder is an efficient cleaning mixture. It may be necessary to use a solvent such as acetone or a caustic solution to remove adhering substances.

After washing, the glass must be thoroughly rinsed in clean water, with a final cleansing in distilled water. For this purpose two adjoining sinks are an advantage. The glassware may then be left in an inverted position to drain and dry. The water tends to cling in small items such as small test tubes and these may be dried in an oven at a temperature of about 100 °C. Drying cupboards are useful for the larger items of glassware.

If the apparatus which has been washed is required for immediate use it may be flushed out with methylated spirits followed by a little ether. It is finally dried on an electric hot air dryer such as has been described by the author[6] or simply by blowing air from a compressor through a heated copper tube into the vessel. To ensure perfect cleanliness the air should be passed through a glass wool filter which retains any small foreign particles and oil droplets from the compressor.

Special

Glassware which has been grossly contaminated may need special cleaning after which it may have to be cleaned by the general method as a final treatment.

Oil and grease. Oil and grease can generally be removed by shaking with a warm detergent solution. It may be necessary to add a light scourer such as small pieces of paper or sawdust. Afterwards the vessel is washed out with water followed by concentrated hydrochloric acid and lastly with distilled water.

The removal of silicone grease is a more difficult problem but several ways of removing silicone stopcock grease and silicone high-vacuum grease have been recommended by a supplier of high-vacuum equipment[7]. The glassware should be first rinsed with paraffin, or other suitable hydrocarbon solvent, then cleaned with a warm solution of chromic acid.

Alternatively, after the solvent has been used, the glassware may be cleaned with a warm solution of one of the following:

1. 5 g sodium perborate in 100 ml 10% sodium hydroxide solution.
2. 10 g sodium hydroxide and 5 g borax dissolved in 100 ml water.
3. 10–15 ml 50% potassium hydroxide solution in 100 ml industrial methylated spirit. (The solution should not be used for a period of more than 10 minutes.)

Alternative methods, depending on the circumstances and nature of the contaminant, have been suggested by Edwards[8].

4. Soaking in fuming sulphuric acid.
5. Cleaning with carbon tetrachloride.
6. A mixture of a detergent such as Lissapol or Teepol 10 parts, water 4 parts, paraffin 45 parts, well stirred, heated to boiling, and used hot.

Tar. Tar can usually be removed by benzene or by other suitable solvents, followed by cleaning with an abrasive powder.

Carbon. Carbon is difficult to remove. Soaking in caustic soda solution is usually effective, but it may be necessary to resort to a chromic acid cleaning solution. When the carbon is particularly obstinate a small quantity of the chromic acid may have to be gently heated in the flask. A steam-heated acid cleaning bath has been described by Hancock[9]. Edwards[3] has recommended the use of a mixture of 2 parts of tri-sodium phosphate to 1 part of sodium oleate in 16 parts of water as an effective solution for removing carbon. He also suggests heating a few grammes of solid sodium sulphate in the flask over a Bunsen. This treatment helps to loosen the deposit.

Stains. Stained glassware may be cleaned by a saturated solution of ferrous sulphate in dilute sulphuric acid. Permanganate stains may be effectively removed by sulphurous acid. Hydrochloric acid diluted with its own volume of water is suitable for removing iron stains.

Other hard deposits. If the nature of the hard deposit can be recognised, the choice of the removing agent is simplified. Deposits such as chalk, for example, which adhere to the sides of glass vessels after boiling, are easily removed by dilute hydrochloric acid. To remove quantities of hard substances which adhere tenaciously to the sides of vessels, it is necessary to use, in addition to the cleaning agent, something which scours the inside of the vessel. Various scouring agents are used, such as paper, sawdust, lead shot, glass beads, or sand. The choice of these depends on the degree of abrasion necessary to remove the adhering substance. Sand should be used only as a last resort, since it scratches glassware. When the nature of the deposit is not known and it does not respond to normal washing methods, chromic acid cleaning mixture, which is probably the nearest approach to a universal cleaner, should be used. So-called chromic acid is made by dissolving 25 g sodium dichromate in 25 ml water and increasing the volume to one litre by carefully adding concentrated sulphuric acid.

Another so-called universal reagent for cleaning glassware has been suggested by Crawley[10]. The mixture consists of a cold solution of

5% hydrofluoric acid
33% nitric acid
2% Teepol
60% water

and is particularly effective for grease, carbon, and mercury contamination. Prolonged treatment should be avoided and the reagent should not be used for cleaning volumetric glassware.

Decon 75, a proprietary brand of cleaning concentrate, available from Medical Pharmaceutical Developments Ltd, Ellen Street, Portslade, Brighton, may be used in place of chromic acid cleaning mixture. It is a non-foaming surface-active agent and is non-corrosive and non-toxic. It can be rinsed off the material being cleaned without leaving traces of detergent film.

Volumetric apparatus

Volumetric glassware should be perfectly clean and entirely free from all grease. Greasy glassware causes any remaining aqueous solution to collect in drops on the wall of the vessel when it is drained, whereas this does not occur if the vessel is chemically clean.

To clean the glassware, a little detergent in water is sufficient for general purposes, and for burettes a brisk cleaning with a brush dipped into this solution is usually sufficient. Whenever a detergent is used several rinsings with water are necessary, and all glassware must be finally rinsed with distilled water.

Abrasives, hydrofluoric acid mixtures, and strong solutions of alkalis should not be used, and caution should also be exercised when soaking graduated glassware in strong acid solutions because unprotected fillers used for graduating some glassware may be affected.

For accurate work, it is essential that volumetric glassware be cleaned with chromic acid cleaning solution. The apparatus should be filled with the solution and left to stand for at least twelve hours before it is rinsed and dried. Pipettes should stand immersed in a tall polythene measuring cylinder or other suitable container. The chromic acid is kept for future use and should be stored in a strong vessel which cannot be easily broken or in a polythene bottle.

Volumetric glassware should not be heated or washed in hot water since the glass takes a considerable time to contract to its original volume. This is particularly important with accurate volumetric ware.

The proper care and cleaning of volumetric glassware has been competently described in a leaflet by Elliott, which also describes the current method for cleaning and greasing stopcocks[11]. To prevent stopcock keys from binding or freezing (e.g. in burettes) they should first be thoroughly cleaned and rinsed in ether. The key should then be warmed slightly by indirect heat and a small quantity of grease carefully smeared down one side of the key. The key is then reinserted in the barrel and rotated until the grease forms an even film. To prevent keys from sticking when strong alkaline solutions are used, the key should be cleaned and the ground surface rubbed with a graphite pencil until completely blackened. The key is then greased in the normal way. Lubricants which are said to be suitable are (a) pure Vaseline, (b) a mixture of Vaseline and resin cerate, (c) a mixture consisting of 1 part of beeswax to 3 parts pure Vaseline, or (d) a hard grease consisting of 1 part of soft black rubber added in small pieces to mixture (c) and heated to 140–150 °C and stirred continuously until thoroughly incorporated. For stopcock grease mixtures and for details of many other useful laboratory aids a technician's handbook is now available[12].

However carefully burette keys are greased, small pieces of the lubricant invariably find their way into the jet. To clean the fine jets is normally very troublesome but if they are cleaned as suggested by Michels[13] little difficulty will be experienced.

A piece of thin copper wire 450–500 mm long is inserted through the burette tip and taken out of one side of the stopcock barrel until about 300 mm is protruding. The wire is then wound around a piece of 3-mm glass rod to form a small coil. The glass rod is removed and the coiled wire is pulled back by means of the short end of wire left protruding from the tip. The rotating motion of the wire as the coil unravels cleans the sides of the tip as the wire is withdrawn.

Special items of glassware

Fragile or intricately shaped items of glassware may be cleaned ultrasonically. The cleaning fluid is contained in a tank to which transducers are attached. A generator causes ultrasonic impulses to be applied to the transducers which convert the electrical energy to mechanical energy. Vibrations from the transducers cause cavitation in the cleaning fluid which sets up a scrubbing action.

Sintered glass discs

To prolong the life of sintered glassware the following points of general care should be observed.

1. They should not be heated above 200 °C.
2. Deposits on the face of the sinter should be cleaned off immediately after use.
3. Strong solutions of alkalis and acid fluoride solutions should not be filtered.
4. The correct working pressure should not be exceeded. For large discs the maximum pressure differential is 380 mm of mercury and for discs below 40 mm 600 mm of mercury.
5. Care should be taken to dry crucibles before placing them in ovens or autoclaves. The oven or autoclave should be cool and the temperature then raised. The temperature should be allowed to fall almost to room temperature before the crucibles are removed.

Cleaning. When they are new glass sinters should be washed with distilled water before being used. Hot dilute hydrochloric acid should then be passed through them and finally they should be washed again with distilled water.

After use the cleaning of glass sinters may present considerable difficulty. In the case of crucibles which have been used for gravimetric analysis, a dilute solution of the di-sodium salt of E.D.T.A. is suitable for removing many precipitates.

Should the sintered glass discs in crucibles and other sintered ware become discoloured and prove difficult to clean, more vigorous methods may be necessary. The crucible or other sinters may be boiled in aqua regia in a large glass beaker but the operation should be carried out in a fume cupboard. Strong alkalis should not be used, because they attack the glass sinters.

An alternative cleaning method is to soak the discs for twelve hours in a solution prepared by adding concentrated nitric acid to sodium dichromate crystals. The sinters should afterwards be thoroughly washed with distilled water.

Another vigorous cleaning method involves sucking a 1% solution of potassium permanganate through the sinters. A few drops of concentrated sulphuric acid are then placed evenly on the discs to produce a gentle heat. The resultant discoloration is then removed by sucking the discs dry and passing a hydrogen peroxide solution acidified with a little sulphuric acid slowly through them. The discs are then washed in distilled water. The procedure is repeated passing the solution through the discs in the reverse direction. Care! The reaction between hydrogen peroxide and permanganic acid is violent. Carry out the operation with great caution.

As a general guide for the removal of particular substances from Pyrex sintered discs the following cleaning methods have been recommended[14, 15].

Fats or greases: carbon tetrachloride or suitable organic solvent.

Organic substances: immerse the disc in warm chromo-sulphuric acid or concentrated sulphuric acid containing a little potassium nitrate or perchlorate (0.5% wt./vol. will suffice) and allow to stand overnight.

Albumen: hydrochloric acid or warm ammonia.

Cuprous oxide or iron stains: hot concentrated hydrochloric acid with potassium chlorate.

Barium sulphate: concentrated sulphuric acid at 100 °C.

Mercury residues: hot concentrated nitric acid.

Mercury sulphide: hot aqua regia.

Silver chloride: sodium hyposulphite.

Stannic oxide: boiling sulphuric acid (leave the disc in the acid to cool).

Microscope slides

New slides should be washed in hot water containing a small quantity of detergent. They should then be thoroughly rinsed and left in a dichromate cleaning solution for several hours. After rinsing with tapwater and again with distilled water, they should be stored under 95% alcohol in a covered dish until required for use.

If the slides are old, they should be soaked in xylol to effect separation and then left in alcohol for several days. They may then be cleaned in the same way as new slides.

Syringes

To clean syringes the following method is recommended by Prunty, McSwiney, and Hawkins[16]. Rinse with water immediately after use to remove blood. Remove paraffin oil by washing in chloroform, then alcohol, and finally ether. Draw air into the syringe and expel it. Repeat until plunger is free in the barrel. Plug ends with non-absorbent cotton wool, put on the cap, and place in its case. Sterilise by heating for one hour in an oven at 160 °C and allow to cool in the oven.

Needles. Squirt water through the needles immediately after use. Sharpen the point of the needle under a lens and wash in alcohol, then ether. Draw air through the needle until it is dry, place in its glass case. Plug with absorbent cotton wool and heat in oven for one hour at 160 °C.

Cleanliness of glass apparatus containing mercury

The mercury in levelling tubes becomes dirty, resulting in poor appearance and tailing, so that the meniscus cannot be clearly seen. Edwards[3] has recommended that about 13 mm of syrupy phosphoric acid be allowed to float above the top of the mercury, which cleans the wall of the tube.

Archard[17] has described how mercury in contact with aluminium forms a thin surface compound which prevents it sticking to glass. He recommends mercury treated in this way for mercury indices in glass tubes. One hour's contact with aluminium foil was found sufficient to condition 1 ml of mercury and samples of mercury so treated retained this property after a period of four months. The property can be removed by squeezing the mercury through a linen handkerchief.

Porcelain

Porcelain crucibles may be cleaned by boiling in aqua regia and in some instances by boiling in caustic soda solution. Evaporating basins usually respond to cleaning with hot water and cleaning powder.

SILICA WARE

To clean silica (Vitreosil) apparatus, the methods advocated by the manufacturers should be followed[18]. Light stains should be removed by fusing sodium bisulphate in the vessel, which, after pouring off the molten salt, should be washed with water. Badly contaminated apparatus should be treated with the following solution:

Hydrofluoric acid (40%)	2 parts by volume
Nitric acid (concentrated)	7 parts by volume
Water	7 parts by volume

Affected vessels should be filled with or immersed in the above solution and allowed to soak for a period of about 30 minutes depending on the degree of contamination. They should then be washed with tapwater, followed by distilled water, and finally dried on a clean glass cloth.

Note that whenever hydrofluoric acid solutions are used the hands should be suitably protected by the use of a barrier cream and rubber gloves. Eyeshields should also be worn.

Transparent silica absorption cells for spectrometric work

The manufacturers of Vitreosil ware also recommend that transparent absorption cells should be cleaned by soaking for 24 hours in a concentrated chromic acid solution or by boiling for one hour in concentrated nitric acid. They should then be washed in distilled water and their surfaces rubbed with cotton wool soaked in alcohol. Finally, dry clean cotton wool should be used. During the process of cleaning and drying, the cells should not be touched by hand. As far as possible the cells should be kept permanently clean by removing any contamination immediately after use and by immersing them in distilled water until they are again required.

Other manufacturers are opposed to immersing cells for long periods in chromo-sulphuric acid mixtures and recommend that solidly fused glass or silica cells may be cleaned occasionally in chromic acid at about 35 °C for periods not exceeding an hour. The complete method recommended by Hellma (Hellma GmbH, Müllheim-Baden, W. Germany) is as follows:

1. Wash the cell under a water jet.
2. Dry the cell with tissue paper or by suction pump and nozzle.
3. Immerse in the acid bath for one hour.
4. Invert the cell and allow to stand a while.
5. Wash under a water jet.
6. Rinse in distilled water.
7. Dry the outside windows with tissue or filter paper and then with special optics cleaning paper. Allow the inside of the cell to dry out in a dust-free atmosphere. If the cell will not be used again for some time, wrap it in a clean piece of optics paper and place in a container.

For normal cleaning when the degree of contamination does not call for the acid bath treatment given above, the following cleaning techniques are considered adequate.

1. *After use with water-soluble samples.* Thoroughly wash with a jet of water. Rinse in distilled water and dry as in (7) above. It may be necessary occasionally to wash first in a good soap solution and with diluted hydrochloric acid (3–5%).
2. *After use with aqueous protein solutions.* Soak overnight in approx. 0.5% hydrochloric acid with the addition of pepsin (1 g/100 ml). Thereafter as in (1).
3. *After use with miscible organic solvents.* First rinse in the pure solvent concerned with a small addition of a base or acid. Thereafter as in (1).
4. *To remove heavy metal traces.* Wash in water. Immerse in concentrated nitric acid (in special cases aqua regia). Rinse in distilled water. Dry as in (7) above.

A further method which has been recommended as satisfactory consists of first soaking the cells in water for a few minutes. If the contamination is not removed a mild sulfonic solution may be used or, if a stronger cleaning agent is required, the cells may be soaked in acid–alcohol solution: 3N HCl and 50% alcohol.

Cells should *never* be cleaned with hot concentrated alkalis, abrasives, or any other agent likely to mark the polished optical surfaces.

TRANSPARENT SILICA MATERIALS

Few people are fully aware of the necessity for the utmost cleanliness when working transparent Vitreosil in the blowpipe flame. Certain substances form chemical compounds with Vitreosil at its fusing temperature and contamination by some other substances, particularly the alkaline metals, causes the material to devitrify. Thus, traces of these substances on the transparent surface may give rise to points of devitrification which spread when the silica is heated. This can be avoided by the thorough cleaning of the material prior to heating. Dust and fingermarks may also give rise to devitrification points on the material when it is worked in the flame.

A wash with water removes settled dust and non-greasy marks on the material and it should afterwards be washed with distilled water. Both the inside and outside surfaces should then be rubbed dry with clean boiled rag which should be free from soap and soda, or with clean cotton wool. After washing and drying the surfaces should be wiped with cotton wool dipped in alcohol. Tubing may be cleaned by pulling alcohol-dipped cotton wool through the bore.

Whenever Vitreosil is heavily contaminated or has been ground with abrasive materials, it is further recommended that the parts be immersed for a few minutes in 10–20% hydrofluoric acid which should be followed by washing with water. A 40% solution (10–15 minutes) is necessary if there is any possibility of chemical attack at high temperatures. Similar cleaning, followed by heating in the blowpipe flame to just below the melting point, is also advised if heavy deposits of volatilised silica are found to have condensed on the cooler parts of the apparatus after working in the blowpipe flame.

PLATINUM WARE

Care and cleaning

Owing to its chemical advantages, platinum is superior to porcelain, silica, or nickel for certain laboratory operations. This is particularly true quantitative work, but because of the high cost of this metal it must be used with the greatest care.

In the presence of many metals including arsenic, antimony, bismuth, zinc, lead, and tin, and especially at high temperatures, platinum forms alloys. It should, therefore, never be heated in contact with these metals, or with their oxides, or with any of their salts which are easily reducible.

Platinum is also affected by selenium, tellurium, silica, sulphur, and phosphorus. Of these, phosphorus is the most dangerous and is destructive even if present in very small quantities. Phosphates and sulphides should also not be heated in platinum vessels.

Halogens and halogen compounds also attack platinum and this is especially true of free chlorine. The metal should therefore never be cleaned with aqua regia. Hydrochloric acid can be used provided that it is free from any oxidising substance. Among the other substances which attack platinum are the oxides, peroxides, nitrates, and nitrites or cyanides of alkali metals. Unknown mixtures should, of course, never be heated in platinum ware.

When platinum ware is heated, the base of the vessel should be above the blue cone of the flame, otherwise the metal becomes brittle. Crucibles should be supported on platinum triangles but silica ones are satisfactory. If pipeclay triangles are used these should be clean and bare portions of wire should not be exposed. Platinum must not be heated in contact with other metals.

To clean platinum crucibles which are slightly soiled or stained, sodium carbonate or borax should be fused in them. If this method fails, potassium bisulphate should be used but this very slightly attacks the metal. When it is molten, the cleaning agent may be brought into contact with the whole of the interior face of the vessel by gently tilting it with platinum-tipped tongs. Nickel tongs may be used if pieces of platinum foil are wound round the tips and secured with platinum wire. The contents of the vessel are then tipped out on to a piece of iron. Water is boiled in the crucible to remove any remaining salt. Final cleaning is effected by immersing the vessel in concentrated hydrochloric acid and heating.

Sodium amalgam has also been recommended for cleaning platinum[19]. The amalgam is carefully wiped over the surface with a soft cloth. The amalgam of mercury with any base metals is then wiped off after first moistening the crucible with water. The crucible is then heated to redness to drive off any remaining mercury.

Platinum ware in time loses its polished appearance and this can be restored by polishing it on a wood former with sand which passes a 100-mesh sieve and which consists of rounded grains. Fine carborundum powder or precipitated chalk is also suitable for polishing purposes.

PLASTICS LABORATORY WARE

Warm soapy water or detergent solution should be used to clean plastics laboratory ware but it may be necessary to us chromic acid cleaning solution or dilute nitric acid in some cases. Abrasives or scraping should be avoided but warm alcohol may be tried for obstinate contaminating materials. Plastics laboratory ware should never be heated over a flame or on a hotplate.

REFERENCES

1. Ansley, A. J., *An Introduction to Laboratory Technique*, 2nd edn, Macmillan, London (1952)
2. Guy, K., 'A Method for Removing Stuck Stopcocks', *S.T.A. Bull.*, **4**, 1, 7 (1954)
3. Edwards, J. A., 'Cleaning, Care, and Maintenance of Glass, Metal and Plastic Apparatus', *Chem. Age, Lond.*, **70**, 1433 (1954)
4. Northall-Laurie, D., 'Making Thermometer Graduation Marks Permanent', *J. scient. Instrum.*, **9**, 96 (1934)
5. Jobling, *Pyrex Laboratory Glass*, bulletin No. 14 issued by James A. Jobling Co. Ltd, Sunderland (1963)
6. Guy, K., 'An Inexpensive Hot Air Drier', *J. Sci. Technol.*, **2**, 1, 29 (1956)
7. Edwards High Vacuum, *Methods for Removing All Traces of Silicone Stopcock Grease or Silicone High Vacuum Grease from Glassware*, pamphlet issued by Edwards High Vacuum Ltd, Crawley (n.d.)
8. Edwards, J. A., *Laboratory Management and Techniques*, Butterworths, London (1960)
9. Hancock, C. K., 'Steam Heated Acid Cleaning Bath', *J. chem. Educ.*, **27**, 621 (1950)
10. Crawley, R. H. A., 'A Universal Reagent for Cleaning Glassware', *Lab. Pract.*, **3**, 462 (1954)
11. Elliott, *E-Mil Notes on Care and Cleanliness of Volumetric Glass Apparatus*, pamphlet issued by H. J. Elliott Ltd, Pontypridd (n.d.)
12. Institute of Science Technology, *Technician's Handbook*, London (n.d.)
13. Michels, W., "A Simple Method for Cleaning Burette Tips', *J. Chem. Educ.*, **38**, 548 (1961)
14. Jobling, *Pyrex Glass*, bulletin No. 3 issued by James A. Jobling Co. Ltd, Sunderland (1961)
15. Baird and Tatlock, *Notes on the Care of B.T.L. Sintered Glassware*, pamphlet No. N.270 issued by Baird and Tatlock (London) Ltd (n.d.)
16. Prunty, F. T. G., McSwiney, R. R., and Hawkins, J. B., *Lab Manual of Chemical Pathology*, Pergamon, London (1959)
17. Archard, G. D., 'A Method of Preventing Mercury from Sticking to Glass', *J. scient. Instrum.*, **29**, 412 (1952)
18. Thermal Syndicate, *About Vitreosil*, reference handbook published by The Thermal Syndicate Ltd, Wallsend, Northumberland (1958)
19. Grant, J. (ed.), *Clowes and Coleman's Quantitative Chemical Analysis*, 15th edn, Churchill, London (1944)

6
Safety in laboratory and workshop

GENERAL

When the experienced laboratory worker recalls his early days in the laboratory he may wonder how in spite of his inexperience he managed to avoid a serious accident. As in all arts and crafts it is the younger worker who has the most need of careful guidance, training, and supervision from the senior personnel. A sound spirit of co-operation and unselfishness on the part of all the laboratory staff is the best safeguard against accidents.

The chief responsibility for the promotion of this kind of spirit rests with the head of the section. He is in the best position to give the necessary lead to his staff and should maintain close contact with everyone under his control. Occasionally accidents caused through negligence call for admonishment, but where accidents are conscientiously reported, undue criticism of the responsible individual may induce an attitude whereby future accidents are covered up and the steps necessary for their prevention are therefore never put into operation.

The most important steps to safety are cleanliness and tidiness, which in themselves promote an accident-free environment. Cluttered benches not only cause accidents but clearly indicate muddled working on the part of the person responsible. Inadequate locker and storage arrangements may contribute to this state of affairs.

Equipment should always be well maintained. Adequate notice boards should be provided so that safety rules may be prominently displayed. The fire, hospital, and ambulance facilities available, and their telephone numbers, should be displayed on notice boards. It is recognised that verbal instruction is far more effective than a printed notice and provision should be made for the instruction of all the staff by regular meetings and discussion on matters of safety. The services of Inspectors of Factories may be obtained for this purpose on request.

It is desirable, too, that all establishments nominate their own safety officer, particularly research and educational institutions not subject to the Factories Acts. The duties of this officer should include the dissemination of safety literature, organisation of first aid facilities both central and local, and promotion of safety lectures.

To supplement the work of the safety officer, all personnel should receive training in first aid and in this connection the various Red Cross Organisations are usually willing to lend assistance. Every encouragement should be given to staff to see potential hazards and to report on them. An accident book should be maintained, for only recording accidents can assure their prevention.

DANGERS FROM GLASS

For accidents, glass is the laboratory worker's worst enemy. Glass apparatus should be erected methodically and deliberately and never in a hurry since this is the chief cause of accidents.

CUTTING GLASS TUBE AND ROD

To cut the smaller diameters of glass tubing or rod into shorter lengths, first mark it with a glass-knife or triangular file. Place the thumbs one each side of the mark and close to it. Then pull the glass slightly towards its ends and at the same time break it away from the body in the one motion. It is not advisable to attempt this with tubing above 15-mm bore and when breaking any size of tubing the hands should be protected with a piece of cloth. The ends of cut tube and rod should always be flame-polished before being used in apparatus.

CUTTING GLASS SHEET

The safe and successful cutting of glass sheet requires an intimate knowledge of the material, which is acquired through practice in handling it. To cut the glass a sharp tool is essential and of the several types available a diamond cutter is the best for use in the laboratory.

The glass sheet must be cut on a level surface such as a good bench top or on a table reserved for the purpose. If it is possible to provide a special table it is advisable to affix to its top two wooden stops against which a straight edge can be firmly held while the glass sheets are being cut. One of the stops is screwed to the table close to the front edge about 450 mm from one end and the other is fixed in a similar position at the rear edge.

When the glass sheet is to be cut it is laid on the table between the wooden stops and the position of the cut is found by measurement. Small identifying marks are then made by means of a grease pencil. The straight edge is then held steady against the stops in line with the pencil marks and, using the lightest possible pressure, a light scratch is drawn across the surface of the glass. Never try to deepen the scratch by retracing the original scratch mark, as this will result in a ragged edge. An unsuccessful first attempt means that a further scratch should be made in a new position. Note that new glass always cuts better and more easily than old; the chances of successfully cutting old glass are more remote.

When the scratch has been made two battens approximately 50 mm wide and 6 mm thick, the length of which exceeds the width of glass being cut, are placed beneath the glass sheet to keep it level.

One of the battens is placed close to the scratch and parallel to it. With a folded duster held in the hand a sharp pressure is then brought to bear on the piece of glass which overhangs the batten while the other hand is used to steady the main sheet.

Sheet glass has an infinite number of uses in the laboratory and various shapes and sizes are in constant demand. A sound knowledge of the safe cutting, drilling, grinding, polishing, and marking of this material is essential and these operations, including sheet glass construction work, have been explained at length by Charlett[1].

BENDING GLASS TUBE

It is essential when bending glass that the bends be well rounded so that the tubing retains its original diameter. Flat bends should be discarded. Good bends not only assist the function of the apparatus, but are also a safeguard against accidents during assembly. The sharp ends of the tubing must always be flame-polished.

HEATING GLASS

Depending upon its composition and its thickness, glass is susceptible to stresses and strains and can withstand only limited thermal and physical shock. Heat-resistant glassware must always be used for heating liquids. It should also be used when heat is likely to result through diluting or dissolving a substance and these operations should never be carried out in nonresistant glass vessels such as Winchester bottles.

CARRYING GLASS

To avoid accidents when glassware is transported, nonslip floors are necessary in the laboratory and sensible shoes should always be worn. The route should be free from obstructions and spilled liquids should be immediately wiped up. Never hurry! Many major accidents have occurred through persons falling, or being struck by a door, when carrying glass apparatus or dangerous chemicals in glass containers. Winchester bottles should not, therefore, be carried by the neck, or cradled in the arms, and trolleys or Winchester carriers should always be used.

ISSUE OF GLASSWARE

Damaged glassware should not be used, nor should it be issued to other people. If it is badly damaged, it should be discarded and placed in a metal receptacle. If repairable, it should be put in the repairable glassware box to await attention.

BORING CORKS TO TAKE GLASS TUBING

Corks or bungs should always be held by their sides between forefinger and thumb and bored on a piece of scrap wood. Bore from the narrow end by rotating the borer in one direction only. The size of hole should be compatible with the size of the glass tubing to be inserted so that no undue forcing in assembly is necessary. For most purposes it is essential that the cork or bung grip the tube reasonably tightly, and therefore when inserting the tube it should be held well down close to the point of entry. A little lubricating medium such as glycerol assists the operation. Rubber bungs which have been in contact with glass tube for long periods adhere to the glass. A safe way of removing the bung is either to cut it off or to select a cork-borer slightly larger than the outside diameter of the tube and insert it into the bung using the glass tube as a guide. Similarly, a cork-borer may also be used as a guide when inserting glass into corks and bungs.

Alternatively, a special tool may be used which consists of a steel tube set in a handle. The tube is fitted with a bullet nosecap which is removed when the tool has been pushed through the rubber bung, thus allowing the insertion of glass tubing via the hollow tube. If the tool is used to free adhering glass tubing, the nosecap is removed prior to the insertion of the tool.

GLASSWARE FOR VACUUM OR PRESSURE WORK

Under pressure or vacuum, glassware which is merely scratched can be extremely dangerous. If used for such purposes it should be of heavy wall and protected by suitable wire-mesh guards. The use of goggles or screens during the experimental period is to be encouraged. All corks and bungs must be especially well fitted and must not protrude too far into the necks of vessels. After evacuation, air must be admitted to the apparatus slowly if organic substances are involved. When glassware containing volatile substances is shaken, the pressure which builds up must be released from time to time.

BROKEN GLASS

Glass fragments, particularly if hot, must always be deposited in a special metal waste receptacle and never mixed with general rubbish. If glassware is broken in a sink the pieces should be immediately removed and small fragments, which are invariably trapped in the grille of the waste outlet, should be extracted with tongs. If broken glass becomes mixed with ice, the whole must be discarded.

OTHER DANGERS FROM GLASS

Carboys and other large glass containers should not be used unless well protected by a basket or other means. Glass tubing stored in racks should not protrude beyond the end of the racks. Round-bottom glass vessels should always stand in cork ring supports.

DANGERS FROM GENERAL LABORATORY OPERATIONS

For general operations in the laboratory, the efficient design of laboratory furniture and the regular overhaul of working equipment are the main factors governing safe laboratory conditions.

FURNITURE

All laboratory furniture, whether of wood or of metal, must be of the highest quality and should have a nonabsorbent finish. The benches must be the correct height in accordance with the work to be done on them and should incorporate sufficient sinks to allow any person working at the bench speedy access to one of them in an emergency.

The shelves, on or above the benches, should not be too high and should have a light beading affixed to the front of them so that bottles are not easily knocked off.

It is also important that the bench service controls be so positioned that there is no necessity to reach over benches. All service pipe lines should be painted with the appropriate service colour which allows them to be more easily identified. Suitable types of rubbish boxes in sufficient numbers should be provided. Broken glass or chemical refuse, which would be dangerous if deposited in the normal waste bins, requires separate bins.

EQUIPMENT

When checking equipment, particular attention should be paid to temperature-regulating devices and to gauges and valves on pressure equipment.

Protective items such as respirators, goggles, and gloves, if openly displayed, are far more likely to be regularly used than if hidden away in cupboards. Glass-fronted wall cabinets may be used to ensure that the appliances are seen and protected from dust. Wall racks, for Winchester carriers and bottle baskets, show at a glance when such items are missing and help to ensure that they are kept together in their proper place. Respirators must be regularly disinfected and rubber gloves kept in good condition by occasional powdering. Laboratory goggles should be comfortable and lightweight. These are worn for any operation which constitutes a risk to the eyes such as work involving acids, bromine, ammonia, cutting of sodium, and chipping and grinding operations. Persons should not be allowed to enter laboratories where eye dangers exist unless they are wearing eye protection.

ACIDS

More accidents are said to occur in the laboratory through mineral acids than from any other liquids. Concentrated sulphuric acid is the most dangerous. When acids are diluted the acid is poured into the water and never water into the acid. During the process of pouring acids and other corrosive liquids, the sensible precaution of wearing rubber gloves and goggles should be adopted. It is sometimes found that the stoppers in concentrated acid bottles have stuck. To deal with this occurrence, protective clothing should be worn and the bottle placed in a pneumatic trough which may be stood in a large sink. The stopper and the neck of the bottle are covered with a cloth and the stopper is then gently tapped. If these measures fail to free the stopper, it may be necessary to mark the bottle around the middle of the neck with a file. A point of hot glass is then applied to the file mark. Such methods, however, are best left to an experienced laboratory worker.

When pouring acids or other corrosive liquids, the stopper of the bottle should not be placed on the bench but with practice can be withdrawn and held in the crook of the little finger. The acid should be poured slowly. If the bottle is tilted too much, air locks will be caused and the acid will discharge in spurts. The bottle should be held with the hands placed on opposite sides so as to avoid contact with drips which may run down. A plastics bottle pourer simplifies this operation. When acid is being poured, or any other hazardous operation is being performed, it is dangerous to distract the attention of the operator. After pouring, and before replacing the bottle, it should be flushed on the outside with water. Should the skin be splashed with acid or other corrosive, it must be flushed immediately with plenty of water and afterwards a solution of sodium bicarbonate should be applied. If acid is splashed on the bench it must be wiped up at once. Any vessel which has contained acid should be rinsed out with water and not left to be washed up with acid dregs remaining in it. Whenever acid is discarded plenty of water should be run into the sink and at the same time the acid should be slowly poured away.

Acids soon destroy paper labels on bottles and for this reason etched labels are best, but a satisfactory alternative is to coat the label with paraffin wax. It is a dangerous procedure to stick a paper label over an etched label and then fill the bottle with a different substance from that denoted by the etched label.

Large quantities of acid should not be put into thin-walled vessels such as flasks. Concentrated sulphuric acid in Dreschel bottles or desiccators must be discarded by the user before these are put out for washing up.

Never put concentrated acids and alkalis adjacent to each other on shelves.

AMPOULES

Extreme care is necessary when sealed ampoules containing dangerous or low-boiling-point liquids are opened. Gaston[2] recommends cooling the ampoule in ice water or if necessary progressively cooling it to lower temperatures by means of ice water, ice salt, or solid carbon dioxide and solvent. A scratch is made with an ampoule file close to the top of the neck and the crack is then touched with a point of hot glass.

SODIUM

Only a limited amount of sodium, which should be cut into small pieces, should be kept in the laboratory reagent bottles. When larger pieces and quantities are required, they should be drawn from stores and the drawer held responsible for their use. Reagent bottles should be inspected frequently to ensure that the sodium is well covered with naphtha. Before putting out vessels for washing up, every trace of sodium must be removed. Old sodium is disposed of by adding it carefully, in small quantities, to alcohol.

Lassaigne's method for sodium fusions is somewhat hazardous, particularly in crowded conditions in teaching laboratories. Other tests described by Middleton which do not involve the use of alkali metal have been recommended by Lawrence[3] as being safer.

PIPETTING

Mouth pipettes should not be used for poisonous, corrosive, or volatile liquids. A rubber bulb pipette, safety pipette, or burette must be used. The use of the rubber bulb pipette is to be recommended for all liquids and is more hygienic. Rubber bulbs should be issued in all student laboratories. After use, pipettes should never be laid down with the end protruding beyond the edge of the bench.

SULPHURETTED HYDROGEN

Although this extremely deadly gas, which forms explosive mixtures with air, has a characteristic odour of rotten eggs, it has been called 'the killer that may not stink'[4]. When the concentration of this gas is high, the sense of smell is lost. This gas, like all other toxic gases, must always be used in a fume cupboard, and gas generators should be recharged in the open air. The dangers from sulphuretted hydrogen should be brought to the notice of all persons using the laboratory. Should large quantities of the gas escape in the laboratory, then, as for any other highly toxic gas, such as chlorine, carbon monoxide, and nitrogen oxides, the room must be vacated and entry prohibited until the gas has dispersed. When the concentration of the gas has diminished, and it is safe to enter the area using a self-contained breathing apparatus or a suitable respirator, all windows should be opened.

PERCHLORIC ACID

Protective clothing and eyeshields should always be worn when working with perchloric acid. It should not be used on wooden tables or benches or be allowed to come into contact with organic matter. If contaminated, the acid should be discarded by mixing it with a large volume of water and pouring it into a sink with the water taps discharging. It is advisable to keep any bottles containing the acid on a tray, which if the bottle is broken will contain the contents.

GENERAL PRECAUTIONS AND REQUIREMENTS

For picking up hot pieces of apparatus, appropriately shaped tongs are necessary. Bumping may occur when materials are heated in test tubes and they must not be pointed at other people, nor should the operator himself look down at the tube while it is being shaken or heated. When a quantity of chemical has been taken from a bottle for use, any surplus should not be returned to the bottle. When wash bottles are used to contain liquids other than water, they should be labelled in accordance with their contents. No one should use a wash bottle other than his own. Before liquids are heated they must be well mixed, and fast-boiling liquids should not be heated in narrow-neck vessels. Care should be taken to ensure that condensers or delivery tubes do not become choked during distillation or thermal decomposition experiments. The pressure which builds up when volatile liquids are shaken in separating funnels should be released frequently. Bottles containing chemicals such as aluminium chloride, which are liable to exert pressure in the bottle, must not be opened without taking due precautions. Dry ice or liquid air should not be handled with the bare hands.

Ventilation

Good ventilation, which keeps the laboratory free from injurious toxic gases, is absolutely essential and such systems must be regularly inspected to ensure their efficiency does not diminish. The cumulative effects of small concentrations of harmful gases or dusts may long pass unnoticed until a worker is taken seriously ill.

Personnel facilities

Proper washing up facilities, good arrangements for personal ablution, and a supply of drinking water, are all necessary to sustain the good health of the laboratory staff.

First aid personnel

A suitable number of persons should be trained in first aid and should carry on their person a key to the first aid facilities. A trained person should always be available and the name of such persons should be known to every laboratory worker. First aid cabinets should be installed in every laboratory. Refresher courses in first aid should be held often and may be made more interesting by showing occasional films on safety measures or by inviting speakers.

In order to introduce personnel to the various safety devices, practices should be held from time to time and are vital to the efficiency of the first aid arrangements. All accidents, even if trivial, should be reported and recorded, and if it is considered helpful the nature of the accident should be publicised in the department. A regular inspection of the record book will allow preventative methods to be adopted for the future.

Laboratory rules

Laboratory regulations must be posted in prominent places. When laboratories are not in use the door should be locked. This prevents work being done without proper supervision and the carrying out of unauthorised experiments. No one should work alone at night. When laboratories are in use, all doors and other exits should be unlocked and free from obstruction.

DANGERS FROM POISONS

Apart from the scheduled poisons which should invariably be kept under lock and key, many other substances handled daily in the laboratory, such as oxalic acid, are also poisonous. For complete safety it is advisable that all chemicals be regarded as poisonous. Poisons are not necessarily taken into the system by way of the mouth, but may also be inhaled, or absorbed through the skin. Certain basic rules are, therefore, to be observed in handling chemicals, and if these are adhered to the probability of poisoning is remote.

Chemicals must never be tasted or unnecessarily smelled. To smell a substance waft the odour to the nostrils by waving the hand gently across the mouth of the container. Chemicals should not be handled and in particular substances such as phosphorus, which apart from being toxic may cause lasting burns. Full use should be made of spatulas when dispensing chemicals and if necessary rubber gloves should be worn or a suitable barrier cream applied.

Food should not be eaten in the laboratory nor should laboratory vessels be used for drinking purposes. On leaving the laboratory the hands must be washed with soap and water before taking food. The use of organic solvents for cleaning the hands is not advised since these tend to dry the skin and cause dermatitis.

For any substance which gives off toxic fumes, the fume cupboard must always be used. All experiments giving off copious fumes, even if the fumes are not particularly dangerous, should also be performed in the fume cupboard. This keeps the laboratory atmosphere clean.

When working with cylinders of dangerous gases such as chlorine, sulphuretted hydrogen, or carbon monoxide, a respirator should be to hand and it should be ascertained beyond doubt that it is a suitable one and contains the proper absorbent.

Vapours of organic solvents are toxic and whenever a choice of solvent exists it is wise to choose the one which is least harmful.

SAFETY IN LABORATORY AND WORKSHOP 247

Apparatus which has been used for containing poisons should be washed up by the person concerned immediately after use.

MERCURY

Mercury is a cumulative poison and its soluble compounds are therefore dangerous. Its vapours, even in low concentrations, are

Figure 6.1. The mercury collector (Courtesy H. J. Elliott Ltd)

also poisonous and mercury distillations should therefore always be performed in a fume cupboard. The whole of the distillation apparatus should stand on a tray. In the same way, whenever mercury is used on the bench it should always be handled over a tray, which in the event of a spillage prevents losses. If mercury is spilled on the bench, or on the floor, it must be recovered at once by mercury tongs or by a seek bottle attached to a water pump.

The most efficient way of quickly picking up spilled mercury, however, is by means of a portable mercury collector. This collector does away with the difficulties associated with the pump and seek bottle and consists simply of a stout rubber bulb attached to a mercury pipette (*Figure 6.1*). If the spillage enters inaccessible cracks, it must be made inactive by applying flowers of sulphur or by a solution of iodine in potassium iodide.

A trap, such as the Vulcathene universal type which unscrews and allows mercury spilled in sinks to be recovered, should be fitted to sinks in laboratories where mercury is used.

DANGERS FROM FIRE

INFLAMMABLE SOLVENTS

Handling

The daily handling of inflammable solvents breeds a contempt for these dangerous materials. Solvents with low boiling points, such as carbon disulphide and ether, are the most dangerous and the degree of danger increases as the boiling point of the solvent becomes lower. At the flash point of the solvent, vapours are given off which form inflammable, and sometimes explosive, mixtures with air. These vapours are capable of travelling considerable distances and when ignited flash back with great rapidity to their source.

The handling of inflammable solvents in bulk should be carried out in fireproof rooms designed for this purpose and provided with alternative exits. In all areas where the nature of the work is such that inflammable vapours, gases, or dusts might be produced, naked lights, smoking, and the carrying of matches should be strictly prohibited. Notices to this effect should be prominently displayed. All possible sources of ignition such as electric switches, fuse boxes, sparking tools, lamps, and other electrical equipment should be well guarded and if possible excluded from the danger area. For general lighting purposes, vapour-tight fittings should be installed.

Use in the laboratory

Containers

In a general laboratory the control of the use of smaller quantities of inflammable solvents and the introduction of the essential safety measures becomes a more difficult matter. The maximum quantity to be kept in the laboratory at any time must be prescribed and stated in laboratory regulations. Proper containers bearing the words 'no naked lights' should be used. The containers should be distinctively coloured or otherwise marked so as to be easily identified. Glass containers should be restricted to a safe size and where it is technically necessary to use wide-mouth ones they should have loose fitted covers. For greater safety containers should be placed on a special sand tray, which if an emergency arises may be quickly removed from the room. As an added precaution it should be ruled that work with solvents be restricted to a specific area within the laboratory and steps must be taken to ensure that the prescribed areas are free from direct sunlight and other possible sources of ignition. Hazardous work should be carried out in a fume cupboard so that if a fire occurs it can be confined and more easily extinguished.

Heating

The best method of heating low-boiling-point solvents is by steam. Heating mantles can be used but there is an element of risk involved in that should the heating vessel fracture the contents might seep through to the heating element of the mantle. Webster[5], however, has pointed out that the heating mantle, because of its large heating surface, is an advantage over naked flames since the actual temperature of the mantle may be maintained at a lower level than the ignition temperature of the liquid. For medium-boiling-point liquids, water baths are permissible if electrically heated or used in conjunction with a safety burner.

Distillation and extraction

For operations involving distillation or extraction, the apparatus is best erected on a metal tray containing a layer of sand. Ground-glass joints are preferable to corks or rubber bungs which may deteriorate and leak in hot vapours. Distillation vessels should never be filled to more than one-half of their capacity and should be filled only when cool. To prevent bumping, bumping stones should be added to the solvent. The stones should never be added when the solvent is hot, since it may be superheated and the addition of the bumping stones would cause a sudden ebullition.

Disposal

The disposal of inflammable solvents by pouring them into a sink is extremely dangerous. This is prohibited by health regulations because of the toxicity of some solvents, and in addition a serious explosion hazard results. Carbon disulphide, for example, is particularly dangerous in this respect and in the presence of oxygen forms an explosive mixture readily ignited by flame, shock, or catalytic agents such as rust which may be present in the drains. To dispose of inflammable solvents they may be spread in suitable quantities over a reasonably wide area on waste ground and allowed to evaporate or burnt in small quantities in open metal trays in a disposal area. The disposal should be carried out by an experienced person and precautions should be taken to prevent other people entering the area until it is safe to do so.

Absorbent materials such as filter paper or extraction thimbles and the solid residues from flasks, if soaked in inflammable solvents, must never be deposited in the rubbish boxes in the laboratory.

Storage

Bulk quantities of inflammable materials should be stored in a properly constructed store built of nonflammable materials and situated at a safe distance from the main building. In built-up areas, it is usually permissible to erect these on the flat roofs of buildings. The store, which should preferably be constructed of concrete or brick and concrete, must also have a heavy fireproof door. The store should be ventilated but the vents should be protected so that no danger of fire from an outside agency is possible. The floor should be sunken or the sill raised so that in the event of fire the complete contents of the store may be retained and flaming liquids are not likely to run out and spread the fire to neighbouring buildings. Light switches are situated outside the store and vapour-proof lights should be fitted inside. Some method of automatic fire control by a piped dry chemical system is desirable.

CAUSE AND PREVENTION OF FIRE

Where laboratory fires are concerned, prevention is certainly better than the cure. The success of a fire brigade is judged not by the number of fires overcome but by the scarcity of fire outbreaks in its district. The desirability of maintaining close relationship with the local fire brigade cannot be overstressed. Occasionally the brigade personnel should be invited to visit the laboratories to assess the fire risks and to familiarise themselves with the layout of the buildings.

Hazardous materials should not be stored unnecessarily in the laboratory and all hazardous operations must be carried out in specific locations. If fireproof rooms are not available, the use of fireproof partitions should be considered, since these serve to prevent the spread of fire. Sprinkler systems should be installed in suitable locations.

The prevention of fire, too, depends on good lab-keeping. This involves tidiness in all laboratories and storerooms, the constant surveillance and maintenance of laboratory services, and the efficiency of the firefighting equipment. All laboratory staff should receive instructions as to their function in the event of fire and a good warning system should be installed. Note that accumulations of rubbish, such as paper, rags, and wooden material, constitute a major fire hazard in the laboratory.

Spilled solvents must be immediately wiped up and the use of electrical apparatus, such as hair dryers, to increase the rate of evaporation of inflammable solvents, should be avoided.

Inflammable liquids should not be stored in household-type refrigerators unless the controls have been competently placed outside the cabinet.

FIREFIGHTING EQUIPMENT

The hand-size and larger sizes of fire extinguishers should be available in all laboratories, and, in addition, large extinguishers should be available in corridors. The position, number, and size of extinguishers depend upon the shape and size of the laboratory and upon particular local hazards within it. Generally speaking, one extinguisher should be positioned at each end of the laboratory but not directly within the area where the fire risk is greatest.

A water hose is necessary at a central point and its length should be sufficient to reach the extremities of the building. In larger buildings several hose points may be necessary.

The mere provision of fire appliances is not enough and only regular familiarisation with them ensures complete safety. Regular fire drills should be held and full use made of the various appliances during the practice. The equipment should be regularly checked and extinguishers refilled. When a fire appliance has been discharged, a report should be made immediately to the officer responsible for fire equipment.

Safety showers which can be used in an emergency should be installed above the exits or in other carefully selected places in chemical laboratories. Showers of this type are designed to give a cascade of water and can be used by persons with their clothing on fire or by those who may have been splashed with acid or other corrosive substances. The showers are usually operated by means of a hanging chain placed at the correct height. Since some difficulty might be experienced in grasping the large ring pull in an emergency, a new pattern, which is self-operating when the victim stands on a sunken mat beneath it, is best. This may eventually be replaced by another type, now in an experimental stage, which is operated by a magic-eye ray.

Although showers are preferable to fire blankets, the provision of both is desirable. Fire blankets may be strategically placed around the laboratory for quick local action.

Classes of fire risks

When using firefighting appliances the type of risk involved must be considered. The four main classes of fire risks are:

Class A: Carbonaceous or general fire risks, e.g. paper, wood, fabrics, general rubbish.

Class B: Inflammable liquids, e.g. petrol, benzene, paint, tar, oils.

Class C: Electrical fires, e.g. electrical apparatus or machinery.

Class D: Metal fires, e.g. aluminium, magnesium, sodium, potassium.

Extinguishers and their use

Soda acid and water–CO_2. (Suitable for Class A fire risks)

Soda acid or water–CO_2 extinguishers are intended only for use on ordinary combustible materials but may also be used on inflammable liquids which are soluble in water. They must not be used on electrical fires, on imflammable solvents immiscible in water, or on sodium or potassium fires.

SAFETY IN LABORATORY AND WORKSHOP

Soda acid extinguishers consist of a metal cylinder containing a solution of sodium bicarbonate. Situated inside the cylinder, below the plunger, is a vessel containing sulphuric acid. Striking the plunger releases the acid to mix with the bicarbonate solution and the pressure created by the resultant gas causes the liquid to be forced from a jet. The jet may be directed at the fire from a distance of up to about 30 feet.

The action of the water–CO_2-type extinguisher is similar. The pressure gas in this case is released from a small CO_2 cartridge situated below the striking plunger.

Foam. (Suitable for Class B fire risks)

Foam-type extinguishers are suitable for fires involving inflammable liquids, but should not be used on electrical fires as the foam is an electrical conductor. The foam is contained in a metal cylinder and is ejected when pressure from a CO_2 high-pressure cartridge situated below the striking knob is released. The minute bubbles of CO_2 which constitute the foam blanket out the flames. Some types of foam extinguishers employ a premixed foam compound which enables the extinguisher to be quickly recharged.

B.C.F. (Bromochlorodifluromethane). (Suitable for Class B and C fire risks)

B.C.F. extinguishers, known as high vaporising liquid extinguishers, may be used on inflammable liquid fires and on electrical fires. These extinguishers should not be used in confined spaces.

An advantage of this type of extinguisher, which is fast replacing the carbon tetrachloride (C.T.C.) extinguisher, is that it does not produce toxic fumes when the liquid comes into contact with the fire. Also, the low boiling point and rapid volatilisation properties of B.C.F. make it an excellent medium for firefighting. Some extinguishers of the smaller variety have an expendable body, so that after the extinguisher has been used a new body is simply screwed on to the head and the extinguisher is then ready for further use.

Carbon dioxide. (Suitable for Class B and C fires)

The CO_2 cylinder is a very popular laboratory extinguisher and can be obtained in sizes ranging from a 50-lb (22.7-kg) mobile unit down to small $2\frac{1}{2}$-lb (1.3-kg) hand-size cylinders suitable for small bench fires. CO_2 extinguishers can be used for electrical fires and those involving materials immiscible with water. They can also be used on a person whose clothing has caught alight but care must be taken that the extinguisher is not discharged into the person's face.

From large cylinders the CO_2 is released by a screw valve and from smaller ones by trigger action. The gas extinguishes the fire by 'blanket' action. This type of extinguisher is especially useful in laboratories because of its clean action which is an essential feature where expensive apparatus is concerned. During use the discharge from the cylinder can be terminated immediately at any time and if the cylinder is not completely emptied (the contents can be checked by weight) it can be re-used.

Dry chemical extinguishers. (Suitable for Class B, C, and D fires)

As well as being suitable for use on electrical fires and fires involving inflammable liquids, dry chemical extinguishers may be used for fires involving metals. The sizes of these extinguishers range from 20 lb (9 kg) down to the small varieties weighing only 3 lb (1.36 kg) which can be operated with one hand and are very suitable for small laboratory fires.

The dry chemical extinguisher consists of a metal container filled with a dry powder which is non-hygroscopic, non-conducting and non-toxic. The powder is released when the CO_2 cartridge situated inside the cylinder is caused to release its gas pressure and after use is quickly replaceable.

Sand

Sand in fire buckets must be kept clean and dry. This material is very useful for sodium, potassium, and lithium hydride fires where the use of water, carbon dioxide, or carbon tetrachloride is dangerous.

Asbestos blankets

The use of asbestos blankets for bench fires is not recommended. The blankets are heavy and may damage apparatus unnecessarily and so spread the fire. They may be used, however, to smother fires which may occur in large wide-mouth vessels.

Water

For fires of a general nature involving ordinary combustible materials, water is still an important firefighting aid and an adequate supply of pressure water is essential. It may be used on inflammable liquids which are soluble in it.

Other extinguishing agents

Dry soda ash, sodium bicarbonate, and salt are all useful fire-extinguishing materials and may be kept ready for use in boxes around the laboratory. These serve a useful function in fighting sodium or potassium fires. Soda ash and sodium bicarbonate serve a dual purpose since they are also a standby for neutralising acid spillages.

FIREFIGHTING

Most bench fires can be tackled in a simple way such as by smothering with a damp bench swab and by closing the mouth of a flask or the lid of a container. Some materials, such as explosives or incendiaries, because they are self-sufficient in fuel, will burn to completion, but with normal fires the exclusion of fuel or air will prevent combustion.

When an outbreak of a fire cannot be controlled by hand appliances and a major blaze results, the room must be evacuated and the fire brigade summoned. The gas and electricity must be shut off at the mains and doors and windows closed. Neighbouring inflammable materials and any valuable records or apparatus should be removed from the vicinity of the fire, if this can be accomplished without danger to persons.

GAS

Any fuel gas supply is a constant danger and requires careful control. The treacle tin experiment of our schooldays vividly illustrates that such gases are, when mixed with air or oxygen in the right proportions, violently explosive. Petrol gas and propane–butane gas are particularly hazardous and piped installations for these gases require expert fitting. The whole installation must be rigorously tested for leaks before the system is used. Since such gases tend to accumulate in closed spaces to an even greater extent than coal gas, the slightest leak which may develop in these systems requires immediate attention. Whenever an accumulation of gas is suspected, naked lights should be extinguished, the gas shut off at the main valve, and doors and windows opened. To trace gas leaks only soap and water should be used and never naked flames.

When closing laboratories at night all burners and local gas cocks must be off and only then should the supply be closed at the main valve. As a further safeguard against leaks, all electrical apparatus should also be turned off at night. If for any reason it is absolutely necessary to leave any gas, water, or electrical appliance on overnight, it should bear a card signed by the person in charge of the section and the caretaker should be informed.

Each room should have its own gas valve, the position of which should be clearly marked and the ventilation in the room should be adequate for the number of gas burners involved or unhealthy conditions will prevail.

Whenever gas is passed through apparatus and is ignited at an outlet, a test tube sample should be first tested to ensure that the gas burns quietly and that all the air has been expelled.

Rubber tubing on gas burners must be regularly inspected and replaced if old or damaged. Pilot lights on gas appliances should always be lit before the main supply, and if, as is sometimes the case, no pilot light is fitted, a light should be held to the burners before the gas is slowly turned on.

When gas rings, Meker burners, and similar appliances are used, the reflected heat is liable to burn wooden bench tops and a piece of thick asbestos should be used to protect these. It is very convenient to have in the laboratory special 'hot' benches the tops of which are covered with a heat-resisting material such as Sindanyo.

PRECAUTIONS AGAINST FIRE

An adequate number of rubbish bins, marked in accordance with the type of rubbish to be deposited in them, are necessary in laboratories and the type with self-closing lids are an advantage. The bins should be emptied daily and the contents burned at a safe distance from the laboratory. Special bins, which should never be kept near radiators or other heat sources, must be provided for oil and paint rags, which are very hazardous materials. Rags which have been used for mopping up spilt acids may also easily take fire and these, too, should not be discarded in bins containing general rubbish.

Smoking in the laboratory should be prohibited, since the careless disposal of matches and cigarette ends has been responsible for many outbreaks of fire, and for the same reason flint lighters ensure greater safety.

At times when the laboratory is not in use, care must be exercised to ensure that wash bottles and other liquid-filled glass containers are not left in the path of direct sunlight.

Hot tripods and charcoal blocks should be allowed to cool before they are put into wooden cupboards.

Sodium residues, even when apparently inactive, should be properly treated with alcohol before disposal and must never be put into the sink.

The design of furnaces and ovens should be such that apparatus may be easily manipulated within them.

Hot crucibles and other hot objects must not be put directly on to the bench but should be placed on an asbestos sheet kept close by for this purpose. In student laboratories a piece of hard asbestos about 13 mm thick and 300 mm square should be issued to each student.

The use of inflammable lagging or insulating material, on vessels likely to be heated, is to be deplored and inert materials such as asbestos string are best for this purpose.

HAZARDOUS OPERATIONS

Persons should not sit at a bench when performing hazardous tasks such as heating vessels filled with acid or oil, since this is inviting disaster. The hot or flaming liquid may be spilt into the lap. For such operations bench trays, or raised rims to the bench tops, are worthwhile safeguards. The oil in oil baths must be free from water or it may froth from the vessel on to the heating medium and take fire. Electrically heated mantles may be used in preference to oil baths and in many cases Duralumin blocks are an advantage.

DANGERS FROM EXPLOSION

LABORATORIES HANDLING EXPLOSIVES

Buildings in which special risks are prevalent owing to the handling or storage of explosives should be so spaced that any building is at a safe distance from its neighbour. Protective walls or earth mounds may separate the buildings.

The building in which laboratory tests involving the handling of explosives are carried out is ideally a single storey of fireproof construction. The building should be well ventilated and have wooden floors and wooden benches. Special sparkproof electrical fittings and the other safety precautions necessary for areas involving fire hazards should be adopted. The facilities in the laboratories should include fume cupboards with good draught, the provision of appliances for personal protection including ample protective clothing, and good arrangements for personal hygiene. The normal services must be provided and these should include steam. Dermatitis has to be carefully guarded against in explosives laboratories because of the handling of nitro bodies such as T.N.T.

Bulk inflammable solvents and explosive samples must be kept in special stores situated away from the main building.

The first rule in the safe handling of explosives is that they must not be subjected to friction or shock. For this reason, smooth bottles of good quality, closed with soft rubber stoppers, are necessary. For dangerous initiatory materials, paper containers and thin paper tools are used. Rooms in which dangerous dusts are produced should be constructed in a way which allows the dust to be easily and regularly removed.

EXPLOSIONS DUE TO INFLAMMABLE SOLVENTS

Volatile inflammable liquids give off inflammable vapours. If the concentration of the vapour reaches certain limits and is mixed in the right proportions with air, ignition causes combustion which proceeds at high speed and with great violence. The sudden expansion which accompanies the combustion constitutes an explosion. Explosions generally occur when the vapour concentration is low. It is for this reason that empty inflammable solvent containers, owing to an accumulation of air and vapour mixtures within them, can be more dangerous than when full.

EXPLOSIONS DUE TO LIQUID AIR

Simple materials such as cotton wool burn explosively if contaminated with liquid air. Liquid air also forms explosive mixtures with reducing agents and is dangerous in the presence of hydrocarbons. A robust trap should be employed when liquid oxygen is used as a coolant. It is much safer to use liquid nitrogen. The investigations carried out by McCarty and Balis[6] show that liquid nitrogen is not appreciably contaminated by the oxygen in the air but that oxygen may be present in liquid nitrogen as an impurity. The oxygen content of impure liquid nitrogen tends to become higher, owing to evaporation of nitrogen, which boils at a lower temperature than oxygen.

EXPLOSIONS DUE TO DUST

A substance when dispersed in the air in a finely divided form is known as dust. Dust from combustible substances can, if sufficiently dense and when mixed with air, ignite and explode violently. This is due to its rapid combustion and the subsequent expansion.

The formation of explosive, irritant, or toxic dusts must therefore be prevented. The best means of doing this is by enclosing the source and exhausting the dust locally. Dangers arising from the fouling of the exhaust system are overcome by keeping it clean and regularly testing its efficiency. In areas where the total prevention of dust is impossible, all likely sources of ignition must be excluded and for personal protection it may be necessary to use dust masks or respirators. Some operations may need to be conducted in an inert atmosphere and for others wet methods of grinding may be possible.

EXPLOSIONS DUE TO GASES

Many gases are also explosive when mixed with oxygen or air, and pieces of apparatus which have been used for containing or transporting such gases should be made safe after use by filling with water or by blowing air through them.

Hydrogen

The explosive properties of hydrogen when mixed with oxygen are vividly illustrated by the soda water bottle experiment. In this particular demonstration experiment, care must be taken to ensure the safety of the onlookers by wrapping the bottle in dusters or by enclosing it in wire netting of fine mesh.

Precautions must be taken with all experiments involving hydrogen, and when a jet of this gas is to be ignited it must always first be sampled to see that all the air has been expelled from the gas-producing apparatus. Open flames should be kept well away from gas generators, especially when air is being displaced from them. For class experiments involving the reduction of oxides, it is safer to use coal gas than hydrogen, provided the local coal gas supply is known to contain sufficient hydrogen for this purpose. Iron tubes are safer than glass ones for the preparation of the gas by the action of steam on metals. The skin of sodium peroxide should be cut off sodium metal pieces when making hydrogen by the action of the metal on water. Hydrogen is a light gas which easily diffuses through small apertures into lines or vessels to form explosive mixtures.

Hydrogen sulphide

Sulphuretted hydrogen is inflammable and when mixed with air is explosive. The partial displacement of air from apparatus containing this gas must be guarded against, and when refilling gas generators, the operation should not be performed in the presence of an open flame.

Fuel gas

Fuel gas mixed with air forms an extremely dangerous mixture. If a gas leak is detected all doors and windows should be opened immediately and precautions taken to ensure that no lights are struck.

Acetylene

Acetylene is a dangerous endothermic compound and at elevated temperatures and pressures is liable to spontaneous decomposition.

EXPLOSIONS DUE TO CHEMICALS

Many chemicals in common use in the laboratory are explosive in nature or may form explosive compounds. Particular care must be taken with all experiments involving substances such as concentrated acids, metallic sodium, potassium, phosphorus, nitrates, concentrated hydrogen peroxide, persulphates, perchloric acid, chlorates, and other strong oxidising agents. Ammonium nitrate, which decomposes exothermically to give gaseous products, is, in common with other oxidising agents, dangerous if mixed with organic material. Experiments involving the decomposition of this salt should be performed with great care as it may explode on heating. The thermal decomposition of solids involves considerable risk and care must be taken to see that outlet tubes are not blocked.

Peroxides

Most peroxides, even in storage, can be dangerous. Hydrogen peroxide, although fairly stable when pure, may decompose explosively, particularly in contact with dust or finely divided metals. It is safe to store when diluted, and like all peroxides it requires cool storage conditions. In contact with the skin, hydrogen peroxide causes white blisters.

Ether peroxides

Ether peroxides are formed by the oxidation of di-ethyl ether, isopropyl ether, and higher ethers, particularly in the presence of ultra-violet rays. For this reason ether should be stored over a spiral of bright copper, active carbon, or aluminium oxide or in an atmosphere of nitrogen which prevents the formation of the peroxides. Ether should be kept in bottles wrapped in black paper. Brown bottles are not suitable. The oxidation of the ether proceeds more quickly when the bottle is only partially filled and the liquid surface is in contact with air.

Because the peroxides have a higher boiling point than the ether from which they are formed, towards the end of a distillation they become concentrated in the vessel and may explode violently. Ether and ether residues must always be tested before distillation and if peroxides are present these should be destroyed by the addition of acidified ferrous sulphate solution.

GENERAL PRECAUTIONS WITH POTENTIALLY EXPLOSIVE MATERIALS

Whenever experiments involving the use of chemical substances and gases are performed, there will always be an element of risk, but by obeying a few simple rules these may be more safely conducted.

Whenever it is possible to do so, experiments involving the use of hazardous materials should be conducted in a specially protected area, in a fume cupboard, or in the open air.

Before a hazardous experiment is conducted permission should be obtained. Experiments with volatile solvents should not be carried out in the presence of flames.

If the behaviour of any substance is in doubt under the conditions in which it is to be used, a small quantity should be used in the first instance. This should be heated on a water bath and not on an open flame. The usual precautions for personal protection should be adopted such as the use of goggles or a safety mask.

For all experiments in which the glass apparatus is subjected to pressure or vacuum, the correct shape and thickness of vessel should be used and the apparatus must be shielded by an appropriate guard or safety screen.

The taking of chemicals from the laboratory must be strictly forbidden. Many accidents have occurred through chemicals being used for pyrotechnics and unauthorised home experiments. These occurrences commonly involve young people of the 15–17 age group. In school laboratories potentially dangerous materials such as chlorates, magnesium, and poisons should be under lock and key.

When demonstration experiments are conducted, such as the thermite experiment or any similar one likely to be dangerous, a safety screen should be used to protect the class and the demonstrator. Remember that an exothermic reaction may be hazardous even when heat is not applied. Before treating sodium peroxide with water to prepare oxygen, it should be ascertained beyond doubt that the peroxide is pure.

SOME COMMON CAUSES OF EXPLOSIONS

It is impossible within the limits of this book to list the many chemical substances or combinations of substances which could present an explosive hazard. The few examples which are given, however, may serve to illustrate the variety of these. Explosions may occur owing to:

The action of concentrated sulphuric acid on potassium permanganate and chlorates.

The storage of anhydrous aluminium chloride in unvented bottles with screw-down lids.

The washing of vessels which have contained sodium or potassium.

The storage of ammoniacal silver solution.

The action of water on calcium carbide.

The improper preparation of mixtures for analysis such as the inclusion of chlorates or the mixing of powdered metals with oxidising agents.

The action of perchloric acid with alcohols, dehydrating agents, and combustible materials.

The heating of nitrates with sodium thiosulphate or stannous chloride.

PERCHLORIC ACID

The normal strength of perchloric acid solution used in the laboratory is 60–72% $HClO_4$. If the solution comes into contact with strong dehydrating agents, however, the anhydrous perchloric acid may be formed which is explosive. If the anhydrous acid, which is a volatile colourless liquid, is prepared, it may explode after standing for a few days. A drop of the acid in contact with wood or paper causes an immediate conflagration and with charcoal explodes. For these reasons organic matter should never come into contact with perchloric acid or with a solution of the acid.

Special fume cupboards are necessary to wash perchloric acid vapours and accumulations of dust, or the residues in the ducts of fume cupboards in which perchloric acid is used, may lead to an explosion.

PRESSURE VESSELS

Autoclaves

Autoclaves are best situated behind a solid wall which provides complete protection for the operator. The construction of the autoclave should be such that it has an efficient safety valve and regulator, and the working of these must be regularly checked. The safety valve should be set at a pressure not exceeding two-thirds of the test pressure.

When choosing a pressure vessel, both the temperature and the pressure must be considered in determining whether the vessel is suitable for the conditions it must withstand. The material of which the vessel is made must be suitable for use at the working temperature.

After use, pressure vessels should be immediately cleaned and all working parts checked over; the information gained by such inspections and tests should be recorded for future reference by any user.

Whenever there is any doubt about the working capabilities of the vessel under conditions which may be expected to arise during the course of new work, a small-scale experiment should first be performed from which useful information may be gained.

DANGERS IN STORAGE

PLANNING CHEMICAL STORES

A well-designed store minimises the hazards associated with the storage of chemicals, liquid reagents, and gas cylinders. It must be well ventilated and spacious enough to allow good clearance in the aisles. The store should be situated, if possible, on a ground floor with a fireproof door opening to the outside. This allows direct access to the open air in an emergency and the convenient and safe delivery of packages. The store should also be constructed in such a way that a moderate and even temperature can be maintained. Separate rooms are required for the safe storage of acids, ammonia, and gas cylinders. Safety measures such as adequate fire extinguishers, protective clothing, and vapour-proof light fittings, as are normally provided in locations where hazardous materials are involved, should be installed. In addition a good supply of sand, a shovel, and an all-purpose respirator are essential. Sprinkler systems are recommended only if the materials stored are safe when water is used.

GENERAL PRECAUTIONS FOR STORES

Smoking

Smoking in the stores should be prohibited. Unauthorised persons should not be admitted or allowed to dispense or carry stores materials.

Shelving

Any shelving must be strong and able to easily withstand the weight of materials stored upon it. Every precaution should be adopted to ensure that accidents arising from the leakage or spillage of any material are prevented. For this reason, a beading which protrudes above the front edge of the shelving should be affixed.

The use of chairs and stools for reaching materials on the shelves should not be allowed and a sensible pair of non-slip steps or a ladder should be provided. Heavy items should be stored on lower shelves.

Trolleys and carriers

Four-wheeled trolleys are an advantage for the safe transportation of materials and safe hand carriers for Winchester and other bottles are necessary.

Inflammable materials

All inflammable materials should be kept in a separate inflammable store. When boxes have been unpacked the wood wool and other packing material should be taken outside to prevent an accumulation of combustible material in the store.

ACID STORES

The floor of the acid store should be made of cement and sloped towards a drain in one corner. The shelves, too, are best constructed of inert material, such as precast slabs made with acid-resisting cement, and these should not extend too high up the walls. A supply of water for washing down the floor in the event of spillage and for emergency use in the event of personal contamination must be available. A large container of soda ash or sodium bicarbonate should also be kept ready for use and irrigation devices for the eyes, such as eyewash bottles, are also necessary. Safety siphons and carboy tilters must be kept in the store for dispensing acids and other dangerous liquids. The store requires a strong ventilating fan.

AMMONIA STORES

Where the quantity of ammonia used is considerable, a separate store is desirable and its construction should be similar to that of the acid store.

STORAGE OF CHEMICALS

HAZARDOUS COMBINATIONS OF CHEMICALS

Hazardous combinations of chemicals must be avoided to reduce the risk of fire, explosion, and fumes. Strong oxidising agents and easily oxidisable material should not be kept close together, and other chemicals liable to react vigorously must also be stored apart. Dangerous chemicals which resemble each other in appearance, and particularly where the media in which they are stored are dangerous one to the other, must also be separated. An example of this is sodium which is kept in closed containers under naphtha and white phosphorus which is stored under water. Sodium is often received in tins which in the corrosive atmosphere of the store rapidly deteriorate. If this occurs the contents should be transferred to a strong sealed-glass container. Bottles containing substances which require to be vented should be kept accessible for regular inspection.

VOLATILE LIQUIDS

Volatile liquids like ether which may exert a high pressure, and others such as nitric acid which are likely to exert pressure, should be kept in a cool place away from direct sunlight, hot water pipes, and other sources of heat. The containers for liquids of these kinds must never be completely filled and should be opened with care. It is advisable to wear goggles. Inflammable liquids must not be stored on the open shelving, and empty inflammable liquid containers must be closed securely or may be kept filled with water.

CHEMICAL CONTAINERS

Bottles containing dangerous substances, such as bromine, must not be kept on high shelves. Any bottle containing a substance liable to give off toxic fumes, or highly objectionable odours if broken, should also be stored in a safe place. An accident involving these may cause no personal harm but may render the store uninhabitable for days. Other chemicals, such as sodium peroxide and phosphorus pentoxide, need to be kept tightly sealed and when opened should be rewaxed around the stopper after use. Where the spillage of a substance may result in dangerous conditions, it may be kept on the shelf in some larger inert container which will contain it in the event of breakage.

Labelling of containers

All containers should be labelled prior to filling and must never be filled with substances other than that detailed on the label.

DANGERS FROM THE DISPOSAL OF CHEMICALS

A quantity of unwanted chemicals usually accumulates in chemical laboratories and stores owing to the deterioration of labels, through contamination, or as residues. Such chemicals may be of known or unknown character, and in either case their disposal may involve a considerable element of danger. Perhaps it is for this reason that the task is shunned by everyone until the accumulation of material is such that the task can no longer be ignored.

All disposals should be carried out by a responsible person and if any doubt exists in his mind concerning the handling of a particular chemical someone else with special knowledge of the chemical should be consulted. Persons responsible for disposing of chemicals normally evolve their own safe methods and in the past little written information on this important subject was available. A publication by Gaston[2] has, however, rectified this omission and some of the techniques recommended by him for the disposal of the more common chemicals are briefly summarised in *Table 6.1*.

In order to accurately assess the degree of danger involved in the disposal of a chemical, a sound knowledge of its properties is necessary. If the nature of the chemical is unknown, extra care is required and *very small quantities* of the substance should be tested before the disposal is attempted.

If the substance is a solid, its appearance, odour, or inflammability may suggest the best method of disposal, and similarly in the case of liquids, miscibility, inflammability, volatility, or odour may prove to be important indications.

During disposal all the necessary safety precautions should be

Table 6.1 TECHNIQUES FOR THE DISPOSAL OF COMMON CHEMICALS

Chemical	Method of disposal
Acid anhydrides	
Acetic anhydride, Acetyl chloride, Thionyl chloride	hose down drain from safe distance; very small quantities may be poured into water with stirring
Chromium trioxide (or chromic acid)	add carefully to plenty of water then wash down drain
Phosphorus pentoxide	spread on an open dish and leave exposed to air in a well-ventilated spot until it takes up sufficient moisture from the air; warning label required; wash small portions at a time of the hydrolysed material down the sink with plenty of water
Sulphur trioxide	break container in a dry pit and hose with water from a safe distance
Concentrated acids	
Aqua regia, Formic, Acetic, Hydrochloric, Nitric, Sulphuric, Perchloric	dilute by pouring carefully into plenty of water and then wash down sink with tap discharging
Chlorosulphonic	break the container from safe distance and hose down in disposal pit
Hydrofluoric	neutralise carefully with powdered calcium carbonate. Leave to stand and after several hours pour down drain with plenty of water
Oleum	surround container with plenty of sodium carbonate and break container from safe distance. Hose down with water
Explosives and other violently reacting substances	
Diazonium salts, Picric acid	dissolve in hot water and wash down drain
Ammonium nitrate	wash down drain
Chlorate, perchlorates	hose away in disposal pit
Hydrides: calcium hydride, lithium aluminium hydride, sodium hydride	very small quantities can be carefully added to a large volume of water OR spread on a dish in a flamefree and well-ventilated room until they take up sufficient moisture from the air. Wash down drain. Warning label required
Peroxides: benzoyl peroxide	add in small portions to strong alkali solutions (sodium hydroxide) and agitate to prevent settling. If mixture becomes thick add water
hydrogen peroxides, peroxides miscible with water	dilute well with water and pour into drains

Table 6.1 TECHNIQUES FOR THE DISPOSAL OF COMMON CHEMICALS—*continued*

Chemical	Method of disposal
slow-burning peroxides (solid)	spread safe quantity over a small pile of wood shavings and kindle from distance with a fuse such as cotton wool soaked in alcohol on metal rod
slow-burning peroxides (liquid)	burn in a pit. Ignite safe quantities with a fuse
sodium oxide and peroxide / potassium oxide and peroxide	dissolve in plenty of water and wash down sink with taps discharging

Inflammable substances
Solids:

magnesium (turnings or powder)	add to 5% hydrochloric acid in small portions in a beaker away from flames. Stir and cool if necessary
white phosphorus	burn in open and keep well away
potassium	cover with glycerine. After some time when metal has dissolved and little or none remains, *cautiously* add small quantities of ethyl alcohol. When no further reaction occurs dilute with water and dispose down drain
sodium	*very small quantities* may be added to ethyl alcohol in a beaker. Carry out operation in a fume cupboard away from flames
Liquids (solvents): benzene, ether, carbon disulphide, ethyl alcohol, methyl alcohol, dioxan, acetaldehyde, acetone	burn in open in shallow metal trays let in the earth. Ignite small quantities (500 ml maximum) with a fuse such as cotton waste soaked in alcohol on end of a metal rod OR small quantities can be spread over a safe area in the open and allowed to evaporate
Other solvents: acetylene tetrachloride trichloroethylene carbon tetrachloride	allow to weather on open ground

Poisons

Mercuric chloride	treat solution with sulphur dioxide gas or sodium bisulphite. Non-poisonous mercurous chloride is then precipitated
Beryllium and its salts	dissolve in hydrochloric acid. Wash down sink with plenty of water
Dimethyl sulphate	treat cautiously with solution of ammonia. Wash down sink
Oxalates	allow to react with excess potassium permanganate in hot aqueous solution (above 60 °C). This is done in acid solution, e.g. in dilute sulphuric acid

taken. Rubber gloves and goggles should always be worn. If substances are to be put down the drain it should first be ensured that this method of disposal is in accordance with the regulations of the local authority. It is safer to dispose of chemicals one at a time and they should never be mixed indiscriminately.

DANGERS FROM GAS CYLINDERS

STORAGE

The storage room for gas cylinders should be unheated but protected from extreme heat or cold and direct sunlight. The room also requires good top and bottom ventilation. Smoking, or the use of naked lights, should be prohibited inside the store. The light fittings should be the vapour-proof type and the light switch should be positioned outside the room. The best position for the store is close to an outside exit. Separate storage for empty and full cylinders is advisable. Outside storage is necessary for cylinders containing poison gases but these must be adequately protected from ice and snow and the direct rays of the sun.

Cylinders should be stored in an upright position and secured to the walls with their protective metal caps in place. Acetylene cylinders must always be stored upright and kept apart from oxygen cylinders. If a cylinder suffers damage of any kind, it should be returned to the suppliers accompanied by a statement giving the exact nature of the damage.

TESTING FOR LEAKS

Soap solution is used for testing for gas leaks and never naked flames.

COLOUR CODE

The contents of a gas cylinder are denoted by the colour of the cylinder. *Table 6.2* shows the colours used for the more common gases used in laboratories. Full colour codes for industrial[7] and medical[8] cylinders are available.

The valve on all combustible gas cylinders has a left-hand thread and those filled with noncombustible gases a right-hand thread. This is a safety measure to prevent the use of a wrong cylinder, since the paint on cylinders may deteriorate through bad storage and handling. It also prevents the interchange of cylinder fittings. On no account must the colouring or markings on cylinders be altered nor should an attempt be made to fill a cylinder with gas. They should be filled only by the gas suppliers.

Table 6.2 COLOUR CODE FOR GAS CYLINDERS

Gas	Cylinder colour	Shoulder colour
Industrial		
Acetylene	maroon	—
Air	grey	—
Ammonia	black	yellow/red
Argon	blue	—
Carbon dioxide	black	—
Carbon monoxide	red	yellow
Chlorine	yellow	—
Coal gas	red	—
Helium	brown	—
Hydrogen	red	—
Methane	red	—
Nitrogen	grey	—
Oxygen	black	—
Sulphur dioxide	green	yellow
Medical		
Air	grey	white and black
Carbon dioxide	grey	—
Helium	brown	—
Nitrogen	grey	black
Nitrous oxide	blue	—
Oxygen	black	white
Oxygen–carbon dioxide mixture	black	white and grey

TRANSPORTATION

Cylinders must never be rolled, dragged, or banged together but should be transported singly with the valve closed and all fittings and regulators detached. It is dangerous for any person to attempt to lift a heavy cylinder on his own since this may result in serious internal injury. Poison gas cylinders should always be handled by two persons.

VALVES AND FITTINGS

The correct pressure regulators must be used in accordance with the type of gas contained in the cylinder. Regulators are usually painted the same colour as the cylinders for which they are suitable, or alternatively the name of the gas is marked on them. Regulators do not need to be hammered tight in the valve, and hand pressure and the correct key are sufficient.

The valve should be turned on slowly since a sudden release of pressure is dangerous and may damage the pressure regulator. Care should be taken to avoid over-tightening the valve when the gas is shut off and nothing should be used to obtain extra leverage on the key, as this will damage the valve spindle.

Valves and fittings do not require lubricating. The valve threads must not be tampered with and should be kept free from dirt, grit, water, oil, and grease.

Cylinders should never be connected to glass apparatus before first testing the valve so as to ascertain the sensitivity of adjustment. The sudden release of gas from a cylinder has caused many bad accidents.

On no account should oil and grease be allowed to come into contact with oxygen or acetylene valves, since these materials may ignite violently or cause an explosion. Oxygen valves should not be handled with greasy hands, greasy rags, or when wearing greasy or oily clothes.

CYLINDERS IN THE LABORATORY

Cylinders must never be laid on the laboratory floor but should be secured in a stand or strapped to a bench. After use, or when the laboratory is to be closed, they must be disconnected from the apparatus or plant with which they have been used.

Acetylene

With copper, silver, and copper alloys containing more than 70% copper, acetylene may form explosive compounds. Copper piping or joints for this reason are not used with this gas. Acetylene cylinders may become hot owing to backfire from faulty equipment or from other accidental heating. If this occurs the cylinder valve should be closed and the cylinder taken outside into the open air and cooled with water. The valve is then fully opened, the cylinder kept cool, and its contents allowed to escape.

Chlorine

Chlorine cylinders are fitted with special needle valves, but even these corrode and become jammed. In some cases they break off. These cylinders should not be stored for lengthy periods and it is expedient to return them to the suppliers, even when they still contain gas, rather than risk the possibility of a stuck valve.

CYLINDER FIRES

If a cylinder takes fire the flame should be extinguished and the valve closed. The cylinder is then cooled by spraying it with water.

DANGERS FROM ELECTRICITY

The passage of small electric currents as low as 25 milliamperes a.c. or 80 milliamperes d.c. through the human body may be sufficient to cause death by the failure of the heart or respiration. It is not always high voltages which are responsible for fatalities, since the degree of danger to life by shock depends on the particular conditions prevailing at the time of contact. The resistance of the body varies according to the circumstances and the area of contact. Wet or damp conditions are particularly dangerous. Fatal accidents have occurred at 60 V.

HIGH TENSION

The growth in recent years of the amount of electrical equipment used in the laboratory has also increased the need for safety measures. In much of the apparatus, voltages above 500 V are used and special precautions are necessary.

In those departments where a great deal of work with high voltages is undertaken, it is desirable to provide specially protected h.t. enclosures. An enclosure of this type has been fully described by Cornwell[9]. It is constructed in a way which ensures perfect safety for qualified staff carrying out experiments involving the use of h.t.

The enclosure described, which is efficiently earthed, is of permanent metal construction and completely encloses any apparatus which can become 'live' at a dangerous voltage. When the door or any other aperture into the cage is opened, the main supplies are disconnected. The accidental closing of the gate is impossible since any person inside the enclosure can be easily seen by others having a reason to close it.

The main supplies to the enclosure are taken through removable fuses, so that whenever repairs or alterations to equipment become necessary the enclosure can be made 'dead'. Only authorised experimental staff may enter the enclosure without special permission, and when it is necessary for other technical staff to enter the officer responsible for the enclosure takes custody of the fuses. The position and number of the fuses are clearly shown on a detailed wiring diagram fixed to the cage. The fuses are not replaced until the visiting technician has vacated the enclosure.

Many other precautions are taken to ensure that the enclosure is safe and included among them is the provision of earthed metal covers over terminals fed direct from external supplies. Electrical leads and pipes are protected so that they cannot conduct high voltages outside the cage.

SPECTROGRAPHIC EQUIPMENT

To comply with the regulations governing the control of high-frequency apparatus, spectrographic equipment should be housed in an efficiently earthed screened room or enclosure. A method of screening which satisfies Post Office requirements has been described by Smith and Walsh[10]. The sparking apparatus used for the spectrographic analysis of metals and alloys is completely enclosed in a wire cage which has a counterbalanced front door. The electrical circuit is broken on opening or closing the door of the cage. This avoids the accidental use of an unscreened spark and protects the operator.

TYPES OF ENCLOSURES

Fully protected

Fully protected enclosures are those made as safe as possible for operations carried out by unskilled persons.

Semi-protected

In semi-protected enclosures, although safety arrangements are made, they may not be completely safe because the output is taken to experimental circuits.

Unprotected

Unprotected enclosures are those in which the nature of the experimental work precludes the full use of safety devices. On this type of enclosure a notice should be displayed to the effect that it is unprotected and entrance to unauthorised persons is prohibited.

GENERAL LABORATORY WORK WITH H.T.

If h.t. work is carried out on open benches in the laboratory, warning notices should be prominently placed on the bench. A light wooden screen at a convenient height should be erected around the edges of the bench. This draws the attention of passers-by to the dangerous possibilities of the work and also restricts apparatus and electrical leads to the confines of the bench. The floor of the laboratory should be covered with good insulating material.

The h.t. bench should be made of wood and any metal pipes should be well hidden and in such a position that they cannot be accidentally touched or contacted by wires. The bench top must be well insulated by covering it with ebony Sindanyo or some other similar material.

Electrical switches, sockets, and gas taps should preferably be situated in recesses in the front of the bench in a position convenient to the operator. This avoids the necessity for reaching over the bench to adjust controls. It should be possible to make the whole apparatus 'dead' by the operation of a single switch which should be clearly marked. H.T. switches should be placed so that a conscious effort is necessary to operate them.

It should not be permitted for any one person to work alone in a room when using h.t. H.T. benches should not be left by the responsible person, even for a few seconds, when the apparatus is 'live'. When the laboratory is closed all apparatus must be made 'dead'.

H.T. terminals should be protected and the use of naked wiring avoided. Adjustments to h.t. should be made, whenever possible, with one hand and never when the light is poor. The circuit should be completely 'dead' when making connections. In circumstances where a small current is required in the h.t. circuit, a current-limiting device may be incorporated to control any high current.

GENERAL INSTALLATION

All electric plant and wiring should be installed by a competent electrician and periodically inspected by him. As much wiring as possible should be on a permanent basis and installed according to the I.E.E. Regulations. Each section of the wiring should be safe for the current it must carry and the conductors should be enclosed in good insulating material of a specified quality and of high resistance. Conduit or metal sheathing should be used to protect the building in the event of the cables overheating and igniting. Fuses, or trip-off devices, must be provided to isolate the current if a fault should cause an excessive current to flow. Equipment, such as transformers and condensers, should be stood on a metal tray large enough to contain their contents in the event of fire.

OVERLOADING

It is a common fault in laboratories that more and more electrical equipment is added without regard for the overloading of supply cables. The main feeders to the switch distribution box may have been designed to carry the load safely when the building was erected but become inadequate with the addition of extra current-consuming apparatus. A general overload for the building results. For similar reasons, before any local extension of existing wiring is carried out, the extra load should be assessed. The total wattage of the additional loading must be carefully computed since the existing wiring in that locality may not be the nearest safe point at which to connect it. The fuse in the distribution box which protects the wiring in the area must also be considered, and the error of fitting heavier fuse wire to carry the load is to be avoided at all costs, because this may overload the supply wiring and give rise to a dangerous situation.

EXPERIMENTAL WIRING

Experimental set-ups involving electrical apparatus are much safer if the method of wiring is sound. All leads should be kept as short as possible and well-insulated rubber-covered or asbestos-covered flexible leads should be used in preference to the twin-braided flex. For joining wires, porcelain or other insulated connectors should be used. The wiring should be kept away from hot surfaces or metal contact and excess wire should be neatly coiled to avoid dangling leads.

Double-pole switches are safer than single-pole switches because they open both sides of the circuit at the same time. If single-pole switches are used they should break the live wire.

Whenever a low voltage is suitable for the particular work to be done on the bench, it is safer to use it rather than the mains. For this purpose a transformer can be used and should be kept in a safe housing off the bench. The secondary circuit of the transformer should be protected by a suitable fuse. When any adjustments to circuits are made, the whole apparatus should be disconnected by taking out the feed plug from the supply socket, or if necessary by withdrawing the main fuses.

EARTH CONNECTIONS

Earth connections should offer the minimum resistance to any faulty current. The earth lead, therefore, must be heavy enough to carry a current which will blow the fuse and isolate faulty sections. The earth connection must be sound, and, since soldered joints may break without the fault being noticed, should also be mechanically secured by a clip or screw. Earth connections may be made to the conduit or lead sheathing which protects the supply leads. Alternatively, they may be connected to water supply pipes or to the main earth wire if this is provided in the laboratory. All paint must be scraped from the metal to ensure a good earth connection. Gas or hot water pipes must not be used for earth connections. Socket outlets of the three-pin type should always be used in preference to local earths for portable apparatus, which should be wired with three-core flexible leads.

PRECAUTIONS WITH INSTRUMENTS AND PORTABLE TOOLS

All electrical apparatus and instruments should be inspected for faults by a competent technician before being installed in the laboratory. With the possible exception of heavy immovable equipment, these should not generally be connected to local earths but should be fitted with a three-core cable and plugged into a three-pin socket. This ensures that if any part of the equipment becomes live the earth cable is connected through the frame of the instrument to the earth pin of the socket.

Most electrical troubles in the laboratory arise through the ageing and consequent breakdown of insulation. This is particularly prevalent in old or badly damaged hotplates, ovens, and furnaces, etc., owing to the insulation being heated and becoming brittle and finally threadbare. Constant inspection is therefore necessary and old or worn flexible leads should be replaced as a routine matter. In good hotplates and other appliances, precautions are taken to protect the insulation on the leads as far as possible, and for this reason it is uneconomical to purchase cheap electrical gear.

Appliances such as portable electrical drills which involve a two-handed grip in their application are the most dangerous. Apparatus which gives the slightest shock must immediately be taken out of use, checked, and repaired. Damaged leads must be replaced and never patched. Work should always be done with insulated tools.

Flexible leads sometimes require to be extended. If for this purpose a plug-and-socket connection is used, the female side, where the connections are covered, should be live and never the male side, for in this case the live pins would be exposed.

SPECIALLY DANGEROUS CIRCUMSTANCES

In certain conditions the shock hazard is more dangerous. Such conditions may exist in darkrooms, damp basement rooms, rooms with concrete or stone floors, or rooms in which steam is used. Special attention must be paid to wiring in this type of room and the need for good earthing will be apparent. Never handle electrical switches and connections with wet hands or when standing in water.

DANGERS FROM RADIATION

In the course of their duties, laboratory workers may be subjected to harmful ionising radiation from radioisotopes or from X-rays produced by certain types of equipment. To overcome any possible harmful effects, safe working methods have been devised, but since they are constantly being improved laboratory workers should keep themselves up to date with current safety measures to ensure maximum safety in the laboratory. Before considering the particular hazards and the present methods used to overcome them, the origin and the nature of the radiations should be considered.

Owing to the rearrangement of unstable nuclei in the atoms, radio-active substances release energy as electromagnetic waves or as particles which have mass and momentum and this energy is measured in electronvolts (eV).

TYPES OF RADIATION

The three forms of radiation associated with the disintegration of radioactive isotopes, and known as α, β, and γ, have different penetrating and ionising powers.

α-particles

α-particles are helium nuclei and accordingly are positively charged. They have little penetrating power but great powers of ionisation.

β-particles

β-particles are negatively charged particles and are identical with electrons. They are given off by most disintegrating radioactive isotopes and have moderate powers of penetration and ionisation.

γ-rays and X-rays

These are electromagnetic waves. γ-rays are X-rays of short wavelength. X-rays are also emitted by radiographic, fluoroscopic, and similar equipment.

UNITS OF RADIATION

The amount of radiation delivered to a specific absorber is known as the 'dose' and the unit dose of radiation for X-rays and γ-rays is called a röntgen (R). One röntgen produces 2.08×10^9 ion pairs per cm^3 of air at s.t.p. Since the röntgen is related only to X-rays and γ-rays another unit, the rep (röntgen equivalent physical), was adopted as the dose unit irrespective of the radiation. It was also expedient to be able to measure the dose in terms of its biological effectiveness and another unit, the rem (röntgen equivalent man) was introduced. Finally, the rad, a unit of absorbed dose irrespective of the nature of the irradiated material, was also adopted. The dose in rems is equal to the dose in rads multiplied by the appropriate r.b.e. (relative biological effectiveness). The rad is defined as the amount of energy imparted to matter by ionising particles per unit mass of irradiated material at the place of interest.

QUANTITY OF RADIATION

To define the quantity of radioactivity a unit called the curie (Ci) is used. This unit is defined as the quantity of radioactive nuclide in which the number of disintegrations per second is 3.70×10^{10}. Smaller quantities are denoted by the millicurie (mCi; 1/1000 curie) and the microcurie (μCi; 1/1000000 curie).

SEALED AND UNSEALED SOURCES

Sources of ionising radiation, which are firmly sealed in strong containers of sufficient strength to exclude the possibility of contact with the source and to prevent the dispersal of the radioactive material, are called 'sealed sources'. Sources of ionising radiation not confined this way are known as 'unsealed sources'.

Figure 6.2. Isotope containers. These are used for the storage and transportation of radioactive isotopes (Courtesy E.R.D. Engineering Co. Ltd)

The containers for sealed sources should be undamaged and uncorroded, and to avoid the necessity for any person to expose himself to radiation risks by physically examining the contents the containers should be clearly marked with the nature of their contents. The standard isotope containers, which are usually made of stainless steel or aluminium, may be transported in strong outer containers as illustrated in *Figure 6.2*.

EFFECTS OF RADIATION

The dangers associated with the exposure of the body to ionising radiations are intensified by the fact that the senses give no warning of the presence of radiation. Irreparable damage may therefore have already been sustained when the effects of the radiation manifest themselves. The nature of the symptoms associated with exposure to radiation depend on several factors including the nature of the radiation and the amount absorbed.

Because harmful radiation destroys the body cells, workers may be required to undergo blood counts before beginning experimental work and at regular intervals thereafter. Regular medical examination may also be necessary.

Radioactive materials may enter the body by inhalation, ingestion, or absorption through the skin. α-particles, which cannot penetrate the skin, are nevertheless particularly dangerous if inhaled or ingested because of the deleterious effects they have on the radiosensitive organs of the body. Similarly, β- or γ-emitting substances are also dangerous if taken into the body. Because of their powers of penetration β- and γ-emitters are also dangerous from the point of view of their external radiation. γ-rays are particularly dangerous in this respect, since they are more difficult to screen than the less penetrating β-particles.

Strict protective measures are adopted to ensure that working areas, and the atmosphere and equipment within them, are not 'contaminated' by unwanted radioactive materials. These measures include the provision of adequate shielding, special methods of ventilation, and clean and efficient methods of working.

MAXIMUM PERMISSIBLE DOSE

The radiation dose received by an individual may be due to a single exposure or may be accumulated over a period of time. The maximum permissible doses have been specified. They are of such magnitude that, in the light of present knowledge, only negligible injuries over the period of their lifetime would be caused to persons subjected to them.

MEASUREMENT OF RADIATION

In laboratories where radioactive materials are handled, frequent measurements of the radiation to which the occupants are exposed are necessary. The frequency with which the measurements are made depends upon the type of work being undertaken. The monitoring instruments employed may be used for the determination of the personal dose received by an individual worker over a period of time, for the systematic measurement of the radiation in the laboratory, including the level of contamination, and for specific purposes.

Film badges

A film badge may be used for measuring the dose of radiation received by a worker over a period of time. This should not exceed 300 mrem per week over the whole body. The badge consists of a photographic film secured in a small holder and is worn in an unscreened position on the laboratory coat. The sensitivity of the badge varies with the type of film used in it. At regular periods the badge is exchanged for a new one and the results obtained from processing exposed badges are recorded. In this way, a complete record of the dose of radiation received by any individual is maintained. If it is considered that the radiation to which the hands are subjected is high in comparison with that received by the body, badges of similar type may be worn on the wrist or as a ring on the finger.

Pocket dosemeters

Another type of personal monitor, called a dosemeter, takes the form of a small ionisation chamber and is approximately the same size and shape as a fountain pen. These may be used to measure more immediate radiation to which a worker may be subjected in connection with specific tasks. The ionisation chamber method of measurement is also used in some other monitoring instruments. In the case of the pocket dosemeter, the system consists of a quartz fibre electrometer and a roll condenser. The quartz fibre is situated at the centre of the chamber and is insulated from the outer case. By removing the end cap of the chamber, the electrode system may be charged by putting a high-tension battery across the electrodes. When the ionisation chamber is ionised by X- or γ-radiation the condenser gradually discharges. This alters the position of the quartz fibre in relation to a small scale incorporated in the dosemeter from which the number of dose units may be read. A charger unit which consists of a transistorised printed circuit fed by a 1.5-V single cell is used to charge the direct-reading dosemeter. An internal lamp illuminates the dosemeter scale during charging (*Figure 6.3*).

Another instrument, called a personal radiation monitor, can be

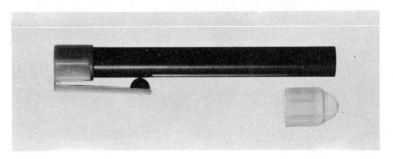

Figure 6.3. Pocket dosemeter (Courtesy Landis & Gyr, S.A.)

carried in a breast pocket. When the safety level of radiation in the vicinity of the wearer is exceeded, an audible warning is heard.

Monitors

Various types of monitors may be used for the systematic measurement of the general level of activity in the laboratory. These may be used as fixed or portable instruments for monitoring working conditions and laboratory operations. Portable survey instruments may also be used for the measurement of contamination in working areas and in surrounding inactive areas.

Basically, the monitors consist of a detector unit such as an ionisation chamber, a Geiger–Müller tube, or a scintillation counter. These are used in conjunction with a power supply and a counting rate meter. In some cases the radiations may also be acoustically detected. Detectors may take the form of a probe head, in which case they are usually attached to the main body of the instrument by a flexible connecting cable. In the case of survey instruments, the detector is usually incorporated in the body of the instrument itself (*Figure 6.4*).

Probe units vary in design according to the nature of the radiation which they are required to detect. Some are sensitive to both β- and γ-radiations and are fitted with a small window, which when open allows β- and γ-radiations to be detected, and, if shielded, monitors only γ-radiation. The difference between the γ-reading and the β-plus-γ-reading provides a measure of the β-activity.

Another type of probe, called a scintillation probe, is used for

284 SAFETY IN LABORATORY AND WORKSHOP

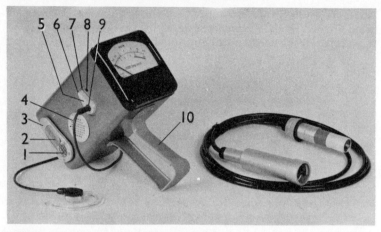

1 Window of measuring head   2 Shutter to absorb beta radiation   3 Lever for unscrewing measuring head   4 Loudspeaker   5 Switch   6 Switch position 'Operation'   7 Switch position 'On'   8 Switch position 'Battery Check'   9 Socket for connecting earphones   10 Pistol grip

Figure 6.4. Radiation survey meter. The earphone can be used for aural detection. The measuring head can be taken out and connected on to the end of a probe cable which is seen lying at the side of the instrument (Courtesy Landis & Gyr, S.A.)

α-particles and has a very thin window which allows α-particles to pass through and to strike a lead sulphide screen. This causes light scintillations which are detected by a photomultiplier, the output from which is registered on a counting rate meter.

Stationary monitors

Stationary monitors are used in fixed positions for specific purposes such as measuring the contamination on the hands and feet of workers leaving active areas. They are also used for monitoring whole buildings, in which case they may be incorporated in a general alarm system. Similar monitors are used for the continuous monitoring of the atmosphere and for monitoring liquid flow such as waste water. These, too, may be coupled to an alarm system.

MEASUREMENT OF CONTAMINATION

Smear tests

Smear tests are often employed to assess the level of removable activity on laboratory surfaces. A small piece of filter paper is lightly rubbed over a known surface area and the level of removable contamination can be ascertained by means of a counting unit.

Air sampling

Sampling devices are employed to measure the amount of radioactive material in the air. The air is drawn through a filtering medium which is afterwards monitored.

PROTECTIVE MEASURES AGAINST RADIATION

Shielding

Shielding materials are used to protect personnel from penetrating rays. Since the range of α-particles is only about 25 mm in air and there is no possibility of them penetrating the skin, special shielding from these particles is unnecessary. Nevertheless, for perfect safety, rubber gloves and a light mask covering the nose and mouth should be worn. The active material should, as far as possible, be handled behind a glass screen. β-particles, which have a much longer range and can penetrate tissue, may be stopped by suitable thicknesses of shielding material, such as glass, aluminium, or Perspex. The penetrating nature of γ- and X-rays, on the other hand, is such that they cannot be entirely stopped by shielding, but their intensity can be greatly reduced if dense materials, such as lead or suitable thicknesses of concrete, are employed. The thickness of the material to be used can be calculated from available information according to the type of radiation and the radiation energies of the particular isotopes. It is always advisable, however, to check such calculations by direct measurement.

Shielding with glass or plastics materials as protection against β-emitters presents little difficulty, compared with the heavy lead or concrete shielding used for γ- and X-rays. Lead shields are usually made up with lead bricks. To avoid the possibility of the radiation penetrating the shield at the joints, the bricks should be interlocking. The shields should also be constructed in such a way that radiations cannot escape at places where the manipulating rods pass through them. Further protection from radiation may be afforded by lead-impregnated wearing-apparel such as gloves, but these are heavy and clumsy to wear. Concrete is not generally as suitable as lead for shielding purposes because of the extra thicknesses involved. Because lead and concrete shields are very heavy, the supports for these must be very strong.

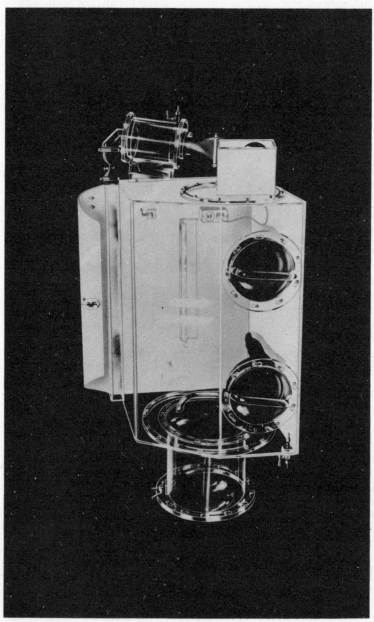

Figure 6.5. Dry box. This one-piece plastics model has an interchangeable air lock and a sealed interchangeable circulating and filter system. The flow from the circulating system is uniformly distributed by means of a graded manifold. Positive 'O' ring seals are used so that the (Courtesy, The Emil Greiner Co.)

The interposition of opaque shields between the operator and the source presents considerable manipulative difficulties and remote handling techniques must be devised. Mirrors and periscopes are often used for observation purposes.

In high-level laboratories much of the work is conducted in 'dry boxes' (*Figure 6.5*). These are extremely useful for work involving dry powders and especially for work involving α-active material.

Since the intensity of the radiation decreases as the square of the distance from the source, remote working methods, involving special handling devices, often provide cheaper methods of protection than expensive shielding.

PERSONAL PROTECTION

The dangers from ionising radiation may be further decreased by adequate personal protection. Laboratory coats, preferably made of nylon, or surgical-type gowns without pockets, should be worn. Rubber gloves should also be used at all times and a light mask should be worn over the mouth and nose. Clothing which has been used in active areas should never be worn in inactive areas and, similarly, that which has been worn in high-level laboratories should not be worn in laboratories of lower levels. To ensure that these rules are observed, the clothing may be coloured in accordance with the class of laboratory to which it belongs.

Change-rooms with separate places for laboratory clothing and normal walking out clothing should also be provided. The change-room also acts as a boundary between active and nonactive areas and should be geographically sited accordingly. A boundary line, or a low division of some kind down the centre of the change-room, defines closely the limits of the two areas.

A good supply of paper towels and paper handkerchiefs should be provided in working areas. The handkerchiefs, which may be used for many purposes, should be interposed between the rubber gloves worn on the hands and laboratory service controls or switches when these are operated. After one use handkerchiefs and towels should be placed in separate waste containers and disposed of as radioactive waste.

When laboratory workers leave the working areas the hands should be washed for several minutes with soap and water until no significant activity remains. Showers should be provided.

ACTIVE AREAS

Work involving radioactive material is conducted in active areas such as special laboratories. Counting apparatus is normally confined to completely separate rooms in which there is no background activity. It is important that these rooms are not contaminated by radioactive materials. Active areas should be clearly marked and only authorised personnel, the numbers of whom should be kept to a minimum, should be allowed to enter them. All persons entering active areas should be acquainted with and subject to the rules applicable to the areas.

Active laboratories should be maintained in a scrupulously clean condition and the workers themselves should assume responsibility for the cleanliness of their own working area.

Special procedures should be adopted and rigorously applied to prevent the spread of contamination from active areas. The apparatus and equipment used in them should be distinctively marked and should not be removed without permission. If, in special circumstances, equipment has to be transferred, it must first be monitored and certified free from contamination. Even slight contamination may be sufficient to upset work in other laboratories.

SAFETY PERSONNEL

The overall responsibility for the organisation of the general safety arrangements and the conduct of the work in laboratories and adjacent buildings should be entrusted to a suitably qualified person. In large establishments this is a task of considerable magnitude, necessitating the appointment of a health physicist. The duties of this officer include the inspection and certification of all laboratories where radioactive work is carried out. All accidents should be reported to him and first aid treatment and medical facilities should be readily available.

METHODS OF WORKING

The radioactive materials used in laboratories should be limited in toxicity and activity within the bounds of the experimental work. The work should, as far as possible, be carried out in closed systems. Wet materials should be used in preference to dry ones. It is also advisable, before any work involving the use of hazardous materials is undertaken, that practice runs be made with materials of low activity. In this way, suitable and safe experimental techniques may be perfected and precautionary measures devised to deal with possible accidents.

Containers of radioactive materials should be opened in a fume cupboard or in a dry box. A special vice or other safe means of holding the container while it is being opened should be employed. Sealed sources should be handled with tongs and not with the hands. Protective clothing should be worn at all times, and if dangers to the eyes exist face shields should be used. Persons with open wounds below the wrist should not be allowed to work in active areas and on no account should any apparatus come into contact with the mouth. Glassblowing should not be permitted in active areas. Full use should be made of the fume cupboards, especially if the experiments give rise to dangerous fumes or sprays or other forms of airborne contamination. When pipettes and similar apparatus are used for handling materials, these should not be placed directly on the bench. If small accidental spills do occur, they should be immediately wiped from the bench top with a paper handkerchief, cotton wool, or other suitable absorbent material held in a pair of tongs. The spill should never be allowed to dry on the bench. Material used for cleaning up spills should be consigned to the radioactive waste bin.

In low-level laboratories the bench tops may be protected with kraft paper or with plastic sheeting which if contaminated may be discarded as radioactive waste. A recent product marketed under the name of Benchkote is excellent for bench protection. It consists of an absorbent paper coated on one side with polythene, is supplied in sheets or reels, and is obtainable from H. Reeve Angel & Co Ltd. Further bench top protection may be provided in all laboratories by the use of trays lined with absorbent paper.

DECONTAMINATION

Decontamination of all apparatus and tools is undertaken immediately following the contamination, which may then be more easily removed. The method of cleaning depends on the nature of the equipment or apparatus. Glassware may be cleaned by normal cleaning agents, including chromic acid, and detergents may be used for tools. Some items are difficult to decontaminate and may have to be kept in special containers until the contamination has 'decayed'. These containers should be clearly labelled with the contents, and the date upon which the radioactivity will have decayed to a safe level should be stated.

When accidents occur which involve a serious spillage, or the release of radioactive powders or aerosols into the atmosphere, emergency procedures should be employed. These are followed by rigorous decontamination of the area. If an inhalation hazard is involved, the room should be evacuated. Persons involved in spillages should remove all contaminated clothing and should wash thoroughly all contaminated parts of the body until no activity remains. They should be monitored before being allowed to leave the area. The general safety procedure includes informing the safety officer, warning persons in the vicinity, and taking steps to avoid spread of contamination. If the area cannot be effectively decontaminated, it should be sealed off and notices displayed prohibiting entry.

Personal decontamination

Normal ablutions, using soap and water, are usually sufficient to decontaminate the skin, but care should be taken to avoid spreading the contamination and to protect the eyes. Should the eyes be affected they must be immediately irrigated with water. Inaccessible parts of the body should be given special attention, especially the finger nails. A soft brush may be used for obstinate contamination, provided care is taken not to abrade the skin.

Decontamination of working areas

Large surface areas, such as bench tops, may usually be decontaminated with soap and water. In extreme cases the surface may have to be removed by solvents, paint removers, or other means.

WASTE DISPOSAL

For the normal laboratory waste materials, pedal-operated rubbish bins, lined with waxed paper or plastics bags, should be provided. The bins should be marked RADIOACTIVE WASTE. The methods of disposing of waste material are determined by the nature of the material and by its degree of toxicity.

Solid waste

Bags containing solid waste should be tied at the neck and if the material has a short half-life they may be stored until the activity reaches a safe level. The waste may then be disposed of through the normal refuse channels. Specially hazardous materials, and materials with a long half-life, are placed in safe storage until a competent authority gives permission for them to be disposed of at sea in special containers, or by other suitable means.

Animal carcases and some other solid materials are incinerated, provided special precautions are taken in respect of the flue gases and the resultant ash.

Liquid waste

Liquid waste involving α-active material must not be disposed of down the sink. β- or γ-active solutions with short half-lives may be put into safe storage until the activity has dropped to a safe level. These may then, with permission, be disposed of into sewers. The disposal is usually accompanied by dilution with large quantities of water or by the addition of inactive carriers.

Small quantities of active liquids are generally kept in stoppered polythene bottles. These may be stored in outer containers of sufficient size and strength to retain the contents of the bottle in case of breakage. Larger quantities of liquids are sometimes kept in special tanks until the activity has decayed.

Gases

The laboratory ventilating arrangements should be so arranged that airborne effluent is discharged well clear of all buildings.

STORAGE OF RADIOACTIVE SOURCES

Safe storage of radioactive sources is essential and the materials stored should be removed only by authorised persons. High-activity sources should be stored in a special storeroom which should preferably be situated in an outside building and provided with adequate shielding and fire protection. The storeroom should be adequately ventilated, and if gases or vapours are likely to be evolved from the solid materials mechanical ventilation to the outside atmosphere is necessary.

The room should be regularly monitored to ensure the outside level of radiation does not exceed 10 mrem per week and the level inside does not exceed 25 mrem per hour.

Glass vessels containing active residues at tracer level should be stoppered with cork, polythene, or rubber stoppers and not with glass or screw-on stoppers. Vented containers may be necessary if unstable solutions are involved and solutions which are thermally unstable and contain radioactive materials should always be stored in vented containers. All the stored sources should be fully labelled and records kept of their purchase, issue, and receipt should be kept.

Sources of low-level activity may be kept in the laboratory in a locked cupboard which, if the activity of the materials demand it, should be shielded.

TRANSPORTATION OF RADIOACTIVE MATERIALS IN BUILDINGS

To limit the effect of accidents during the transport of radioactive material within a building, emergency routines with which the staff should be acquainted must be established. A label bearing full details of the contents should be attached to the containers. The containers themselves should be enclosed in outer containers, which if the inner containers should break will safely hold their contents.

DESIGN OF LABORATORIES FOR RADIOACTIVE WORK

The geographical arrangement of the laboratories in buildings in which radioactive work is undertaken should allow a gradual transition from the lower-level laboratories to those of higher level. The building itself should, if possible, be quite separate from laboratories unconnected with radioactive work.

The laboratories are classed as type A, B, and C according to the radiotoxicity and the radioactive quantity of the materials used in them. Type A laboratories should be specially designed. Type B are used for work involving an intermediate level of activity and Type C for low-level work. Ordinary laboratories are sometimes converted to Type C laboratories by providing suitable floor covering, painting the interior of the laboratory with hard-gloss paint, and other minor alterations.

Laboratories generally should be certified as suitable for the class of work it is proposed to carry out in them. In the following discussion the general principles mentioned are applicable to all laboratories, although some of the special precautions may be tempered to suit low-level laboratories.

To avoid the accumulation of dust, the interior of the laboratories, including their fitments, should be free from crevices, ledges, and projections. Sharp corners should be avoided. Walls and ceilings should be faced with hard, smooth impervious materials and finished with strippable hard-gloss paint. Well-waxed linoleum is suitable for floors provided the joints are suitably filled. Vinyl tiles may also be used.

The junctions between the floor and walls should be coved and steel window frames and flush automatically closing doors should be fitted. The service lines should be enclosed in the walls and access may be provided by flush-fitting panels. The light fittings should fit flush to the ceiling.

The floor space per person should be not less than 100 ft^2 (9.4 m^2), and better ventilation than is provided in normal laboratories is necessary. The higher the level of the work the greater should be the number of air changes. The plenum system of ventilation should be used and the air intake should be less than that of the extraction rate to give a reduced pressure in the laboratory and so prevent the outflow of contaminated air into other areas. The incoming air should preferably be filtered and should enter at openings set at a higher level than those for the extracted air. Recirculation of the air should not be practised. Whatever the system of ventilation used, the air flow should always be from the area of low contamination to that of the greater contamination.

Although it might be desirable to use only the fume cupboards for ventilating purposes, this is usually impracticable because of their working position and the insufficiency of their air intake. They may, however, materially assist the general ventilation.

The fume cupboard ducts should be made of some impervious material and should be easy to clean. Fume exhaust and vacuum systems should be independent of systems in neighbouring laboratories. The ejector fan system is best for ventilating the cupboards. The stacks from the cupboards should be high enough to allow the natural turbulence of the air to dilute the effluent. This avoids the possibility of harmful effects on persons at ground level or in neighbouring buildings. Above certain limits of contamination, airborne effluent must be filtered. The interior of fume cupboards for radioactive work should be finished in nonabsorbent materials. Alternatively, they should be painted with high-gloss paint, which may be stripped off if necessary. Plenty of fume cupboard space should be available and the rate of extraction through the open fronts of the cupboard should not be less than 100 ft^3 (2.83 m^3)/min.

It is important that the fume cupboards be well designed so that leakages of contaminated air into the laboratory, by eddy currents or through other causes, do not occur. The cupboards should, therefore, be smoke-tested after installation. Good top lighting above the fume cupboard should be provided.

The laboratory should be well lit with natural and artificial light and the floor should be free from obstacles.

The plumbing should be given most careful consideration. All sinks should be of the best quality and should have no imperfections in their glazing. Open channels should be avoided and all drainage pipes, which should be made of nonabsorbent material, should have a good fall. The joints, traps, and bends must be easy to inspect and clean, particularly where blockages in the drainage system could cause flooding in nonactive areas. The drainage system should, however, not pass through nonactive areas. Direct discharge of liquid effluent into sewers may be undesirable unless the activity is very small. Special drainage may be necessary from A-level laboratories, and from fume cupboards in other lower-level laboratories or from any other place where specially active effluent is discharged. The effluent may be passed to tanks where its activity can decay. Before the storage tanks are finally discharged into the sewers, they should be monitored.

Laboratory furniture should be easy to clean and should be designed so that radioactive dust cannot easily penetrate into it. The bench tops may be made of stainless steel or other similar nonabsorbent materials suitable to the work of the laboratory. In many cases thoroughly waxed hardwood tops may be suitable. Bench tops and fume cupboard bases should be strongly supported to take the weight of any shielding material which may be placed on them. Frontal service controls are very desirable, particularly on fume cupboards.

DANGERS FROM BIOLOGICAL HAZARDS

In biological laboratories diseases may be transmitted from man to animals or from animals to man. Indeed in many cases the spread of a disease has been attributed to humans or animals who from outward appearances were apparently quite healthy. For this and other reasons, unauthorised persons should be excluded from possible danger areas in animal laboratories and special precautions should be observed by bona fide staff employed in them. If these measures appear unduly severe, they may be fully justified in that many recorded human deaths have been directly related to laboratory-acquired infections. Therefore, in laboratories where special infection rules exist or where experimental or diagnostic animals, especially monkeys, are used, the laboratory staff should be immunised by protective vaccination.

The organisms responsible for infection may be passed to a susceptible host by respiratory process. They may also be transmitted by excretions from the intestinal or urinary tracts, or by discharge from open sores or from the shedding of skin or hair. The suck or bite of insects is of course another well-known source of infection.

The primary methods employed to prevent the spread of disease may involve the confinement of the infection at its source by the use of microbiological cabinets or closed ventilation systems, or by the confinement of infectious materials by other local means.

Other secondary methods depend for their success on good laboratory design. Such methods include the complete isolation of infectious areas and the extensive use of impervious and easily cleaned materials for the construction of floors, walls, and ceilings. Ultra-violet air locks and door barriers, supplemented by adequate change and washrooms for the use of laboratory personnel, are further lines of defence.

Possibly the most important way of combating infection is by the conscientious use of correct working techniques by the laboratory staff, whose awareness depends to a large extent on competent instruction in laboratory procedures. A special warning should be given to trainees concerning the bacteriological hazards most likely to be encountered in the areas in which they will be employed.

The greatest risks involved are those arising from the handling of animals and correct handling techniques are therefore most important. Cleanliness is also of great importance, as is the use of protective clothing such as rubber gloves, gowns, and rubber boots. Safety clothing appropriate to the work in hand should always be worn. At the close of work and after animals have been handled, full use should be made of the washing facilities.

One of the chief dangers associated with the handling of animals is due to bites or scratches and persons suffering wounds of this nature should report them immediately and receive proper treatment. All wounds in the skin should be completely covered before work with animals is undertaken.

Other elementary laboratory precautions should be observed, especially correct pipetting procedures. Bulb pipettes should always be used and liquids should never be pipetted by mouth. Smoking, drinking, or eating in infectious areas should be strictly forbidden.

In laboratories where radioactive materials are used, precautions against possible contamination are essential and protective clothing, such as gowns, rubber boots, and disposable plastics gloves should be worn. The wearing apparel should be monitored before work commences and again when the work is complete.

Radioactive laboratories should be completely isolated from all others and in their construction should comply with the safety regulations. In particular, all the laboratory surfaces should consist of nonabsorbent materials and where necessary floor gulleys should be provided to allow for washing down.

The disposal of radioactive waste should be carefully carried out and the use of hay or sawdust in animal cages and in radioactive areas should be avoided.

Waste from cages and other solid matter must be carefully collected in approved vessels and burnt in an incinerator. Precautions should be taken during this process of destruction so that any gaseous products given off do not contaminate surrounding buildings. The disposals should be carried out under the supervision of the appointed safety officer.

SAFETY IN THE WORKSHOP

Modern science has promoted the need for more and better machinery to serve laboratories. Consequently, the size and character of science workshops has changed, and in order to centralise expensive machinery it is now common practice to have one workshop to serve the needs of a whole institution. In spite of this, the laboratory or local workshop continues to exist and many scientific staff agree it should. It is in the heart of the laboratory areas that the scientific staff and the workshop technician can really work together to satisfy the on-the-spot requirements of experimental work. It is not intended, however, to argue the merits of local workshops as opposed to central ones, but merely to emphasise that the two types do exist, and to point out that the following rules of workshop safety may apply to either but generally to both.

LIGHTING

Good natural and artificial lighting must be available at the working plane in the workshop. Adequate lighting is the chief safeguard against accidents and avoids the eyestrain often associated with the accurate and close work involved in workshop techniques.

Natural light

The windows must be designed to make the most of the natural light and machines, such as lathes, which involve close work, should be positioned where the light intensity is greatest.

Artificial light

The general artificial lighting arrangements should be such that no shadows are cast on the work and special lighting may be required at certain work places. If strip lighting is used this may give rise to stroboscopic effects and it is advisable to arrange the lamps to run on more than one phase of a multiphase supply. This has the effect that all the lamps are not extinguished at the same time.

VENTILATION

Natural ventilation must be good and in many types of workshops additional mechanical ventilation is necessary.

HEATING

A comfortable working temperature should be maintained. Low temperatures result in loss of tool control and induce the technician to wear heavy clothing. This increases the possibility of accidents.

FLOORS

Non-slip durable flooring is desirable. It should not absorb, or be affected by, grease and oil and should not depend upon slippery dressings, such as floor polish, for its upkeep. The floor should be on one level.

GENERAL FINISH

The workshop should have smooth walls with no projections. The walls should be finished in a cheerful colour, which should have a good light reflectance value to promote the best seeing conditions.

Drab workshops are things of the past, since they encourage untidiness, disorder, and fatigue. The colour scheme can also embrace a safety colour code. The code colours are used to draw attention to the various workshop hazards which exist. Red may be used for danger points, green for safety points such as first aid boxes. Yellow indicates caution and orange may be used to distinguish electrical hazards.

Doors should open outwards as a precaution against fire.

SPACING

The workshop should be spacious so that overcrowding is avoided. The minimum cubic space per person should be 400 ft^3 (11.3 m^3). Gangways must be adequate and kept free from obstructions. Machines should be carefully grouped so that traffic at working places is reduced to a minimum.

STORAGE

Ample storage space is essential for all workshops. The extent of the store should be such that all raw materials, tools, and ladders not in use, finished work, and other items are kept quite clear of the workshop. Such materials must never be stored in gangways or near to working places. The store must be readily accessible and should not encroach on the working area in any way.

TIDINESS

Tidiness of the workshop helps to reduce the possibility of accidents and everything should be returned after use to the store or quickly dispatched to its destination. Suitable rubbish boxes should be available and special bins provided for the disposal of cotton waste. Waste materials should be regularly disposed of. Grease, oil, or any other substance which is spilled and likely to make the floor slippery must be wiped up at once. A dirty workshop is the result of slack supervision and is a primary cause of many workshop accidents.

PERSONAL CLEANLINESS

Washing facilities for personal hygiene are necessary in the workshop. Oil and grease block the pores of the skin and promote infection. The slightest cut or abrasion merits treatment as a protection against dermatitis and the possibility of an infected wound. The use of solvents such as petrol or benzene for cleaning the hands is inadvisable. Such solvents dissolve the natural skin oils and, like strong soaps and alkalis, may cause dermatitis.

CARE OF THE EYES

The workshop should be spacious so that overcrowding is avoided. The minimum cubic space per person should be 400 ft^3 (11.3 m^3). by hand tools and abrasive wheels. Small splashes of molten metal are particularly dangerous. Welding processes subject the eyes to penetrating rays.

To protect the eyes goggles should always be worn for welding, riveting, chipping, and other dangerous operations. To wear unsuitable goggles may do more harm than good and for a particular operation the correct type of goggles is necessary. The necessity of wearing the right kind of goggles has been adequately emphasised by Park[11]. Goggles should be well-fitting, light, ventilated, and have a non-sweat finish. Transparent guards are essential on all grinding wheels.

Welding processes are especially dangerous and many people are unaware of the dangerous properties of the radiant energy given off from arc and flame sources. Arc welding is the most dangerous because the fusion temperatures are higher. Ultra-violet and infrared rays are given off from welding processes and unless these rays are absorbed or reflected serious damage to the eyes may result. In addition to the protection from light rays which goggles or screens may afford, the glass must also be optically suitable and of good quality. Welding assistants and other workers in the immediate area also need eye protection. If prolonged welding is to be undertaken this should be carried out in a separate place and a welding booth may have to be provided.

Goggles tend to restrict side vision and for this reason the welder, during the course of his operations, should not be spoken to for he may turn around suddenly and point the torch in the direction of the inquirer. It is bad practice to speak to anybody engaged on a hazardous task.

A further safeguard for welding operations is a leather apron which should reach from the neck to a point below the knees.

When drums are to be welded, make certain they are completely empty and contain no dangerous fumes.

PROTECTIVE CLOTHING

Protective items of clothing may be necessary for specific operations. Rubber gloves, for instance, are required for strong alkalis and handling corrosive substances, and leather ones for sharp materials. Asbestos gloves are necessary for hot processes.

HAND TOOLS

The right tool should be selected for the job and the temptation to use something else should be avoided. Tools must be checked before use. Mushroom-headed chisels, hammers with loose handles, and defective wrenches should be scrapped or put aside for repairs.

MACHINERY

New machinery should not be operated until the necessary safeguards have been provided. All electrical work should be installed by a competent workman.

Machine guards need to be carefully designed, securely installed, and easily removed and replaced. They should be constructed of clear material or strong wire mesh and should have no rough edges. They should be fixed in such a way that they do not constitute a tripping hazard. All moving parts of machinery must be securely fenced or guarded but the controls should remain accessible. Open gears should be covered. The guard should not interfere with the oiling or adjustment of a machine but should give protection to persons carrying out these operations.

Overhead gears or belts should be guarded in such a way that should they break the guard will take the weight of the protected part. This is particularly important over gangways.

MAINTENANCE

A regular check and maintenance of all machinery should be undertaken and necessary repairs effected. A high standard is required for machine guards, clutch gear, lifting tackle, and protective and electrical equipment. The lighting facilities may require overhaul and old or faulty electric light bulbs should be replaced. Steam boilers and pressure vessels demand special attention.

GENERAL SAFETY HINTS

Numerous hints for safety apply in the workshop and a few worth practising are as follows.

- Never overreach when on a ladder and always ascend and descend facing the ladder.
- Use wood chisels and other sharp implements in the direction away from the body.
- If work is to be drilled, secure it in a hand vice or bolt it down and never hold it in the hands.
- Drill thin metal sheet on a piece of wood and secure it by wood screws screwed into the wood so that the projections of the heads grip the edges of the metal. Never allow a necktie to dangle when using a lathe, or bend over the work when using a vertical drilling machine.
- When a new grinding wheel has been fitted, stand aside when the machine is first set into motion.
- Use a steel hook for removing swarf from the lathe and never the fingers.
- Never handle brass filings with the fingers since they are sharper than steel and cause resulting wounds to fester.
- When using the bandsaw, rest the hands on the outside edges of the work so that they can be moved quickly outwards should the saw blade break.
- Treat circular wood saws with great caution, and use a piece of scrap wood to push the last few inches of the work being cut past the saw teeth.

FIRST AID IN LABORATORIES AND WORKSHOPS

The laboratory first aid arrangements to be provided in any particular establishment depend upon the number of staff employed. In teaching institutions the number of students using the laboratory may be the deciding factor. The nature of the work and the hazards it involves must also be taken into consideration. In deciding upon the extent of the facilities to be provided, the regulations embraced in the Factories Act may well serve as a guide.

FIRST AID CABINETS

A sufficient number of first aid cabinets should be provided. These should be well positioned and must contain nothing but the prescribed medical appliances and requisites. The boxes should at all times be well stocked and in the charge of a responsible person who should be readily available during working hours. The name of the person should be clearly indicated in every laboratory and workshop.

At all times the telephone number of the local hospital, the names and addresses of several doctors, and their telephone numbers should be attached to the ambulance boxes in the various laboratories and workshops.

Sufficient numbers of staff should be trained in first aid and refresher courses should be held from time to time. If any risk of gassing exists the training should include the recognition of gases, the symptoms of gas poisoning, use of breathing apparatus, oxygen administration, and artificial respiration. Posters outlining the first aid measures to be adopted should also be placed where they can be easily seen.

FIRST AID ROOM

Where the number of people working in the laboratories exceeds 250 a suitable ambulance room should be maintained solely for the purpose of treatment. The room should be in the charge of a qualified person, who should be available during the hours when the laboratories are open, and a record of accidents and sickness cases treated should be kept.

The first aid room should cater for the more elaborate care of patients who may have received preliminary treatment from the laboratory first aid boxes. It provides for the careful treatment of shock cases and serves as a rest room for people before they are dispatched to hospital. Ideally the room should be centrally situated on a ground floor and close to an outside exit. This facilitates the removal of stretcher cases to the ambulance. The room should be quiet and capable of being heated and ventilated without draughts. For ease of cleaning, the walls and floor should be constructed of a hard impervious material and all corners should be rounded to allow for washing and disinfection. Electrical sockets should be provided for boiling water and for the use of a steriliser. Good natural and artificial lighting are necessary.

Equipment for first aid room

The following list of furniture is intended as a guide. Sink with hot and cold water, examination couch with rubber sheet, table for dressings, table with smooth top, desk, chairs, blankets, pail for dirty dressings, instrument cabinet, medicine cabinet, steriliser, oxygen cylinder and administering apparatus, self-contained breathing apparatus, accident report book. The windows should be fitted with blinds. The smaller items of equipment should include bowls, kidney dishes, towels, jugs, medicine glasses, and jars for lotion.

Contents of first aid cupboard

In larger establishments a main first aid box or cupboard will be kept in the first aid room. When the number of persons involved does not necessitate the provision of a first aid room, a main box, in addition to the local first aid boxes, should be situated in a convenient place where hot and cold water and good light are obtainable. The contents of the box should be comprehensive, as in *List 1* below. For chemical laboratories, *List 2* is also applicable.

List 1

Adhesive plaster, 25 mm wide	1 roll
Adhesive (strip) plaster dressing, 65 mm × 90 cm	2 boxes
Applicators, wooden	6
Aspirin tablets	100
Bandages, 25 mm	9
Bandages, 50 mm	9
Bandages, 75 mm	6
Bandages, triangular, 96.5 cm side	9
Boric acid powder	25 g
Burn cream (or jelly)	1 tube
Camel hair brush	1
Castor oil	100 ml
Cetrimide solution, 1%	500 ml
Cotton wool (sterile), 12.5-g pkts	6
Cotton wool dispenser	1
Dried milk	1 tin
Eye bath	1
Eye dressings (sterile)	4
Eye dropper	1
First aid manual*	1
Forceps, dressing	1

* *Laboratory First Aid.*[12]

Forceps, splinter	1
Funnel, plastics	1
Gauze (75 × 90 cm)	3 pkts
Kidney dish, 150 mm	1
Labels (tie-on)	12
Lint, 25 g pkts	3
Measuring cylinder, 50 ml	1
Mustard powder	25 g
Notebook	1
Olive oil	50 g
Oxygen (Lif-o-gen) unit with mask	1
Petroleum jelly	50 g
Safety pins (assorted sizes)	12
Sal volatile	50 ml
Scissors, blunt 125 mm	1
Smelling salts	1
Sodium bicarbonate solution, saturated	500 ml
Sodium bicarbonate solution, 5%	500 ml
Sodium bicarbonate solution, 2%	500 ml
Sodium bicarbonate powder	250 g
Sodium chloride crystals	250 g
Splints (wooden), 450 mm × 100 mm × 12 mm thick	8
Splints (wooden), rectangular type (for elbow, forearm, and wrist)	500 g 2
Tablespoon (stainless steel)	1
Teaspoon (stainless steel)	1
Thermometer, clinical	1
Tourniquet (rubber bandage type)	1
Tow, surgical	500 g
Tumbler (unbreakable)	1
Wound dressings (sterilised), small	12
Wound dressings (sterilised), medium	6
Wound dressings (sterilised), large	6

List 2

Acetic acid, 2%	250 ml
Acetic acid, 5%	250 ml
Alcohol	250 ml
Ammonium carbonate, 5%	250 ml
Amyl nitrite capsules (3 minims)	12
Ammonium hydroxide, 1%	250 ml
Charcoal (medicinal)	100 g

Copper sulphate crystals	25 g
Copper sulphate solution, 1%	250 ml
Egg albumen	50 g
Ether	100 ml
Ferric chloride solution*, 5%	250 ml
Glycerine	100 ml
Magnesium oxide powder	100 g
Magnesium sulphate crystals	250 g
Milk of magnesia (liquid)	200 ml
Mineral oil	100 ml
Peppermint essence	25 ml
Potassium ferrocyanide crystals	25 g
Potassium permanganate crystals	25 g
Sodium thiosulphate crystals	100 g
Sodium thiosulphate solution, 1%	250 ml
Starch (soluble)	50 g
Tannic acid powder	30 g
Universal antidote	60 g

Contents of subsidiary first aid boxes in laboratories and workshops

The first aid boxes situated in the various laboratories should be easily removable from the walls. They should have the following minimum contents.

Acetic acid, 5%	250 ml
Adhesive plaster (strip) dressing, 65 mm × 90 cm	1 box
Bandages, 25 mm	3
Bandages, 50 mm	3
Bandages, 75 mm	2
Bandages, triangular (96.5 cm side)	2
Cetrimide solution, 1%	250 ml
Cotton wool (sterile, 12.5-g pkts	3
Gauze (75 × 90 cm)	1 pkt
Laboratory first aid manual†	1
Lint, 12.5-g pkt	1
Magnesium sulphate crystals	250 g
Milk of magnesia (liquid)	100 ml
Safety pins (assorted)	12

* For preparation of ferric hydroxide mix together equal quantities of ammonium carbonate solution and ferric chloride solution as required. Filtration unnecessary.

† *Laboratory First Aid.*[12]

Sal volatile	30 ml
Scissors, blunt 125 mm	1
Sodium chloride crystals	250 g
Teaspoon (stainless steel)	1
Tourniquet (rubber bandage type)	1
Tumbler (unbreakable)	1
Wound dressings (sterilised), small	3
Wound dressings (sterilised), medium	3
Wound dressings (sterilised), large	3
Universal antidote, dry powder (need only be provided in chemical laboratories)	30 g

TYPES OF INJURIES AND THEIR TREATMENT

Shock

A condition of shock may be caused by most injuries and may arise almost at once or some hours after the accident. If the injury is painful shock may be more severe. The degree of shock depends to a large extent on the patient, and conditions producing severe shock in the case of one person might produce only mild shock in another. Severe shock can result in the collapse of the patient and can be fatal.

Symptoms of shock

1. Shallow respiration.
2. Weak and rapid pulse.
3. Pallor, cold sweat.
4. Subnormal temperature.

Treatment of shock

To treat a patient for shock he must be laid down and completely wrapped in blankets or any other warm material to hand. The head should be kept low by raising the foot of the bed. If the patient is on the ground and there are no injuries to the legs, something should be placed beneath them. Provided he is conscious and no symptoms of internal injury are apparent, hot drinks may be given as a stimulant. A doctor should be summoned as quickly as possible.

Mild shock

In cases of mild shock any tight clothing is loosened. The patient is allowed to get plenty of fresh air but should be kept warm. After reassuring the patient the injury is then treated. Calmness on the part of the person rendering first aid does much to reassure the patient.

Electric shock

The victim must be removed from contact with the live apparatus or the supply shut off. Electric shock may produce asphyxia and in this case artificial respiration must be started at once. The importance of the immediate application of artificial respiration has been stressed by Scott[13], who has pointed out that statistics show that when treatment is begun within one minute of the electrical shock having been received, 90% of the victims recover, whereas after a delay of six minutes only 10% recover.

Asphyxia

Asphyxia may be caused in a number of ways such as drowning, poisoning, insufficient oxygen in the air, or by electrical shock. The effect of any of these is the same, namely a deficiency of oxygen in the blood. In the laboratory the more likely causes of asphyxia are electrocution, suffocation, and the inhalation of poisonous gases.

Treatment for asphyxia

The first aid treatment for asphyxia is to remove the cause, or remove the patient from the cause, and except in cases where the lung tissue has been damaged to commence artificial respiration immediately. Whenever possible, carbon dioxide–oxygen mixture or pure oxygen should be administered while artificial respiration is being applied. For the administration of pure oxygen the disposable type of oxygen cylinder is ideal. (Supplied by Lif-o-Gen Inc., P.O. Box 302, Lumberton, N.J., U.S.A.) It weighs only 225 g and contains a ten-minute supply with the valve kept constantly open. Larger cylinder sizes are available. The oxygen is administered instantly by simply pressing the valve outlet and a light plastics face mask is provided with the unit.

Artificial respiration

Many means of artificial respiration have been used. Of these the Schaefer, Riley, Silvester, Holger Nielsen, and insufflation methods are still in use. The methods most suitable for cases of asphyxia likely to be encountered in laboratories are the Silvester and Holger Nielsen back-pressure-arm-lift and the mouth-to-mouth (or mouth-to-nose) direct insufflation methods.

Silvester method

First lay the casualty on his back.

Place a folded laboratory coat or other clothing beneath his shoulders.

Clear the victim's mouth if necessary and turn the head to one side.

Kneel just behind the casualty's head.

Grasp the wrists and cross them over the lower part of the chest.

Rocking the body forward, press down on the victim's chest.

Release the pressure and draw the victim's arms backwards and outwards as far as possible.

Repeat the procedure twelve times a minute.

Holger Nielsen method (back-pressure-arm-lift)

This may be employed if the patient is suffering from facial injuries which may make direct insufflation impracticable.

Place the patient face downwards. Remove any obstruction to the airway and bring the tongue forward.

Bend the victim's arms at the elbow and turn the head to one side so it rests on the hands with the palm of the hands facing downwards.

Kneel on one knee, which is positioned at the side of the victim's head.

Place the opposite foot close to the patient's elbow.

Stretch out the arms and place the hands, with the fingers spread outwards and downwards, one on each shoulder blade. Press lightly on the lower half of the shoulder blades. For adults the recommended pressure is 9–11 kg.

In order to exert the pressure rock the body forwards, with the arms straight, until it is almost vertical. Exert the pressure for about $2\frac{1}{2}$ seconds and then relax it by rocking backwards. During this motion slide the hands under the victim's elbows and draw his arms upwards until the shoulders are slightly raised and pulled forward. This movement also takes about 2½ seconds.

After the arms are lowered repeat the whole cycle of operations. A method for the simultaneous resuscitation of two victims by an adaptation of the Holger Nielsen method has been described by Bennett[14].

Insufflation method

Because it is simple and easy to apply and causes the operator no fatigue the mouth-to-mouth or mouth-to-nose method is best. The method is as follows. Lay the victim on his back and clear any debris from the mouth and throat.

Tilt the head backwards until fully retracted, and at the same time, to ensure that the airway is unobstructed, hold the mandible forward.

Place yourself on the left-hand side of the victim and use the thumb and forefinger of the right hand to close the victim's nose. The left hand supports the lower jaw.

Breathe in deeply, and placing your lips so as to form a complete seal over the mouth of the victim, blow in air (*Figure 6.6*). More forceful blowing is required for adults than for children.

When the chest of the victim is seen to rise, remove your mouth, whereupon the chest wall of the victim relaxes. Repeat the inflation at the rate of 12–20 times per minute.

The insufflation method may also be applied by a mouth-to-nose technique. This should only be applied when it is not possible to open the victim's mouth sufficiently for the mouth-to-mouth operation to be effectively used. In the mouth-to-nose method, retract the head fully as before, and while the mandible is pushed upwards and forwards close the victim's mouth by the first two fingers of the right hand.

Place your mouth over the nostrils of the victim.

Wounds

A wound is caused when the skin is broken owing to an injury. The degree of seriousness of the wound depends usually on the depth of penetration and the amount of damage suffered by the deeper tissues.

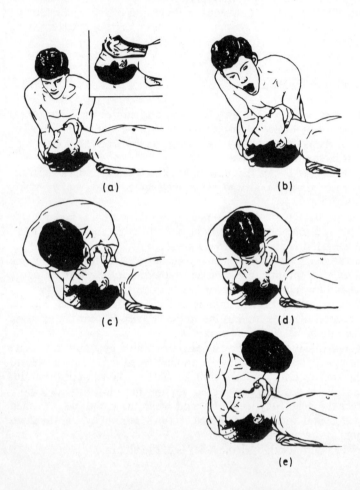

Figure 6.6. The drawings illustrate the mouth-to-nose and mouth-to-mouth methods of artificial respiration. (a) The patient is placed in the supine position with his head tilted back and chin pulled forward. (b) The operator takes a deep breath and opens his mouth wide and (c) seals his lips on the patient's cheeks around the patient's nose. When the mouth-to-mouth method is being employed (d) the operator seals his lips around the patient's open mouth and closes the patient's nostrils by pinching them between his fingers or as shown in the drawing by blocking them with his cheek. He then blows into the patient's mouth until he sees the patient's chest rise whereupon (e) he removes his mouth and takes another breath. He listens for the patient to breathe out and then again inflates the patient's lungs (Courtesy African Oxygen Ltd)

Types of wounds

Incised wounds. Incised wounds are caused by sharp instruments and they have no ragged edges and bruising does not occur. The wound bleeds freely and gapes open. In the laboratory, the most common cause of incised wounds is broken glass.

Although incised wounds are generally the easiest to treat, stab wounds, such as may be caused by the sharp end of a piece of glass tubing, may penetrate deeply and cause bleeding beneath the surface. With this type of wound the risk of infection is far greater. Deep stab wounds may also penetrate internal organs. In cases where doubt exists as to the depth of a wound, seek medical treatment.

If the wound is large and the edges gape, it may require stitching and medical aid is necessary. If tight bandaging is necessary to stop bleeding, the bandage may require to be loosened when the bleeding has stopped.

Lacerated wounds. When the tissues are torn and the edges of the wound are irregular, the injury is called a lacerated wound. Bleeding may not be so severe as with an incised wound but bruising will be seen around its edges.

Shock is usually more severe with lacerated wounds. In the case of large lacerations the haemorrhage could be severe and must be stopped before the wound is treated. This type of wound may be enveloped in a large dressing or a clean towel. Medical assistance should be obtained.

Contusions. In many cases, although the skin may be unbroken, considerable damage and bruising of the tissue may have occurred. Underlying organs may also have been injured. Such damage to the small blood vessels beneath the skin causes the blood to ooze into them and the skin appears red and may later become dark purple. Swelling occurs. This type of wound may be caused by a fall or by heavy objects falling on to the body. For this type of wound, treat first for shock. Apply a bandage soaked in cold water to the affected part. If possible elevate the part and rest it. In severe cases death may result owing to damage to the internal organs, and when the contusions are severe medical aid must be sought.

Infection of wounds

Infection of the wound is to be avoided at all costs. Germs may be introduced into the wound by dirt from the object which caused it or from the skin or clothing of the wounded person. Infection may also be caused by the introduction of germs when the wound is dressed. Every care should be taken, therefore, to ensure the dressings used are sterile and that all water used for cleansing is itself clean or has been boiled. The hands must be thoroughly cleansed. The wound should not be touched during treatment, nor should the sterile part of the dressing be handled when it is being applied.

General treatment of wounds

The general treatment for wounds:
1. Stop the bleeding.
2. Treat the patient for shock.
3. Clean the wound, or in the case of serious wounds, the area around the wound.
4. Rest the wounded part if the wound is serious and get medical aid.

The patient should always be seated or laid down when wounds are being treated.

To stop bleeding. In order to stop bleeding it is necessary to have an understanding of how blood circulates through the body. The blood is pumped from the heart through the arteries under pressure. The waves of pressure are evident when the pulse is felt. If, through an injury, an artery is opened the bright red oxygenated blood issues from it in spurts because of the pumping action of the heart and arterial bleeding can be recognised in this way. Because the blood in the arteries is under pressure, arterial bleeding is the most difficult to stop. Digital pressure, or, in extreme cases, the use of a tourniquet, may be necessary to control it. The pressure is applied on the proximal side of the wound.

The blood from the arteries is eventually distributed at a much lower pressure through very fine tubes called capillaries which nourish the body tissues. Light abrasions of the skin cause damage to these capillaries and blood oozes slowly out. Such bleeding is easily controlled by pressure applied to the wound in the form of a dressing and by the natural clotting process of the blood.

From the capillaries the blood returns at an even lower pressure through the veins and, having lost much of its oxygen, is changed to a purple colour. Venous bleeding can be recognised by the colour of the blood, which flows steadily and not in spurts as does arterial bleeding. Venous bleeding, unless it is a large vein which has been opened, can be stopped by pressure on the wound, but indirect pressure may be necessary. Indirect pressure is applied by tying a narrow bandage round the limb on the distal side of the wound. The bandage is tied tightly enough to collapse the vein but not tightly enough to collapse the artery, which has a thicker wall.

Haemorrhage. Haemorrhage or bleeding may take place internally or externally and may be severe or only slight or moderate.

Slight or moderate bleeding. Slight or moderate bleeding may be stopped by applied pressure in the form of a dressing and a firm bandage.

Severe bleeding. At various points distributed over the body, the arteries pass near to or over a bone. By applying pressure on the artery at these points, the supply of blood to parts beyond the point at which the pressure is applied may be cut off. Digital pressure is applied to the main line of the arteries by pressing with the thumbs, except in the case of the brachial artery, when pressure is applied by the fingers. In the same way, bleeding may be controlled by a tourniquet. A pad and bandage tied firmly on the wound is sufficient to stop bleeding, but in almost every case the application of digital pressure or the use of a tourniquet may be necessary while the dressing is being applied. If the blood oozes through the dressing it should not be removed but a second pad and bandage may be applied over the first and secured more firmly.

A tourniquet should not be used unless it is obvious that the bleeding cannot be controlled by a firmly tied dressing, pad, and bandage, and even then should be used with great care. Tourniquets may be improvised or of a standard pattern. Rubber tubing, stockings, and neckties are examples of improvised tourniquets. Rubber tubing and similar materials have a small surface area and are liable to bite into the skin. If such materials are used as tourniquets, a piece of bandage or other soft material should be placed beneath them to prevent bruising of the artery lining.

The tourniquet is used only on the lower or upper limbs, that is on the femoral or brachial arteries, respectively. Sufficient force is used in tightening just to prevent the flow of blood. Tourniquets should not be used below the knee or below the elbow, except when a limb has been severed. In this case it is necessary to apply the tourniquet as near as possible to the point of severance to prevent loss of blood.

Laboratories and workshops are not usually situated in places far removed from medical assistance, but if help is not forthcoming within fifteen minutes from the time of application, tourniquets should be loosened for a few seconds. If the bleeding continues the tourniquet should be reapplied in a different place but close to the original position. The movement of the tourniquet should be towards the wound. This procedure should be repeated at fifteen-minute intervals until medical help is available or the patient can be moved to hospital. Failure to do this may seriously damage the limb.

Internal haemorrhage. Internal bleeding may be recognised by a rapid pulse, clammy skin, pale lips, shallow breathing, or breathlessness due to air hunger as described by Croton[15]. Bleeding may become visible by the coughing up of red frothy blood, indicating damage to the lungs. Blood from stomach injuries may be vomited and is darker in colour. According to White Knox[16], a cold compress may be applied over the origin of the haemorrhage, provided the first aid worker has correctly diagnosed the condition. The patient is laid down, treated for shock, and conveyed to hospital as quickly as possible.

To clean the wound

Bleeding helps to clean a wound and sometimes as with a pricked finger or similar small wound, may even be encouraged by squeezing. Small foreign bodies should be wiped away from the area in a direction away from the wound, by a swab soaked in antiseptic lotion. Small glass fragments may be removed by tweezers. The wound should be well cleaned with a weak antiseptic lotion such as diluted (1 : 10) Dettol. If the wound is large, only the skin around the wound should be cleaned. Soap and water, saline solution, or clean water may also be used for cleaning wounds. A sterilised dressing is then applied and a pad is usually necessary to press this firmly on to the wound.

Large pieces of glass or other deeply embedded fragments are not removed from wounds but are left for a doctor's attention. When bandages are applied to the wound the area around such fragments is built up with dressings to avoid pressure on the fragment. A ring pad can also be employed for this purpose. Wounds which are bleeding profusely should be tightly dressed.

To rest the wound

The injured part may be supported by a sling or the limb may be tied to another or to some part of the body.

Burns

The most serious danger from burns is due to shock at the time of injury. Deep burns may not necessarily be dangerous to life but if a large area of the body is affected, even by first-degree burns, the injury may prove fatal. The first duty of the person rendering first aid is to observe the condition of the patient and, if necessary, to treat for shock.

First-degree burns

First-degree burns are those which cause the skin to redden and swell.

Second-degree burns

Second-degree burns cause blisters to form.

Third-degree burns

Burns of the third degree involve the destruction of the superficial layer of skin.

Slight burns

When there is no general disturbance apparent in the victim, the burn is known as a slight burn and the first aid treatment applicable to the nature of the wound may be given. This involves:

1. Treatment for shock (if necessary).
2. Relief of the pain.
3. Exclusion of air and infection from the wound area.

Severe burns

A severe burn is one which demands that the victim should receive urgent general treatment for shock rather than the local treatment of his wounds.

Causes of burns

Burns received in the laboratory may be caused by:

1. Flames, hot objects, or electricity.
2. Acids, alkalis, corrosive chemicals.

The type of burn should always be ascertained before treatment is given because the treatment for thermal burns is different from that for corrosive burns.

General treatment for thermal burns

The person is first treated for shock. Before treating any burns the first aid personnel should wash their hands thoroughly and wear a mask or clean handkerchief covering the mouth and nose. It was formerly recommended that the victim's clothing be cut away to disclose the area of the burn. Modern treatment suggests, however, that if the burn is so extensive that it cannot be exposed without removal of clothes, then only the exposed areas should be covered with a sterile bandage, cloth, or towel, and the patient removed to hospital for treatment.

Treatment for slight thermal burns

Slight burns can be more successfully dealt with by first aid methods. They should be first washed with soap and water or with cetrimide solution and then bathed with a saturated solution of sodium bicarbonate. The area should then be covered with a piece of gauze or lint which has been soaked in bicarbonate solution and lightly squeezed out. If the burns are on the face a mask dressing consisting of two halves is made. In the upper half of the mask, holes are left for the eyes and a cutaway portion for the nose. In the lower portion a hole for the mouth is formed. The dressing is soaked in sodium bicarbonate solution and applied. Flemming[17] recommends the use of greasy dressings as being comfortable and proper on the face and fingers for slight burns, provided they are not used too continuously.

Treatment for electrical burns

These are treated as for thermal burns.

Treatment of severe burns

In cases of severe burns the patient is treated for shock immediately. The patient should not be moved and medical help or the ambulance should be summoned at once. Blankets must not come in direct contact with the burns.

Persons with clothing on fire

If a person's clothes catch fire the victim should be laid on the floor, with the burning side uppermost. This prevents the flames spreading and protects the neck and head. The victim's body is then covered quickly with an asbestos blanket, or some other handy extinguishing material, and the flames smothered. The patient is reassured and treated for shock.

Treatment for burns caused by corrosive substances

Acid burns. The treatment for acid burns is simple. Any person at the scene of the accident can render invaluable assistance by applying plenty of water to the affected parts as quickly as possible. The usual laboratory procedure is to get the victim quickly to the nearest sink. When the area of contamination is large this is not satisfactory because of the physical difficulty involved in applying water from the tap. The victim's clothes may have been contaminated, in which case they must be removed at once. This is best done while the victim stands under a discharging shower.

When the affected parts of the body have been well flushed with water, they should be bathed with sodium bicarbonate solution and wet dressings applied as for thermal burns.

Hydrofluoric acid. Hydrofluoric acid burns necessitate special treatment. The affected parts should first be washed well with water and then with a 5% solution of sodium bicarbonate. Finally a paste of glycerine and powdered magnesium oxide is applied.

Alkali burns. Alkali burns should be irrigated immediately with plenty of water as for acid burns. The affected parts may be later bathed with a 5% solution of ammonium chloride, a saturated solution of boric acid, or a 2% solution of acetic acid.

Phosphorus burns. Phosphorus burns are highly dangerous and oily or greasy dressings should not be used. A 1% solution of copper sulphate has been recommended[18]. Another recommended treatment is to immerse the burned part in water immediately. It should then be soaked in a 2% solution of sodium bicarbonate and afterwards swabbed with a 1% copper sulphate solution. This is followed by a further washing with sodium bicarbonate solution. McCartan and Fecitt[19], however, found that treating the burn with copper sulphate ointment, followed by washing with a 2% solution of sodium bicarbonate, gave the best results.

Phenol burns. For phenol burns the area should be well flushed with water and treated as for mineral acid burns.

Bromine. All traces of the bromine must be immediately washed away with plenty of water. Bathe with 5% solution of sodium thiosulphate for 15 minutes. The patient should be taken to hospital for treatment.

Eye injuries

Chemical

Strong acid or alkali in the eyes is extremely dangerous and the sight may be permanently lost unless *immediate* treatment is given. Alkalis are especially dangerous and damage the eyes more extensively and intensively and more rapidly than acids. The eyes must be irrigated *at once* with plenty of water. If he has the presence of mind to do so, the irrigation is best done by the victim himself under the nearest tap. When the eyes are injured, however, it is a natural tendency to close them, and in rendering first aid, it may be necessary to forcibly hold open the eyelids and thoroughly irrigate the eyes with water. A little liquid paraffin or drop of castor oil in the eyes will help to relieve pain. The victim should then be sent to the hospital.

It has been pointed out by Van Arsdell[20] that immediate irrigation with water is the only first aid treatment which should be given, since weak acid or alkali solutions are ineffective and may cause further damage if administered by unqualified persons.

Foreign bodies

Foreign bodies, such as particles of solid substances, which enter the eye should be carefully removed by means of a piece of moist linen or by a camel hair brush, the point of which has been dipped in liquid paraffin. The movement of the brush should be towards the inner corner of the eye, which allows the particle to be more easily removed. To locate the particle the lower eyelid is drawn gently down and away from the eye. If the particle is under the top lid a matchstick is placed horizontally across the lid and the lid is turned back over the match. Meanwhile the patient should be told to look down. If the particle is embedded, continued efforts to remove it may result in further serious eye injury. A drop of liquid paraffin or castor oil should be used to give temporary relief and the eye should then be covered with a pad to check its movement and the victim taken to a doctor.

If the eyes are damaged by direct thermal heat or by hot metal, a few drops of castor oil should be put in the eye. In the case of metal small particles may be removed by careful irrigation.

TYPES OF POISONING AND THEIR TREATMENT

In accordance with the Pharmacy and Poisons Act 1933, persons or institutions concerned with scientific education or research are allowed to purchase poisons for the purposes of their work. Since laboratory workers are afforded this privilege, it behoves the users to take every possible care in the storage and usage of such chemicals.

It is impossible to keep locked up the many other poisons which are in common use in chemical work. Because laboratory workers are above average intelligence and are well trained, cases of poisoning are rare. All laboratory personnel, however, should have a knowledge of the first aid treatment necessary to combat poisoning which may occur by swallowing, inhalation, or absorption through the skin.

Terms used in the treatment of poisoning

Emetic

Emetics are substances given to induce vomiting and to rid the stomach of the poison. They may have to be given forcibly. Examples are as follows:

1. *Mustard.* One tablespoonful (30 g) is added to a glass of warm water. A quarter of this amount is given and this is followed by a glass of warm water. The procedure is repeated at one-minute intervals until all the mustard has been used.
2. *Salt water.* Two tablespoonfuls (60 g) of salt is dissolved in a glass of warm water and administered. This is repeated at one-minute intervals until four glassfuls have been given.

3. *Soap suds.* A piece of mild soap is shaken in warm water until it makes suds. One quarter of a glassful is given and this is followed by a glass of warm water. This is repeated at one-minute intervals until all the soap solution is used up.

The throat may be tickled following an emetic to induce vomiting.

Antidote

An antidote is a substance administered to render the poison harmless or to retard its absorption. Examples are: magnesia (for strong corrosive acid poisoning), vinegar or lemon juice (for strong alkali poisoning).

Demulcent

Demulcents soothe the pain of inflamed membranes. Examples are: milk, barley water, whites of eggs.

Conditions governing the administering of emetics, antidotes, and demulcents

Emetics

Emetics are never administered when the patient is:

1. Unconscious or in convulsions.
2. Unable to swallow.
3. Suffering from the swallowing of a corrosive poison.

Antidotes

Antidotes may themselves be poisonous and following administration may have to be removed from the stomach. When the nature of the poison taken is unknown and cannot be easily ascertained, it is better to give a general antidote to avoid any waste of time. In all cases where the specific antidote is not known, the universal antidote is administered.

Universal antidote. The universal antidote consists of: 2 parts activated charcoal, 1 part magnesium oxide, 1 part tannic acid mixed together. It is best kept dry until required. 15 gram is given in a half-glass of warm water. After the antidote is given, the stomach is washed out, except after corrosive poisoning.

Demulcents

Demulcents are usually administered after the poison has been removed.

Classification of poisons

Corrosive poisons

Corrosive poisons are those which destroy the tissue with which they come into contact. Examples are: strong acids and alkalis.

Irritant poisons

Irritant poisons cause the stomach and the intestines to become irritated and inflamed. Examples are: arsenic compounds, antimony compounds, and phosphorus compounds.

Nerve poisons

Nerve poisons are absorbed into the blood and upset the nervous system. Examples are: morphine, opium, and strychnine.

Recognition of poisoning symptoms

Corrosive or caustic poisons

If corrosive or caustic poisons have been swallowed, the lips are stained, and the lips, tongue, and throat become swollen. The victim retches and vomits. Suffocation may occur and there will be considerable shock.

Irritant poisons

If irritant poisons have been swallowed the lips and mouth are not stained. Any vomiting may be bloodstained. Other symptoms are diarrhoea, nausea, and shock.

Nerve poisons

The nerve poisons may cause the patient to have convulsions (strychnine), to suffer from delirium (belladonna or alcohol), or to become drowsy with the pupils of the eyes contracted and the face flushed (opium, morphine, chloroform, chloral hydrate, barbiturates).

Poisonous gases or fumes

Poisonous gases or fumes may give warning of their presence by their irritating effects or by their smell. Some are odourless, even though they may be present in dangerous concentration. The gas or fumes may cause poisoning by inhalation or by skin contact.

General treatment for gas or fumes poisoning

The affected person should be carried into the fresh air at once and must not be allowed to walk. If breathing has stopped or is poor, artificial respiration is given immediately. The direct insufflation method is best and in cases of poisoning by chlorine, bromine, or phosgene only this method should be used. Oxygen should be administered and is given in spite of the fact that breathing may be good. A mixture of oxygen and carbon dioxide should be given if recommended in the table of treatments. The patient is kept warm and quiet. Some poisonous gases may cause convulsions, which when subsided may be brought on again by noise or by excitement of the patient. Alcohol should not be given unless recommended in the table of treatments and only in a few cases are other stimulants allowed.

Table 6.3 TREATMENT FOR POISONING BY SPECIFIC GASES

Type of gas or fumes	Treatment
Acetylene	Apply general treatment
Ammonia	Apply general treatment. If the patient is unconscious allow him to inhale the fumes of acetic acid. If the eyes are affected, irrigate with water. Wash affected parts of the skin with water. Apply ice pack or cold compresses to the throat to reduce swelling
Arsine	Apply general treatment. Give tea as stimulant

Table 6.3 TREATMENT FOR POISONING BY SPECIFIC GASES—*continued*

Type of gas or fumes	Treatment
Bromine	Apply general treatment. Administer oxygen and if breathing stops apply artificial respiration *by direct insufflation*. Allow the patient to inhale fumes from dilute ammonium hydroxide solution. Treat eyes and skin for bromine burns. Absolute rest is essential
Carbon dioxide	Apply general treatment. Administer pure oxygen *not* CO_2–O_2 mixture
Carbon monoxide	Apply general treatment. Prevent chilling and do not give stimulant. Administer oxygen or 7% CO_2–93% O_2 mixture
Chlorine	Apply general treatment. Administer oxygen and, if breathing stops, apply artificial respiration *by direct insufflation*. Allow to inhale fumes from dilute ammonium hydroxide. Fumes of ethyl alcohol also give relief. 2.5ml ($\frac{1}{2}$ teaspoon) of essence of peppermint in a small glass of water. Absolute rest is essential. Treat eyes and skin for chlorine burns if necessary
Fuel gas	Apply general treatment. Administer oxygen or, if available, 7% CO_2–93% O_2 mixture
Hydrochloric acid gas	Apply general treatment. Allow to inhale the fumes of dilute ammonium hydroxide. Treat eyes and skin for acid burns if necessary. Apply cold compresses to throat to relieve swelling
Hydrofluoric acid gas	Apply general treatment. Allow to inhale the fumes of dilute ammonium hydroxide. Treat the eyes and skin for hydrofluoric acid burns if necessary.
Hydrogen sulphide	Apply general treatment. Administer oxygen or, if available, 7% CO_2–93% O_2 mixture. Irrigate the eyes with water. Give coffee as stimulant
Nitric acid	Apply general treatment. Allow to inhale the fumes from dilute ammonium hydroxide. Administer oxygen if necessary but do not give CO_2–O_2 mixture. If breathing stops apply artificial respiration *by direct insufflation*. Treat the eyes and skin for acid burns if necessary. Send to hospital
Nitrous fumes	Apply general treatment. Allow to inhale fumes from dilute ammonium hydroxide. Administer oxygen if necessary but do not give CO_2–O_2 mixture. If breathing stops apply artificial respiration *by direct insufflation*. Treat eyes and skin for acid burns if necessary. Complete rest and quiet essential. Send to hospital
Phosgene	Apply general treatment. Administer oxygen but not CO_2–O_2 mixture. If breathing stops, apply artificial respiration *by direct insufflation*. Complete rest and quiet are essential. Send to hospital
Smoke	Apply general treatment
Sulphur dioxide	Apply general treatment. Allow to inhale fumes from dilute ammonium hydroxide. Administer oxygen

Arrangements should be made to transport the patient quickly to hospital. Some gases have a delayed action and their symptoms may not become apparent until several hours have passed. All cases of gas poisoning should therefore be kept under observation in hospital for 24 hours. Details of the treatment to be given for particular gases are shown in *Table 6.3*.

General treatment to be given when poisons have been swallowed

In all cases where poisons have been swallowed an assistant should be sent at once for medical assistance. Provided the patient is not unconscious or suffering from convulsions AN EMETIC IS ADMINISTERED AT ONCE. EMETICS ARE NEVER GIVEN IF CORROSIVE POISONS HAVE BEEN SWALLOWED. Vomiting should be induced repeatedly. Mustard in water is the best emetic. In some cases the emetic also acts as an antidote and whenever a particular emetic is specified in the table of treatments (*Table 6.4*) it should be used.

Table 6.4 TREATMENT FOR SPECIFIC POISONS WHICH HAVE BEEN SWALLOWED

Poison	Treatment
Acetone	Apply general treatment. Administer the universal antidote. Keep the patient awake. Give tea or coffee as stimulant
Acids: mineral acids, acetic acid, oxalic acid, phosphoric acid	For concentrated acids do not give an emetic or induce vomiting. Give plenty of water until the antidote is available. Give milk of magnesia or lime water. Follow with milk or whites of eggs (or egg albumen) in cold water. Keep warm and quiet. Give carbonate or bicarbonates in dilute solutions only when other antidotes are unavailable
Alcohol, ethyl, isopropyl	Give large quantities of warm water. Emetic. Cold applications to the head. Give coffee or tea as stimulant. Keep warm. Apply artificial respiration if necessary
Alcohol, methyl	Apply general treatment. Give 60 g (2 tablespoons) magnesium sulphate in 500 ml of water. Give milk or whites of eggs. Apply artificial respiration if necessary
Alkalis, caustic, and concentrated ammonia	Do not give an emetic or induce vomiting. Give plenty of water to drink until the antidote is available. Give 5% acetic acid or 6.5 g (level teaspoon) citric or tartaric acid in 250 ml water or give any citrus fruit juice. Every 15 minutes give a few ml of mineral oil until 4 doses given. Give whites of eggs in cold water, milk, or other demulcent. Keep warm and quiet
Aniline	Apply general treatment. Give 5% solution acetic acid. Give oxygen and artificial respiration if necessary

Table 6.4 TREATMENT FOR SPECIFIC POISONS WHICH HAVE BEEN SWALLOWED—*con.*

Poison	Treatment
Antimony compounds	Apply general treatment. Give 4.0 g (2 teaspoons) tannic acid in 250 ml of warm water or give the universal antidote or strong tea. Give whites of eggs in water, milk, or other demulcent. No oils or fats
Arsenic compounds	Alternate a salt water emetic with doses of 60 g (2 tablespoons) magnesium sulphate in 250 ml water. Repeat several times. Give ferric hydroxide freshly prepared (mix laboratory reagent solution of ferric chloride and ammonium carbonate). The solution may be filtered through a coarse filter, e.g. a handkerchief. Give whites of eggs in milk or cold water. Give strong tea as a stimulant
Aspirin	Give the universal antidote. Administer 0.25 g potassium permanganate in 250 ml water. Give an emetic. Give 500 ml of 5% aqueous solution of sodium bicarbonate.
Atropine	Apply general treatment. Then give 0.5 g tannic acid in water, or give universal antidote or strong tea. Repeat the emetic. Hot strong coffee as stimulant. Apply artificial respiration or give 7% CO_2–93% O_2 mixture if available
Barbiturates	Apply general treatment. The patient must be kept warm. Keep the patient awake by flicking him with a wet towel across the neck, face, and back, and give plenty of black coffee or strong tea. If necessary apply artificial respiration and administer oxygen.
Barium compounds	Apply general treatment. Give 60 g (2 tablespoons) magnesium sulphate in 500 ml water. Repeat the emetic. Give whites of eggs. Milk
Belladonna	As atropine
Bismuth compounds	Apply general treatment. Give whites of eggs in cold water. Milk. Stimulant
Camphor	Apply general treatment. Allow the patient to inhale the fumes of dilute ammonium hydroxide. Repeat emetic. Give hot coffee as stimulant. Apply artificial respiration if necessary
Carbolic acid (phenol)	Apply general treatment. Administer 60 g (2 tablespoons) magnesium sulphate or sodium sulphate in 250 ml water or milk. Give plenty of demulcent drinks
Chloroform	Apply general treatment. Keep patient awake. Apply artificial respiration if necessary. Administer oxygen
Copper compounds	Apply general treatment. Give liberal amounts of milk or whites of eggs in cold water. Give $\frac{1}{3}$ g potassium ferrocyanide in water. Give strong tea as stimulant
Cresols	As carbolic acid (phenol)
Cyanides	Apply general treatment. As an emetic give 1% sodium thiosulphate solution. Assist vomiting by tickling back of throat if necessary. Give brandy, strong coffee or tea, or sal volatile as stimulant

Table 6.4 TREATMENT FOR SPECIFIC POISONS WHICH HAVE BEEN SWALLOWED—*con.*

Poison	Treatment
Fluorides	Give large quantities of water or lime water. Milk of magnesia. Whites of eggs and milk. Stimulant. Artificial respiration if necessary
Formaldehyde	Apply general treatment. Give repeated doses of 1% ammonium hydroxide solution. Give milk of magnesia in 125 ml water. Give whites of eggs or milk. Apply artificial respiration if necessary
Hydrogen peroxide	Apply general treatment. Give 60 g (2 tablespoons) magnesium sulphate in 500 ml of water. Repeat emetic. Give plenty of water
Iodine	Give frequent drinks of soluble starch in water. Give an emetic and demulcent drinks. Black coffee or sal volatile as stimulant
Lead compounds	Apply general treatment. Give 30 g (1 tablespoon) magnesium sulphate or sodium sulphate in 250 ml of warm water. Repeat emetic. Give white of eggs in cold water and milk. Sal volatile or strong coffee or tea as stimulant
Mercury compounds	Apply general treatment. Repeat emetic several times and give large quantities of water. Give whites of eggs in water. Sal volatile as stimulant
Morphine (opium, laudanum, codeine, heroin, paregoric)	Administer 0.7 g potassium permanganate in 500 ml of water followed by an emetic. Repeat. Give coffee as stimulant. Keep the patient awake. Administer 7% CO_2–93% O_2 mixture and give artificial respiration if necessary
Nitrobenzene	Apply general treatment. Give plenty of water. Administer 60 g (2 tablespoons) magnesium sulphate in 250 ml of water
Phenol (carbolic acid)	Apply general treatment. Administer 60 g (2 tablespoons) magnesium sulphate or sodium sulphate in 250 ml of water or milk. Give plenty of demulcent drinks
Phosphorus red	Apply general treatment. Give whites of eggs in water
Phosphorus white	Apply general treatment. Administer 0.2 g copper sulphate in 250 ml warm water. Repeat every 5 minutes until the patient vomits. Give an emetic of mustard and water to rid the stomach of copper sulphate. Give whites of eggs in water
Permanganates	Apply general treatment. Give milk or whites of eggs. Give 5 ml of 3% hydrogen peroxide in 125 ml water made slightly acid with acetic acid
Silver compounds	Give large quantities of salt water as emetic. Give milk or whites of eggs and stimulant
Silver cyanide	Give 20 ml of 3% hydrogen peroxide. Follow with a mustard in water emetic. Give whites of eggs in cold water. Give whisky in water or sal volatile as stimulant

Table 6.4 TREATMENT FOR SPECIFIC POISONS WHICH HAVE BEEN SWALLOWED—*con.*

Poison	Treatment
Strychnine	No emetic. Give 0.6 g potassium permanganate in 250 ml water. Alternatively, 27 g (1 tablespoon) powdered charcoal. Give strong tea. Amyl nitrite may be inhaled to prevent collapse. If convulsions are severe the patient may be allowed to inhale ether or chloroform as an anaesthetic. If necessary give artificial respiration between convulsions. Complete rest and quiet
Thallium compounds	Give salt water emetic. Hot coffee as stimulant. Artificial respiration if necessary
Zinc compounds	Apply general treatment. Give liberal amounts of milk or whites of eggs in cold water. Give $\frac{1}{3}$ g potassium ferrocyanide in water. Strong tea as stimulant

If the poison is known, administer the antidote. If not known administer the universal antidote. Do not give alcohol unless specifically stated in the treatment. Keep the patient warm and quiet.

Note: Egg albumen (1 teaspoonful to a glass of water) may be given in place of whites of eggs.

Cyanide Poisoning

Hydrogen cyanide, the deadly poison gas known as prussic acid, has the smell of bitter almonds, and, being a weak acid, forms salts known as cyanides. Cyanides are also extremely poisonous. Hydrogen cyanide may be swallowed, inhaled, or absorbed through the skin. Very small quantities, if taken by way of the mouth, can cause death owing to the formation of HCN when the cyanide is acted upon by the stomach acids. A concentration of the gas in the atmosphere as low as 1 part in 500 can be fatal and strong concentrations, if breathed, produce instant death.

Symptoms

The characteristic odour of bitter almonds will be noticed on the patient's breath. The skin becomes cold and short convulsive breathing is apparent. The pulse quickly fails and convulsions may occur. There is a rapid loss of consciousness and finally collapse. Prussic acid is an extremely rapid poison which paralyses the respiratory system, and if treatment is to be of use it must be given quickly.

Treatment

A doctor should be summoned immediately. If amyl nitrite is available, apply it by giving the patient inhalations from a pad held close to the nostrils and mouth for about 20 seconds. This can be repeated every 5 minutes for 20 minutes, and care should be taken so that the administrator does not himself inhale the amyl nitrite. If natural breathing ceases, artificial respiration should be commenced and the administration of the amyl nitrite is continued by a helper while this is in progress. Oxygen can also be administered.

If amyl nitrite is not available, ammonia may be used as a substitute.

If a cyanide has been swallowed, and provided the victim is still conscious, an emetic such as salt water, mustard and water, or a 1% sodium thiosulphate solution should be given. This may be followed by brandy, strong coffee, or strong tea as a stimulant.

FACTORIES ACT

The Factories Act was passed in 1937 and consolidated, with amendments, the Factory and Workshop Acts of 1901 and 1929. At the same time, it repealed the whole or part of other enactments which are listed in the fourth schedule to the Act.

IMPLICATIONS OF THE ACT

The Factories Act protects both the employee and the employer and ensures that conditions in factories are such that the welfare and safety of employees are safeguarded.

Claims

Accidents, which may be sustained by a person in the course of his employment, may give rise to:

1. A claim at common law.
2. A claim under the Industrial Injuries Act.

If an employer has provided safe working conditions in his factory, in accordance with the regulations of the Factories Act, it is unlikely that a claim at common law will be lodged against him in as much as a complainant would be required to prove failure on the part of the employer to provide safe working conditions.

In addition to possible claims against the employer, a claim may be lodged involving a third party. If, for instance, while travelling to or from his place of employment, an employee were injured by a vehicle owing to negligence on the part of the owner of the vehicle, then the employee might successfully claim against him.

In the same way, an employee might meet with an accident within the confines of a factory in which he is employed, through a piece of machinery or some other appliance which under the Act was required to be protected by a person other than the factory occupier or owner.

If, in spite of the provision of safeguards by the employer, an employee sustains a personal injury or contracts a disease in the course of his employment, he may claim under the Industrial Injuries Act.

In cases when a person meets with an accident due to his own carelessness or disregard for rules, or through drunkenness it is possible that he may receive no benefit under the Industrial Injuries Act and has no claim against his employer or any other person.

The term 'in the course of his employment' is interpreted in the widest sense and generally takes effect from the time he sets foot on his employer's premises until the time he leaves. The term may also apply indirectly when a person is travelling in the course of his duties.

Although the Act may not apply to certain establishments persons employed in such establishments may still claim at common law or under the Industrial Injuries Act. In such establishments employers generally employ the safeguards as laid down in the Act both for their own protection and for the protection of their employees. Factory inspectors, irrespective of the fact that a particular establishment concerned may not be required to apply the safeguards under the Factories Act, are always willing to advise employers as to the proper safeguards.

APPLICATION OF THE ACT

The application of the Act states that the Act applies to all places of work, whether indoors or in the open air, which are defined by the Act to be factories.

INTERPRETATION OF THE ACT

By the interpretation of the Act, the term factory means premises in which persons are engaged in manual labour on certain types of work. Other premises occupied by the Crown or managed by public authorities may also be classed as factory premises.

ENFORCEMENT OF THE ACT

The Act may be enforced, too, in premises, which are subject to inspection under the authority of any government department, where labour is exercised otherwise than for purposes of instruction. The Secretary of State may arrange that the premises be inspected in respect of matters dealt with by the Act.

Certain power and duties conferred or imposed on district councils are exercised in a county by the county council and in a burgh by the town council. Duties which are to be found in Part II of the third schedule of the Act are carried out in small burghs by the county council of the county in which the burgh is situated, and references in the Act to district councils should be construed accordingly.

If offences are committed against the parts of the Act where the county or town council have the responsibility, the town or county council may prosecute offenders through their appointed officer. The town or county council may therefore institute by-laws with which local factories must comply. Other powers are conferred on county or town councils by the Public Health Act and these powers extend to factories.

ADMINISTRATION OF THE ACT

To administer the Act, inspectors are appointed by the Secretary of State for Employment.

Powers of inspectors

Inspectors have the power to enter factories and to inspect and take copies of registers, certificates, and other documents kept in pursuance of the Act. They may inspect a factory to see that the regulations of the Act are being complied with and may interview any person employed in the factory in respect of matters which come under the Act.

Factory occupiers and their employers are required to assist the inspector and should not delay him or withhold information. Responsible persons in the factory may ask the inspector to produce his identification.

Doctors may also be appointed by the Secretary of State, or by the Chief Inspector of Factories, to act as Appointed Factory Doctors in particular factories.

County councils or district councils and their officers have similar powers as inspectors, provided they can produce written authorisation.

Persons eligible to inspect a factory are not allowed to disclose information acquired in the course of their official duties.

OFFENCES, PENALTIES, AND LEGAL PROCEEDINGS

Under the offences, penalties, and legal proceedings part of the Act, the owner, or occupier, whichever is responsible, may be prosecuted if the Act is contravened. Additionally, the contravening parties may be required to remedy any matter in respect of the contravention. Employees, too, may be guilty of an offence under the Act if they interfere with any appliance or otherwise jeopardise the safety of other employees.

Parents of young persons may likewise be prosecuted if they wilfully allow such persons to be employed in contravention of the Act.

In certain cases where the Act is contravened by an employee, who commits an act for which the owner or occupier is liable, then the employee, too, may be held responsible and may himself suffer a fine as well as his employer.

Offences committed by a person in charge of a machine which has been hired to the factory may make the machine owner liable as if he were the factory occupier.

MISCELLANEOUS

Not less than one month before premises are used as a factory, the occupier must inform the district inspector, and furnish him with particulars about the factory.

Factory occupiers are required to display the prescribed abstract of the Act at the factory entrance and possibly in other parts of the factory. Other information which must be displayed includes the name of the district inspector and the Appointed Factory Doctor.

Employees are entitled to ask the employer for a copy of any notice or document which the Act requires should be posted up.

A register, which must be available for inspection, must also be kept which should contain information relating to young persons employed and other particulars. Periodically, returns may be asked for by the Chief Inspector relating to the number, age, sex, and occupation of persons employed. A notice setting out the hours of employment of young persons and women is also required to be affixed.

GENERAL PROVISIONS OF THE ACT

Health

Factories must be kept clean and reasonable working temperatures must be maintained to provide for the health of factory employees. Overcrowding is not permitted and adequate lighting and sanitary arrangements must prevail.

Where there is no mechanical power used in the factory, the responsibility for the maintenance of these conditions rests mainly on district councils. They are required to keep a register of factories in that area and to take any necessary action to see that factory occupiers put right matters requiring attention. Sanitary accommodation requirements are enforced by the local authority in all factories.

Safety

To comply with the Act, employers must provide adequate safeguards against the various mechanical hazards by the efficient guarding of dangerous machinery, or by other means which will provide adequate protection for the users. Women and young persons are not permitted to clean moving machinery.

Vessels containing dangerous liquids must also be protected.

The mechanical condition of hoists, lifts, lifting tackle, cranes, and other lifting machinery must be sound. They must be examined at prescribed intervals and the results of such examination must be recorded. Lifting machinery must only be loaded within its safe working limits.

Factory floors must be in good condition and handrails must be provided on staircases. Openings in floors should be guarded and ladders must be well constructed and maintained. The access to and the working positions for employees must be safe.

Dangerous fumes

Certain precautions are required to be taken by persons entering chambers, vats, and other confined spaces in which dangerous fumes are liable to be present. The persons entering such spaces must wear a belt with a rope attached while a second person holds the free end. As an alternative, a suitable breathing apparatus must be worn.

Inflammable gases, dusts, vapours, or substances

Dangerous accumulations of dust must be prevented and steps must be taken to minimise the effects of any possible explosions. Explosive or inflammable substances or fumes must be removed from empty drums before the drums are welded or brazed.

Steam boilers

Steam boilers are required to have safety valves, stop valves, and visible pressure and water gauges. Boilers must be examined at regular and prescribed intervals and the results of such examinations recorded.

Air receivers

Air receivers must also be regularly inspected and the results recorded. They must have safety valves, pressure gauges, draining plugs, and also facilities for cleaning their interior.

Water-sealed gas holders

The exterior of water-sealed gas holders must be examined by competent persons at prescribed intervals. In the case of holders over twenty years old, the interior must also be inspected. The results of all inspections must be recorded.

Means of escape in case of fire

District councils are responsible for the examination and certification of certain factories situated within their area in respect of the provision of adequate means of escape from fire. Such provisions must be sufficient for the number of persons employed.

Particulars of the nature of inflammable materials stored must be shown on certificates issued to the factory and any changes in buildings, or in the materials stored, must be notified to the district council by the factory occupier. To enforce their requirements, district councils may apply by-laws but these must be consistent with any regulations laid down by the Secretary of State.

During working hours the escape doors, which should open outwards, must not be locked in such a way that they cannot be easily opened from the inside. Hoists and lifts inside buildings should be completely enclosed with fire-resisting materials, and the means of access to these should be fitted with doors made of fire-resisting materials. Hoistways and liftways must be enclosed at the top only by materials which are easily broken by fire or they should be provided with a vent.

Welfare

Amenities for employees must be provided and should include drinking water and drinking vessels. Washing facilities and accommodation for clothing and the drying of clothing must be provided.

First aid facilities are required to be provided and trained first aid workers must be available during working hours.

Special provisions for health, safety, and welfare

Further regulations for the health, safety, and welfare of employees require that injurious fumes and dusts must be extracted. Persons are not permitted to partake of food or drink in rooms where lead, arsenic, or other poisonous substances are used, or where a process gives rise to siliceous or asbestos dusts. Similarly, persons should not remain in the rooms during meal or rest intervals other than intervals allowed in the course of a spell of continuous employment. A place for eating meals should be provided and situated away from areas where poisonous substances are used. Eye protection devices must be provided where required. Humidity in factories in which atmospheric humidity is artificially produced must be maintained at a proper level. Working in unsuitable underground places which are dangerous to health, or constitute dangerous fire hazards, is not permitted.

The lifting by persons of weights in excess of those prescribed by regulations is not allowed.

Women and young persons may not be employed in certain operations connected with lead manufacture. To protect persons working with lead compounds, protective clothing must be provided and other safeguards applied. A record of the health of such persons must be kept.

Notification of accidents and industrial diseases

Accidents involving more than three days' absence from work, and cases which are treated by medical practitioners involving poisoning by certain substances, should be notified to the Factory Inspector. Formal investigations may be held on cases of accidents or diseases contracted in a factory. Appointed Factory Doctors must report upon cases of death or injury arising in factories through fumes or noxious substances, and upon any case of disease of which they receive notice under the Factories Act.

Employment of women and young persons

The regulations of the Act deal with the hours of work, times of commencement and cessation of duties, rest periods, holidays, and other aspects of the employment of women and young persons. Other regulations also deal with the employment of persons under 16 years of age. Such persons are required to undergo a medical examination certifying them fit for the particular type of employment.

Special applications and extensions

Under the special applications and extensions of the Act, owners, although they may not be occupiers, of factory premises are responsible for the carrying out of certain requirements of the Act in respect of the premises.

REFERENCES

1 Charlett, S. M., 'Sheet Glass in the Laboratory', *Lab. Pract.*, **6**, 447 and 526 (1957)
2 Gaston, P. J., *The Care, Handling and Disposal of Dangerous Chemicals*, Northern Publishers (Aberdeen) Ltd (1971)

3. Lawrence, E. C. H., 'Contribution to the Discussion on the Origins and Prevention of Laboratory Accidents', *R. Inst. Chem. Report* No. 4, 74 (1949)
4. Fawcett, H. H., 'Hydrogen Sulphide, the Killer that May Not Stink', *J. chem. Educ*, **25**, 511 (1948)
5. Webster, A., 'Safe Heating Using Electric Heating Mantles', *Chem. Age, Lond.*, **68**, 961 (1953)
6. McCarty, L. V., and Balis, E. W., 'The Contamination of Liquid Nitrogen by Oxygen of the Air', *Chem. Engng News*, **27**, 2612 (1949)
7. BS 349, *Identification Colours for Gas Cylinders*, British Standards Institution, London (1932)
8. BS 1319, *Medical Gas Cylinders and Anaesthetic Apparatus*, British Standards Institution, London (1955)
9. Cornwell, J. C., 'Prevention of Accidents Arising from Electrical Work', *R. Inst. Chem. Report* No. 4, 24 (1949)
10. Smith, D. M., and Walsh, A., 'The Electrical Screening of Sparking Apparatus for Use in Spectrographic Analysis', *J. scient. Instrum.*, **20**, 63 (1943)
11. Park, B. P., *Industrial Eye Accidents*, pamphlet issued by J. R. Flemming Ltd, London (1956), reprinted from *J. Instn ind. Saf. Offrs*
12. Guy, K., *Laboratory First Aid*, Macmillan, London (1965)
13. Scott, M. M., 'First Aid in Asphyxia', *Practitioner*, **147**, 580 (1941)
14. Bennett, A. L., 'A Manual Method of Artificial Respiration for the Simultaneous Resuscitation of Two Victims', *J. appl. Physiol.*, **8**, 604 (1956)
15. Croton, L. M., 'Wounds and Haemorrhage', *J. Sci. Technol.*, **3**, No. 3, 35 (1957)
16. White Knox, A. C., 'First Aid in Wounds and Haemorrhage', *Practitioner*, **147**, 572 (1941)
17. Flemming, C. W., 'Treatment of Burns. A Plea for Simplicity', *Brit. med. J.*, **2**, 314 (1945)
18. Pieters, H. A. J., and Creyghton, J. W., *Safety in the Chemical Laboratory*, Butterworths, London (1951)
19. McCartan, W., and Fecitt, E., 'First Aid Treatment of Phosphorus Burns', *Brit. med. J.*, **2**, 316 (1945)
20. Van Arsdell, P. M., 'First Aid for Chemical Eye Injuries', *Industr. Med.*, **16**, 188 (1947)

7
Special needs of teaching laboratories

The special needs of teaching laboratories arise from the fact that they are heavily populated by persons undergoing training. Because the students have reached different levels in the instruction, no two teaching laboratories, even though they may be used for teaching the same subject, are exactly alike in design or equipment. For these reasons they differ considerably from non-teaching laboratories. Certain basic needs are, however, common to all teaching laboratories.

DISCIPLINE

For the safety of everyone concerned, a strict discipline must be maintained in the laboratory by the lecturer or demonstrator in charge of the class. A strict compliance with the standing laboratory regulations, as displayed in the laboratory, should be insisted upon.

The technical staff, too, can assist in this by maintaining the laboratories efficiently so that the good order will be reflected by the students in their work and by their co-operation. In order to promote these conditions throughout the department, the senior technician, or other person responsible for the upkeep of the laboratories, can only expect to maintain order if he himself sets an example, and in this way the staff acquires a confidence in his ability and methods. The confidence and the enthusiasm of the staff can only be maintained by the delegation of responsibility to them, and this must be done irrespective of the fact that certain duties may not be quite so well carried out as if the senior person had himself performed the task. It is far more rewarding in the long run to delegate work than to attempt to do everything oneself. When such responsibilities have been given, the decisions and actions of the persons accepting the responsibility must be fully supported by their senior. Only in this way can the full confidence of the staff be gained. The laboratory staff must not be allowed to become the scapegoat for the untidy habits of others and every laboratory worker, whatever his position, must be required to share the responsibility for the state of the laboratories. The person in charge must visit each part of the department regularly so as to be fully aware of the problems and difficulties encountered by his staff in the course of their duties.

If the smaller points of order in laboratories receive adequate attention, this breeds further order into the whole system of the establishment. To illustrate this a few examples of good laboratory discipline are worth consideration.

LABORATORY SERVICES

Laboratory workers are notorious for their wasteful attitude in respect of laboratory services. Abandoned Bunsen burners which are left to burn furiously when not in use, and water filter pumps operated for unnecessary lengths of time, are examples. Oxygen cylinders improperly turned off at the main valve, the use of large hotplates for small vessels, the abuse of distilled water, and neglecting to switch off unwanted lights, all add to the cost of running the laboratory. By adequate supervision much of this kind of waste can be prevented.

LABORATORY CLOTHING

All students and laboratory personnel should wear a laboratory coat. This is a protective measure and is especially important in the case of female persons, whose clothing is the most dangerous in the laboratory. The other advantages of wearing laboratory clothing are obvious if a coat which has had good wear is inspected.

MAINTENANCE OF PERSONAL EQUIPMENT

The student should be taught to maintain his personal equipment and to regard it in the same way as any experienced tradesman does his tools. All glassware should be kept in a clean condition. Students should also provide themselves with the minimum requirements for the course. In biological laboratories, for instance, this consists of a pocket lens, one section razor, two pairs of forceps, section lifters or mounted needles, two scalpels, a camel hair brush, and a Pasteur pipette. Dissecting instruments should be kept sharpened and ready for instant use. In junior classes an occasional locker check should be carried out.

BENCH CLEANLINESS

During practical periods the benches should be kept clean and tidy and all spillages of chemicals and stains should be immediately wiped up to protect both the bench tops and personal clothing. At the close of the class the bench top should be cleaned down with a squeegee and finally with a swab, before the student leaves the laboratory.

In research laboratories, difficulties are encountered and the appearance of the laboratory may be adversely affected by the deplorable attitude of some workers who insist that their bench should never be cleaned. A cluttered bench neither helps their work nor enhances the orderly appearance of the laboratory. On the other hand, unnecessary interference with set-up apparatus could quite obviously be fatal to experimental results and a workable compromise, which allows for general cleaning, can quite easily be worked out.

Waste containers

All students should be trained to use the waste containers which should be provided and wide-mouth bottles or other suitable waste containers should be provided in organic chemistry laboratories for tarry residues. This prevents the permanent fouling, and often the subsequent loss, of flasks and other containers and also prevents sinks from being abused.

PREVENTION OF THEFT

Laboratories throughout the country suffer losses by theft that amount to huge sums of money annually. The technician in charge should see that a careful locking up procedure is adopted when the laboratories are vacated. The responsibility for securing apparatus does not rest entirely with the technician and every laboratory worker must play his or her part by securing expensive portable equipment after use. Note also that however intent a worker may be on the job in hand he must leave the laboratory promptly at the agreed locking up hour. Nothing is more infuriating to the custodian of laboratory apparatus than to be frustrated in the course of his duties by a keen but unthinking enthusiast.

CONTINUOUS EXPERIMENTS

Experiments which require the laboratory services to be left on overnight present a problem, and in fairness to the laboratory staff a procedure applicable to everyone must be laid down. From the point of view of safety, it is desirable that none of the services, particularly gas and water, be left on. In cases where this is unavoidable, the permission of the senior technician should be sought who will inform the night caretaker and make any other necessary arrangements. The responsibility for closing down services, and the locking up of empty laboratories, is one which should be carefully delegated.

REPORTING OF DAMAGE

Accidents which may happen, involving damage to instruments, may not be serious if the damage is reported immediately. No attempt should be made by the student himself to effect repairs to an instrument, for even though he may act with the best of intentions irreparable damage may be done. This rule is particularly important as applied to balance mechanisms and to microscope lens systems. Similarly the temperature settings on ovens, furnaces, or incubators must never be altered without permission. No student should handle any precision instrument unless he has received instruction in its use.

SUPERVISION

At all times experimental work in the laboratory, especially where more junior students are involved, must be adequately supervised. In these days when classes are large, the danger of accidents is quite considerable. Eating and drinking in the laboratory should not be allowed.

REGULATIONS

The disposal of sodium residues and other dangerous substances, and the issue of alcohol and poisons, are some of the matters which should be embraced in standing laboratory regulations. The regulations should be prominently displayed in all laboratories and should be signed by the head of the department concerned.

SUGGESTION BOOK

A suggestion book hung in the laboratory is helpful for everyone and sensible suggestions should be quickly put into operation so as to encourage the further use of the book.

GENERAL REQUIREMENTS

ACCOMMODATION

The laboratory must be large enough to accommodate the students and to allow each person ample room for movement. This is an essential safety measure. The bench space allocated should be sufficient to allow each student to manipulate his apparatus correctly for the particular subject being taught. The student must also have access to the service outlets he needs.

LIGHTING

The room must be well lit by natural and artificial means. This is particularly important for biological and similar laboratories where microscopes are used. The benches for microscopical work should face large window areas. In plant physiology and other laboratories where a great deal of work with microscopes is done, an electrical socket is required for each student so that microscope lamps may be used on dull days.

VENTILATION

In teaching laboratories generally, and chemical laboratories particularly, the ventilation should be sufficient to maintain safe and pleasant working conditions. Adequate numbers of fume cupboards are essential and the need for these varies with the type of work carried out in the laboratory. For qualitative analysis at least one 0.9-m wide cupboard will be required for every ten students.

DRAINAGE

Good drainage is the important requirement for most laboratories. Accessibility to the drainage and to the plumbing services is essential. The regular inspection of service and drainage pipes will save considerable repair costs if steps are taken to deal with any corrosion.

SERVICES

Supplies of gas, water, and electricity on the student benches are necessary in most teaching laboratories and a supply of steam may be required for organic chemistry. Without attempting to describe the extent of the various services which should be provided, it will suffice to say that these must be ample for the present, and sufficient for the future, requirements of the laboratory. The need for more service outlets, and particularly for electrical sockets, increases with time.

Various other piped systems for laboratory services may be installed, but with the exception of gas, water, and possibly steam, the advantages of piped systems is a debatable point. In many laboratories local methods are preferred for providing vacuum, compressed air, and other services. In such cases the service equipment is usually mounted on wheeled trolleys.

DISTILLED WATER

A supply of distilled water is necessary in most teaching laboratories and the means of producing it vary according to the purity of the water required. Water of high purity may be necessary for special work as in physical chemistry laboratories where it is used for conductivity experiments. For such purposes glass or quartz stills are often used, but the water may need to be redistilled a number of times.

For normal laboratory purposes metal stills of the Manesty type, which may be wall mounted, are suitable and popular. These may be heated by paraffin, gas, electricity, or steam, depending on the model. Electrically heated stills are clean and efficient and are fitted with an automatic cut-out and ejection plug which operate if the water supply fails or the still becomes overheated for any other reason. Manesty stills are available in two sizes and yield 1.5 litres or 4.5 litres per hour for a loading of 1500 W or 3000 W, respectively.

Ion exchange columns, known as water deionisers, are now in common use which produce deionised water without the need for a fuel supply. In this respect they are economical, but the initial expense of the resin and its replacement or regeneration may have to be taken into account when comparing costs. Water for general use with a specific resistance of $1.0 \text{ M}\Omega/\text{cm}$ is obtainable for such columns. This compares favourably with water prepared by triple distillation from quartz apparatus.

The storage of distilled water in bulk involves the use of a large container from which it may be tapped off by students for filling wash bottles and for other purposes. Glass or stoneware receptacles with a tubular outlet at the bottom may be used. Because these are vulnerable, a safer container is desirable and for normal laboratory purposes a large copper urn, provided it is tinned on the inside, is suitable.

There is inevitably a certain amount of spillage of water beneath the tap of the distilled water butt, which may have a deleterious effect on the floor, apart from being a nuisance in other respects. To overcome this problem a catchpot can be placed under the tap. This is not entirely suitable and a better method is to let in a metal grille flush with the floor. The grille is situated above a receiver which is connected to the normal drainage arrangements. Alternatively, the spillage is carried through the wall and clear of the building by a small overflow pipe.

WATER PRESSURE

A sufficient and constant water pressure is required in the laboratory for the successful operation of water filter pumps. The pressure may fluctuate because of the alternating nature of the draw off of water in the various laboratories. In this case it may be necessary to install a booster pump and an experienced water engineer should be consulted in this connection.

FURNITURE

The essential items of furniture for teaching laboratories are the student working benches. These vary according to the needs of the particular subject taught. Plant physiology benches, for example, should be wide with a large sink to every two students, whereas laboratories for other plant work may require narrower benches and smaller sinks. Similar variations in bench design are applicable to the various types of chemical and other laboratories.

One of the main requirements for any bench in a teaching laboratory is that it should not be too long, or difficulties in attending on the students are presented. Another feature is that sufficient working space, in accordance with the level of instruction, be afforded to each student. It is reasonable to allow 1.2 m for first-year students and 1.8 m for those more advanced in their studies. For research students a minimum of 2.5 m is necessary. The number of working places on single-sided benches, or on one side of double benches, for first-year work should not exceed four. For more advanced work two places are suitable and for research work only one.

For reasons of space and the sharing of supply services, double-sided benches are the most suitable for chemistry laboratories. This arrangement also allows the contents of a centre bottle rack, placed between the students working on opposite sides of the bench, to be commonly shared. With the advent of small-scale chemistry, however, the need for large centre bottle racks is becoming less apparent. This is advantageous in as much as there is less bench obstruction and the class can be more easily supervised by the instructor.

In physical laboratories the island-type bench, which allows all-round access to the various instruments employed in these laboratories, is still preferred.

For school laboratories the Nuffield Teaching Project has necessitated new bench designs with improved storage capacity. One of these is a diamond-shaped bench capable of accommodating four 'O' level or two 'A' level students (*Figure 7.1*). Special trolleys may be used in conjunction with the benches and when wheeled into position between the benches effect a longer bench top and provide extra working space.

To accommodate the equipment used in the Nuffield project, special thought was given to the storage capacity of the benches. Consequently, four cupboards are provided which are fitted with polypropylene trays and a storage pocket designed to accommodate long pieces of equipment. Among the advantages claimed for this shape of bench are better all-round vision, easier access for both

Figure 7.1. (a) Diamond-shaped bench. The bench illustrated has a polypropylene sink and gas, water, and electricity outlets. Kneehole recesses are provided and writing slides are incorporated for use during tutorial or demonstration periods. (b) Arrangement of the trays used for storage of student sets. The trays are removable so that when necessary their contents may be quickly checked (Courtesy Grundy Equipment Ltd)

students and staff, quicker installation, and more flexible student grouping.

The general requirements for benches are described elsewhere but note that hardwood tops, which contain natural oils, are essential. These should be further protected by suitable treatment.

Wall benches are necessary for items of equipment such as incubators, ovens, and furnaces.

STUDENT LOCKERS

It is generally desirable in teaching laboratories that all students be provided with a set of apparatus and a locker with a key. A master key should be held by the technician responsible for the laboratory and a set of duplicate keys for the lockers should also be kept. There are certain exceptions, such as in the case of physical laboratories, where the personal issue of apparatus to students may not be necessary.

The method of issuing keys to students and the efficient storage and issue of duplicates pose a difficult problem. When large numbers of students are involved, key cupboards or key boards take up considerable space and the keys which are invariably 'lost' from them leave no clue as to their whereabouts. Therefore, a filing and record system whereby each key is kept in a small envelope along with its record card has much to recommend it, and in this way a single-drawer record file of small dimensions can accommodate up to four hundred keys. The details of this method have been described by Otte[1].

Lockers need not be large for first-year classes but should increase in size in accordance with the academic level of the students. The numbers of students tend to decrease in the advanced classes, so that the working space and locker ratio maintain a close relationship. Sufficient room can therefore be found in the bench understructures to provide the required number of lockers.

In elementary laboratories the necessity for lockers may not arise and apparatus may be set out on the benches or on trays for class use and is easily checked when the class period is over.

Arrangement of student lockers

Various arrangements for student lockers have been advocated from time to time and some of these are now described.

The first arrangement is designed for economy where the availability of apparatus is limited. In this case only the centre lockers are issued to the students and end lockers are devoted to the storage of Bunsen burners and similar items of equipment which may be shared by the students using the bench.

Another system allows a certain number of complete sets of apparatus to each bench, corresponding with the number of working places. The same sets of equipment which are kept in the bench lockers and drawers may be used by successive classes. At the end of each class the items, which are kept in marked positions, are checked to ensure that they are all present and intact. The broken or missing items are immediately replaced from the stores. In addition, other smaller lockers are provided for the private use of students for the storage of personal items and for experiments which have not been completed and which are held over until the next class.

The advantages claimed for this system are that students are not tied to one locker or location and the laboratory's needs in terms of area, benches, and apparatus are thereby reduced.

It is my opinion that the best arrangement is to provide every student with a complete set of apparatus and a locker large enough to contain it. The locker should be fitted with an interior drawer and in this way all apparatus is secured with one key by locking the locker door, and becomes the sole responsibility of the student.

Each working space is divided up into locker accommodation. The 1.8-m space, for example, allowed for advanced students, is sufficient for three lockers of suitable size. This allows three changes of classes to use the laboratory, which is normally sufficient. The initial outlay on equipment is high but this is ultimately justified for the following reasons:

1. The student is encouraged to take a personal pride in his equipment.
2. Time and labour are saved by avoiding the setting out, cleaning, and checking of apparatus.
3. The student works with a familiar set of apparatus and retains it for his class examinations.
4. The laboratory is tidy at all times.
5. Broken or missing apparatus can be charged to the student responsible.

OTHER FURNITURE

The other items of furniture needed in the laboratory include a blackboard for discussions and explanations. The blackboard need not be large and the roller type is compact yet offers a large surface area. In advanced laboratories a framed white acrylic sheet attached to the wall is now a very popular writing board for local discussions. A grease pencil may be used for writing on the acrylic sheet which, unlike the chalk board, produces no dust.

In many laboratories a small glassblowing bench is required to introduce this useful art to students and should be situated in a position removed from draughts. The soft asbestos top formerly used for glassblowing benches has lost its popularity because loose fibres tend to be picked up by the glass, making 'pinholes' in the finished ware. A hard black asbestos composition top is better. The wall at the back of the unit should be protected by 3-mm thick ebony Sindanyo or similar material.

For physics, physical chemistry, and other laboratories, where delicate equipment of value is to be stored away from dust, wall cupboards are essential. Plastics dust covers should be used over equipment set out on the benches. Some elementary chemistry laboratories may also require cupboard space when dispensing stores are not provided.

In other laboratories, the main requirement is for wall shelves on which to keep reagent bottles, specimens, and preserved material.

STORES

The laboratory work depends on the prompt and local supply of materials. A dispensing store should be within easy reach of each laboratory to serve student needs. No person, other than the storeman, should be permitted to enter the store even though he may be of higher standing than the storeman and certainly no person should ever be allowed to remove stock.

EQUIPMENT

The equipment needed in teaching laboratories also varies with the subject, but certain items are common to most laboratories.

BALANCES

In the physics laboratory student balances may be kept with reasonable safety in the laboratory. This is not so in many other cases and in biology laboratories separate balance rooms are desirable; for chemistry laboratories they are essential.

Balance rooms should be situated at each end of large laboratories so that working areas are as close as possible to the balances. This arrangement reduces the amount of movement to a minimum and is convenient for everyone concerned.

Up to six students may be assigned to a modern balance during a laboratory period. A card should be kept with each balance with a list of the names of students entitled to use it.

After each class period the balance should be checked. This duty is best shared between several people so that each has only a limited number of balances to deal with. It is also necessary that the balance be serviced at regular intervals by a departmental technician especially skilled in this work, or by the manufacturer's technical representative.

REAGENT BOTTLES

LABELLING

The correct and systematic labelling of bottles adds to the general efficiency of the laboratory and makes life easier for everyone concerned. The label on the bottle must accurately describe the contents, and where a number of laboratories are situated in close proximity the name of the laboratory to which the bottle belongs should also be given. The shelf number and the position of the bottle on the shelf is described. The bottles belonging to the first shelf would thus all bear the number 1 followed by a second number, e.g. 1-1, 1-2, and so on, to define the position of the bottle on the shelf.

This method assists the replacement of bottles, which are liable to stray all over the department, especially when a shortage of a particular commodity occurs. To make labelling an easy matter a rubber stamp may be designed which includes all the necessary headings. The rubber stamp may be used on plain white sticky-backed labels which should afterwards be protected by label varnish or paraffin wax.

Several proprietary brands of label varnish are available but it may also be made in the laboratory by dissolving transparent celluloid in acetone until a syrupy liquid is obtained. Sandblasted labels on bottles are excellent for permanent labels but are expensive and the range of titles is limited. Other labels known as Flexirings consist of rubber rings available in various sizes which are coloured for coding purposes. Many different titles are available and these washable labels may be slipped on to plain reagent bottles. Probably the most popular labelling material is vinyl tape which may be purchased in several widths and in many different colours. The rolls of tape are loaded into a small hand-size tapewriter and by dialling the required lettering by means of a wheel an embossed label is quickly produced. The plastics skin on the back of the lettered tape is then peeled off to expose the adhesive backing and the label is ready to be affixed. In addition to being suitable for bottles these attractive and virtually indestructible labels have a host of other laboratory uses.

SIZE OF BOTTLES

It is desirable for the sake of general appearance that the laboratory bottles conform to a certain size. The 350-ml size is generally convenient, but the selection is largely a matter of personal taste. The deciding factors are the frequency with which the bottles have to be filled and whether the size is convenient for handling.

Smaller sets of bottles with capacities as small as 30 ml are recommended[2] for semi-micro work. These are fitted with a dropper and a rubber teat.

Bulk solutions may be kept in Winchesters or aspirators and it is usually found convenient to use both. Winchesters, which may be stored in side cupboards or on 150-mm shelves recessed into the ends of the benches, are adequate in size to fill those bench reagent bottles which require infrequent attention. Half-Winchesters are suitable containers for stock solutions of indicators.

The bulk solutions for bottles which require to be filled often, such as the dilute acids and sodium hydroxide solutions, should be kept in aspirators. Polythene aspirators, because of their durability and the fact that the outlet taps are little affected by alkalis, are best. As an alternative to polythene aspirators, the glass tap in the normal aspirator can be replaced by rubber tubing and pinch clips for solutions such as sodium hydroxide and sodium carbonate.

The main difficulty attending the storage of all bulk solutions is the drips which fall on to the bench whatever method of tapping the aspirator is used. The author has found it convenient to mount aspirators on a low shelf above, and at the back of, the bench top. The width of the shelf is such that the taps project just clear of the edge of the shelf and are positioned directly above a narrow lead-lined trough in the bench top. A water tap is situated above one end of the trough for flushing purposes. The top of the trough is covered with a plastics grille which is let into and finished flush with the bench top.

PATTERN OF BOTTLES

A uniformity in the pattern of bottles is desirable. Round bottles with narrow mouths and flat stoppers are suitable. The flat stopper can be placed when necessary on the bench without the fear of it rolling off or the possibility of the introduction of foreign matter into the solution. The dustproof type of stopper serves a useful purpose but is more easily chipped when it has to be tapped to free it from a bottle. Tapping can be avoided if rubber or polythene stoppers are used in all bottles which contain encrusting solutions.

For solid reagents the wide-mouth round powder bottles with a plastics screw top are best. These are available in all sizes.

CANADA BALSAM BOTTLES

The drying out of Canada balsam bottles which causes difficulty in biological departments may be overcome by storing the bottles in the manner described by Ford[3]. The balsam bottle is kept inside a jar with a wide mouth and secured in a cavity cut in a cork bung. The bung itself is held in position on the inside of the base of the bottle by packing cotton wool around it. The cotton wool is wetted with xylene to provide the xylene atmosphere.

In a similar way numbers of balsam bottles may be kept in larger airtight containers in a xylene atmosphere and removed just before they are required for practical classes.

CHEMICALS

An ample supply of technical-quality chemicals in the laboratory prevents unnecessary use of the pure grade. For instance, commercial acid is a suitable material for use in Kipp gas generators and for melting point determinations and commercial sodium carbonate for neutralising acid spillages. If these and other technical substitutes are not visibly to hand, then pure and costly materials will be used from the reagent shelves. It is also very important that persons in charge of numbers of people in training should be acutely aware of the prices of chemicals, and should adjust the quantities of materials used for class preparations accordingly. When it is chemically necessary to employ expensive substances for experimental use, it may be possible to restrict either the quantity or the strength of the solution. It is common practice, for instance, for students to use smaller sizes of volumetric glassware when doing silver titrations.

Expensive chemicals are often wasted in the laboratory because the students are unfamiliar with their price.

To draw attention to the fact that a costly item is being handled, a label bearing the word EXPENSIVE should be stuck on to the applicable bottles. A rubber stamp can quickly turn out labels suitable for this purpose. In the instructions given to students the amount of material to be used in the experiment should be stated precisely in terms of weight or volume and the size of the apparatus required should also be indicated. Vague instructions only confuse and result in waste.

RESIDUES BOTTLES

The provision of residues bottles is far from being a new feature in the laboratory but there yet remains to be devised a method to ensure that students make use of them. This can be done only by setting an example and maintaining good discipline. The bottles should be placed at strategic points around the laboratories and their use encouraged by a regular attendance upon them. This imparts to the users the impression that their care in saving residues is matched by the careful recovery of the material. Silver, iodine, mercury, and solvents represent some of the recoverable materials. It is a matter of opinion whether or not in the long run laboratory residues, other than silver residues, are worth saving in view of the time spent in recovering them. A psychological advantage is certainly gained and residues bottles induce tidiness in every laboratory worker. If large quantities of one type of material are used, the recovery of the substance may be profitable.

RECOVERY OF RESIDUES

RECOVERY OF SILVER

The recovery of silver presents little difficulty. A few drops of hydrochloric acid are added to the bulk residues to precipitate the silver as silver chloride. When the precipitate has settled the clear liquid is decanted off and the sludge is filtered through a large Buchner funnel. It is then well washed with water and dried in an oven. Firms which deal in precious metals will accept the silver chloride and give credit for it.

If the metal is required an amount of fusion mixture, twice the bulk of the chloride, and a little glucose to assist the reduction, are added to the dried residues. The whole is ignited in a fireclay crucible placed in a muffle furnace and the silver is then poured into an iron mould and allowed to cool.

RECOVERY OF IODINE

Iodine recoveries should be carried out in a fume cupboard. To the residues potassium dichromate and concentrated sulphuric acid are added in the proportion of 10 g of potassium dichromate and 10 ml of acid to each litre of residues. After one hour the resultant sludge is washed by decantation and can be sublimed from the evaporating dish on to the cold surface of a flask through which water is passed.

Alternatively, the sludge may be heated in a large round-bottom flask in which a large cold finger is suspended. A slight bulge is formed near the top of the cold finger to prevent it slipping through a bored cork inserted into the neck of the flask. A 'V' is cut in the side of the cork to serve as a vent. Periodically the cold finger is carefully removed and the iodine scraped off.

The above methods allow some of the iodine fumes, which have an irritating effect on the eyes and throat, to escape. The best method, therefore, is to heat the sludge in a flask and pass the vapours through a Liebig condenser. To prevent the tube becoming completely blocked, the solidified iodine is removed at intervals by pushing the contents of the tube into a receiver. With the aid of a special still head, which has been described by de Taranto[4], this can be done simply by pushing a polythene plug through the tube. There is no necessity to dismantle any part of the apparatus. The iodine is then dried in a Buchner funnel and finally in a desiccator.

Further methods of recovering iodine have been given by Smithers[5] and Fowles[6].

RECOVERY OF SOLVENTS

Solvents may be recovered by fractional distillation. This should be done in a fume cupboard and all necessary precautions against the attendant fire hazards must be taken.

Prospective users regard recovered solvents with suspicion unless they themselves have effected the recovery, but recovered alcohol and alcohol which is only slightly contaminated may be quite suitable for topping up museum jars in the biology department.

It also has its uses in the workshop for various purposes such as the preparation of shellac or as a cleaning solution.

RECOVERY OF PLATINUM

Small scraps of platinum should be kept until sufficient is obtained for recovery or for sale to dealers. If the platinum is to be recovered the pieces are cleaned by heating them in concentrated nitric acid and washing with water. The platinum is then dissolved in aqua regia and the solution of hydrochloroplatinic acid is evaporated to dryness. The residue is then gently heated. Hot water is added until the solid just dissolves and crystals of platinum tetrachloride form when the solution cools.

Foner[7] also gives a simple method for the recovery of platinum when large amounts have been dissolved by chemical attack. The method consists of acidifying with hydrochloric acid and evaporating nearly to dryness. If nitric acid is present, repeated evaporation with hydrochloric acid is necessary. Large amounts of ammonium salts should be destroyed by preliminary evaporation with concentrated nitric acid on a water bath.

The salts are dissolved in water and a strong solution of ammonium chloride in 50% ethanol is added until no more ammonium platinic chloride is precipitated. Cool and filter. Wash with dilute alcoholic ammonium chloride solution. Char off the paper and ignite at 1000 °C in a porcelain crucible. An oxyacetylene flame can be used to melt the resultant platinum sponge into a bead.

PURIFICATION OF MERCURY

Contaminated mercury may be purified in several ways, depending upon the type of contamination. Superficial dirt is removed by 'filtering' it through a filter paper cone which has a small hole pricked at the apex and which is supported in a funnel in the usual way.

Mercury may be cleaned in a similar way by filtering it through a G3 sintered glass crucible and collecting it in a Buchner flask. These methods have been described by Ansley[8].

Another simple method is to force the mercury by pressure through a piece of chamois leather (*Figure 7.2*). For this purpose the apparatus consists of a piece of glass tubing of about 25-mm bore which is belled at one end and over which is wired a piece of chamois leather. The other end of the tube is closed by a rubber bung and a

Figure 7.2. Simple apparatus for removing superficial dirt from mercury

bicycle valve is inserted through a hole in the centre of the bung. Pressure is introduced by pumping with a bicycle pump.

If mercury is contaminated with amalgams, it is usually cleaned by the familiar method of allowing it to fall in the form of fine globules down a glass tube containing 10% nitric acid. The mercury, reservoir consists of a separating funnel with its stem drawn out to a fine jet. The stem also has a slight bend in it so that the deflection causes the small drops of mercury to rebound from the side of the glass tube and to pursue a zigzag course down the tube. This increases the length of path and assists the cleaning process. As an alternative to this, the stem of the funnel is not drawn out but is slightly constricted close to the open end, and a piece of muslin is tied over it. The funnel is supported so that the muslin is just below the surface level of the nitric acid. This causes the mercury to break up into very fine globules.

Porter[9] has described an apparatus in which the mercury is allowed to fall down a column packed with pieces of glass tubing which is filled with 10% nitric acid. It is claimed that in this way a greater degree of liquid–liquid contact is obtained. A drain cock is situated just above the level of the mercury seal which allows the nitric acid to be run off. The mercury is then washed by passing it again through the tube which is refilled with water. After washing, the mercury is initially shaken in a glass jar with a solution of potassium permanganate and occasionally shaken over a period of 12 hours. The aqueous layer is decanted off and the mercury is again shaken with 25% nitric acid. It is then washed and dried.

Davis[10] has pointed out that all dirt and grease should be removed from the mercury before it is passed through the acid column and suggests that the mercury should be covered with hot water through which a stream of compressed air should be passed. This causes dirt and grease to collect at the surface of the water which can then be decanted off. It would appear better, however, to agitate the mercury with weak sodium hydroxide or detergent solution.

Mercury may also be cleaned in the laboratory by placing it in a large filter flask, so as to expose a large surface area, and agitating it with dilute nitric acid. This is done by drawing air through the contents of the flask by means of a water filter pump (*Figure 7.3*). This method, however, is very slow if the mercury is contaminated by iron because this metal is more difficult to remove than base metals. Two methods of removing iron have been described by Tewari and Agnihotri[11].

The first consists of shaking 400 g of mercury with 100 ml of 10% potassium dichromate solution for 15 minutes. The supernatant liquid is then drained off and the mercury washed by shaking it repeatedly with water until the precipitate is removed. It is then washed three times with 10% nitric acid.

A little mercury is lost when cleaned in this way and a second method is recommended which, although somewhat slower than the first, involves no loss of mercury.

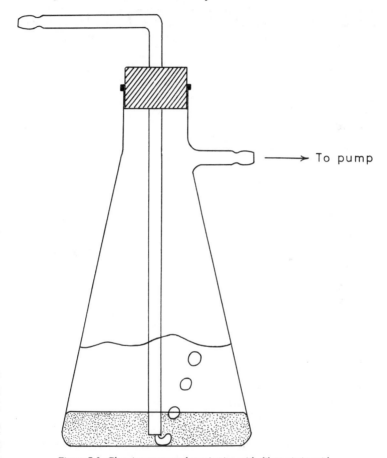

Figure 7.3. Cleaning mercury by agitation with dilute nitric acid

400 g of mercury is covered with 200 ml of water containing 4 to 5 ml of ammonium hydroxide and 5 ml of 3% hydrogen peroxide. The whole is shaken, or may be agitated by drawing air through it, and the cleaning solution is replaced every 30 minutes until the mercury is pure. The mercury is finally washed with 5% nitric acid.

The most effective way to purify mercury and to remove the metalled amalgam completely is to distil it. Various types of stills have been recommended. The type described by Collins[12] has an advantage in that it requires little supervision and stops automatically when the required amount of mercury has distilled over.

For normal laboratory purposes a simple laboratory still which is easily erected is suitable. The apparatus (*Figure 7.4*) consists of a 500-ml round-bottom distillation flask A with a side arm S 20 cm

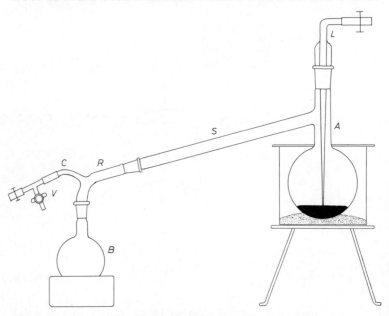

Figure 7.4. Shows a simple mercury still. The still should be erected and used in a fume cupboard

long and of 18-mm diameter sealed into the side of the neck. The end of the arm is connected by means of a receiver adaptor R to a 250-ml flask B. A side outlet C, at the bend of the receiver adaptor, is used for connecting the apparatus to a Hyvac pump. An air leak tube L which is drawn out at one end to a capillary jet passes through the neck of A. This is sealed internally into a short piece of tube above the B 24 cone which closes the neck of the flask. A piece of pressure tubing and a screw clip are used for adjusting the rate of air flow through the air leak tube which is drawn out at the other end to a capillary jet. The end of the jet almost touches the bottom of the flask. The distillation flask is heated in a sand bath which is closed at the top by two pieces of asbestos sheet.

To operate the apparatus, the pump is switched on and the rate of air flow through L is adjusted so that about two bubbles per second pass through the mercury. The initial rate of the distillation is controlled by adjusting a glass vacuum tap V which is attached by pressure tubing to the side arm of R. By regulating the Bunsen flame a steady distillation is obtained.

A method of cleaning mercury introduced in recent years involves the preliminary oxidation of dissolved base metal contaminants to their oxides. The solid oxide precipitates, plus other superficial contaminants such as dirt, grease, and water, are then removed by a decantation process.

The mercury is first agitated inside a glass bowl in an oxifier in which it is broken up into very fine droplets and subjected to repeated contact with air. At the same time the oxide skins are also broken up and carried under and through the bulk of the mercury until oxidation is complete. The whole of the mercury contamination is now in a superficial form and may be removed by a second process.

The second process involves a gold adhesion filter. The mercury is placed in a plastics reservoir with an aperture at the bottom surrounded by a gold ring. Since the mercury does not 'wet' the sides of the container the contaminants are thus able to creep down between the mercury and the sides of the container. They are stopped by the positive seal formed by the 'wetting' of the gold ring by the mercury. This gives rise to an adhesion between the two elements. The clean mercury, drawn from the centre of the mercury, is able to pass through the aperture and is retained in a dustproof container situated beneath it.

REFERENCES

1 Otte, B. J., 'A Filing and Record System for Laboratory Keys', *J. chem. Educ.*, **6**, 518 (1929)
2 Holness, H., *Inorganic Qualitative Analysis Semi-micro Methods*, Pitman, London (1954)
3 Ford, D. F., 'Storage of Canada Balsam', *Sch. Sci. Rev.*, **41**, 517 (1959)
4 de Taranto, M., 'A Simple Still Head Attachment for Use with Solidifying Distillates', *S.T.A. Bull.*, **2**, No. 11, 7 (1952)
5 Smithers, C. W., 'Scheme for the Recovery of Iodine Residues as Potassium Iodide', *S.T.A. Bull.*, **4**, No. 4, 13 (1955)
6 Fowles, G., 'Extraction of Iodine from Iodine Residues', *Sch. Sci. Rev.*, **14**, 373 (1933)
7 Foner, A. H., 'The Use and Care of Platinum Laboratory Apparatus', *Lab. Pract.*, **14**, 944 (1965)
8 Ansley, A. J., *An Introduction to Laboratory Technique*, 2nd edn, Macmillan, London (1952)

9 Porter, M. G., 'The Cleaning of Mercury on a Small Scale', *J. Sci. Technol.*, **5**, No. 1, 25 (1959)
10 Davis, J. J., 'Some Further Views on the Cleaning of Mercury on a Small Scale', *J. Sci. Technol.*, **5**, No. 3, 3 (1959)
11 Tewari, D. H., and Agnihotri, S. K., 'A Note on the Purification of Mercury', *Lab. Pract.*, **14**, 1411 (1965)
12 Collins, T. J., 'An Electric Mercury Still', *S.T.A. Bull.*, **1**, No. 11, 4 (1950)

8
Optical projection

PRINCIPAL METHODS

Images of various objects may be produced on a screen by episcopic or diascopic methods. In the diascopic method light is directed from a source through a transparency such as a lantern slide or film. Episcopic projection involves the production of an image by the reflection of light from the surface of an opaque object. The clarity of the image in any form of projection invariably depends upon the quality of the components making up the optical system, and in particular upon the quality of the condenser and the lens.

DIASCOPIC PROJECTION

Owing to variation in design, the external appearance of diascopic projectors differs considerably. Basically they are all similar. Their essential components are a strong light source, a condenser, and a lens. The condenser concentrates the rays of light which diverge from the light source into a projection lens and the light is further concentrated by placing a concave mirror behind the lamp. *Figure 8.1* shows how the light from the source is utilised to produce on a screen an enlarged inverted image of a transparent object placed in front of the condenser.

The image produced in this way is inverted and laterally reversed and to overcome this, as in the case of the lantern slide, the transparency is inverted and reversed left to right when inserted in the carrier. When the image is finally viewed on the screen, it is seen correctly orientated.

If the slide is to be titled the title should be affixed so that it can be read when the picture appears the right way round. To further assist correct orientation by the operator the slide may also be spotted. There are various ways of doing this but the best is to affix a white spot in the bottom left-hand corner when the slide is

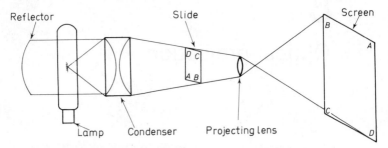

Figure 8.1. Basic components of a diascopic projector. The reflector and the condensing lenses concentrate the light from the lamp on to a transparency

viewed the right way round. When the slide is placed in the carrier or magazine it is inverted so that the spot then appears in the top right-hand corner. If the slide is held between the thumb and forefinger of the right hand with the thumb covering the spot it is then impossible to insert the slide in any but the correct position.

EPISCOPIC PROJECTION

When certain experiments are projected episcopically in lecture demonstrations, the nature of the objects used may be such that the inversion of the specimen is impossible. In these circumstances additional attachments, such as mirrors or prisms, are necessary for the erect projection of the image and such methods of projection have been fully described by Hartung[1].

For the production of images of opaque objects the episcopic method of projection is used. Light from a lamp is concentrated on an object by means of condensers and mirrors and a portion of the light reflected by the object enters a lens to form an image on a screen. The brightness of the image is inferior to that produced by diascopic means and for the best results the room should be completely darkened.

OPTICAL COMPONENTS

CONDENSERS

Condensers consist of two or more component lenses, depending on the type of projector. The cheaper variety are made of moulded glass and the more expensive ones of glass which has been optically worked. For maximum efficiency the glass should be free from spherical and chromatic aberrations. The function of the condenser is to collect and concentrate light from the source on to the transparency and thence to the lens, and upon the effective utilisation of the light depends the ultimate brilliance of the screen image.

LENS

The projection lens is the most important component in the optical system, and if a sharp and undistorted picture is to be obtained it must be of the best quality. Anastigmat lenses are the best and these are corrected for spherical and chromatic aberrations. Good-quality lenses are coated or 'bloomed' to reduce the amount of light lost by reflection. The main function of the lens is to produce a clear, sharp screen image. When filament lamps are used, corrected lenses of large aperture are desirable. The price of such lenses increases with the diameter, and since large light sources require large-diameter lenses to utilise fully the light from the condenser, small but intense light sources are desirable.

Focal length

For a particular projection distance the focal length of the lens determines the picture size. In selecting a projector, therefore, the focal length of the lens should be correctly related to the screen size, to the distance of the screen from the projector, and to the object size. The size of picture may be calculated by applying the formula

$$m = \frac{D-F}{F}$$

but since the focal length of lens F is small when compared with the distance of the projector from the screen D, the magnification m may be more simply calculated from the formula

$$m = \frac{D}{F}$$

The size of picture on a 5-cm slide is 36×24 mm. From the formula it is seen that by projecting the slide from a distance of 10 m using a 100-mm lens, a picture 3.6 m wide is produced. If a 200-mm lens is used for the same distance, the picture will be 1.8 m wide but if used at twice this distance will also produce a 3.6-m wide picture.

LAMPHOUSE

The projector lamp should be firmly supported in a lamphouse. Owing to the high wattage involved, these lamps dissipate a considerable amount of heat and the lamphouse is therefore usually designed with baffled openings at the top and bottom to allow the heat to escape.

The lamphouse should be constructed in such a way that it allows all-round clearance for the lamp and the other internal components. To facilitate the removal of the lamp and for other internal servicing, many instruments are fitted with side doors or, alternatively, the whole of the upper portion of the housing can be folded back after releasing a finger screw. To prevent the operator being burned, the lamphouse on some projectors has a double wall so that the outer wall remains cool.

Projectors with high-wattage lamps are almost invariably fitted with a fan. The fan, which prolongs the life of the lamp and protects the transparencies from damage, should operate quietly without causing vibration. Further protection is provided for film and film strips by means of a glass heat filter which absorbs most of the heat yet allows the light from the visible part of the spectrum to pass through.

ILLUMINATING SOURCES

The illuminating sources for optical projectors are generally tungsten filament lamps, but carbon or tungsten arcs are also used.

Tungsten filament lamps

The quality of a projected picture depends upon the efficiency of the optical system as a whole and therefore the tungsten filament lamp must also be of special design. The shape and the position of the filament are of great importance and it is essential, when new lamps are ordered, to obtain the correct design. The voltage, wattage, and type of cap should be specified.

Halogen lamps

Halogen lamps have replaced tungsten lamps for many purposes including optical projection. These lamps have a small amount of halogen (iodine or bromine) added to the gas filling. The addition of the halogen slows down the disintegration of the filament because the evaporated tungsten actually returns to the filament.

The main advantages of these lamps are the constant and higher luminous flux and a constant colour temperature during the lamp's life.

Because a chemical reaction is involved, the lamp temperature is higher and thus the lamp envelope is made of quartz. The quartz should always be handled through a piece of clean cloth and not touched by the naked hands or its working life-span may be reduced.

The dimensions of the lamp are much smaller yet its optical efficiency is greater. The lamp is therefore ideal for use with projectors where compactness is an essential feature. They are often called Q.I. ('quartz–iodine') lamps.

Design of lamps

To obtain a well-illuminated picture on the screen the wattage of the projector lamp should be high. This also allows the projector to be used in undarkened or partially darkened rooms. For maximum efficiency the light given out by the lamp should be concentrated upon a small area and it follows, therefore, that a small intense point of light is generally the most satisfactory for projection work. Accordingly, lamp manufacturers have produced lamps in which the filament is very closely wound and occupies a very small area. Since it is easier to produce strong, compact filaments for low-voltage lamps, and because the low-voltage lamps provide a greater proportion of effective light, manufacturers have tended towards this design. The use of lower voltages, however, means that a higher current is necessary to operate the lamp, and this entails heavier connections in the projector along with certain other disadvantages. A reasonable compromise is to use a voltage of 110 V, which is now the common voltage for projection instruments. By this means the dual advantage of the compact filament and the high-wattage lamp is gained. To obtain a lower voltage a transformer is used for a.c. instruments, which is a disadvantage in terms of extra weight and expense. Some manufacturers therefore prefer to install a mains voltage lamp in their projectors in spite of the fact that it is not so efficient.

Life of lamps

Although the life of a projector lamp varies, the average life is about fifty hours. As the lamp ages its efficiency decreases owing to a deposit which forms on the glass envelope, so that when a lamp is known to be nearing the end of its working life it is worthwhile changing it for a new one. The life of the lamp can be prolonged up to 10% by burning it at an undervoltage but this reduces the light output. When a lamp is first switched on a sudden heavy surge of current passes through the filament, and to prevent this, a resistance may be wired in series with the lamp. The resistance may be cut out after a moment or two by means of a switch and in some modern projectors this is now done automatically.

Care of lamps

The lamp filament is brittle when cold and soft when hot. It is not advisable, therefore, to move a projector until the lamp is extinguished and the filament is cold and even then it should be moved with great care. If preferred, the lamp may be removed when the projector is transported, but if this is done the lamp should be contained in a box padded with soft material.

Temperature of lamps

In many projectors fans are used to keep the lamps cool. The fan should be designed to keep the temperature of the lamp constant. Each lamp has its own working temperature and for maximum performance this must be correctly maintained.

Arc lamps

The best point source of high-intensity light is obtained by using an arc source but because of the inconvenience associated with arc sources, these are unsuitable for amateur use. They are still widely used for 35-mm projection and also for micro-projection work. Both d.c. and a.c. arcs are available but the direct-current arc gives the best results. The arc is produced by connecting the electrical supply to the two arms of a special holder which supports a hard carbon rod at the end of each arm. By manipulating the holder, the ends of the rods are brought together with the current switched on and this causes the ends of the carbons to glow. The carbons are then mechanically separated by about 3 mm, whereupon an electric arc persists between them.

Direct-current carbon arcs

The direct current travels from the positive to the negative carbon. The positive carbon is supported above the negative carbon and at right angles to it, and to ensure steady burning a resistance is placed in series with the arc. The greater amount of light is given off from the positive carbon and it is this light which is used for illuminating purposes. As the carbons are consumed some means of maintaining a regular gap between them is necessary and this is done by the manual manipulation of the holder or by an automatic mechanism.

Alternating-current carbon arcs

When an alternating current is used for a carbon arc the light is emitted from both carbons. Since these are in fact two light sources, a.c. arcs are not as efficient optically as d.c. Owing to the wandering of the craters formed in the carbon rods, the a.c. arc also tends to travel around the carbons and is somewhat noisier than the d.c. An inductance or choke is put into series with the a.c. arc to stabilise it.

Tungsten arc lamps

Tungsten arc lamps such as the Pointolite give a point source of light which is strongly concentrated and does not flutter. The arc is enclosed in a glass bulb which is filled with gas at low pressure.

THE REFLECTOR

To concentrate the light in the direction of the condenser, curved mirrors made of silvered glass or aluminised metal are used in conjunction with metal filament lamps. The aluminised reflectors are considered better than the glass ones and are not so easily damaged. The reflector is of great value since it increases considerably the amount of effective light. It is essential that the mirror be securely and properly fitted and that the lamp filament be correctly placed in relation to the mirror. The light is reflected back towards the condenser and the spaces between the lamp filament coils are thus filled with the reflections from the mirror. The filament therefore assumes an even more solid aspect. Mirrors are also used in epidiascopes both for concentrating light from the lamp and for the illumination of the opaque object projected.

PROJECTION SCREENS

However perfect the optical system of a projector, the ultimate quality of the projected picture depends upon the effectiveness of the screen. The best position for the screen is that least affected by light from windows or other interfering sources. The optimum screen brightness should be of the order of 10 apparent lux.

For front projection a plain white screen, a beaded screen, or a silver screen may be used, but for rear projection only the translucent screen is suitable. The final choice of screen depends upon the conditions under which the projection is to take place and a number of factors must be considered.

The screen must be capable of presenting a picture which has the necessary degree of brightness, tone, texture, and contrast. If colour transparencies are to be used, the colour reproduction must be up to standard. The screen may also have to be suitable for projection in an undarkened room. The most important consideration is that the screen must be capable of presenting a picture which can be seen by the whole of the audience wherever they may be seated in the room.

PLAIN WHITE SCREENS

The plain white screen is usually made of white opaque fabric which is dressed with a diffusive emulsion compound containing titanium oxide. For fixed screens grainless matt white plastics material is also used. A suitable screen may also be made by painting a plaster wall or other suitable surface with matt white water paint. If space is restricted, the wall behind sliding blackboards may be treated in this way, although it is usually more convenient to treat a side or oblique wall.

For general projection the plain white screen is best, since it is not directional. The reflective surface gives a smooth picture without accentuating the light and dark contrast or the picture grain. This type of screen does not glare. For these reasons, the matt screen is preferred for micro-projection and is excellent for the reproduction of natural colour. Colour reproduction is fair with beaded screens, but with the silver screen it is poor. The diffuse, even reflection from the plain white screen permits a wide viewing angle so that at an angle of 50° from the normal a reasonable picture is still seen. Viewing angles of above 30° are not recommended, because of picture distortion. Persons seated directly in front of a white screen do not, however, see such a bright image as they would seated in the same position facing a beaded or silver screen.

SILVER SCREENS

The silver screen is made by spraying aluminium paint on to a white material. Such a surface has high reflecting powers but reflects the light specularly. Because the reflected light is not diffused, the strength of the picture at angles greater than 20° diminishes and appears dull. At wide viewing angles very little light is reflected at all. This means persons seated in the centre of the hall see an excellent picture, whereas those at the sides see a poor one. This type of screen is therefore suitable only for long narrow rooms.

BEADED SCREENS

The beaded screen is made by spraying very small glass beads on to a white fabric coated with adhesive. This screen is expensive, but if conditions are satisfactory, brilliant pictures may be obtained by its use. The screen, although not so highly directional as the silver screen, is nevertheless best used in long narrow rooms. The audience should, if possible, not be seated at viewing angles greater than 20°, at which angle the brilliance of the picture falls off. Between 20° and 30° the picture produced is inferior to that given by the silver screen but at angles of 30–40° the image is superior. At angles greater than 50° the picture is very poor. The beaded screen is durable but like all screens should be well protected from dust.

TRANSLUCENT SCREENS

A variety of materials including ground glass, good-quality tracing paper, non-glossy tracing paper, Celastoid, and other flexible translucent plastics can be used for making translucent screens. Good-quality tracing linen is better than tracing paper, which is easily damaged. Ground-glass screens are heavy but are usually cheaper than plastics. For best results Celastoid is recommended, provided it is carefully mounted. It is inexpensive and is sold in sheets 132×60 cm. Half-sheets may be purchased. Translucent screens, if made with fine-grain materials, are suitable for microprojection. The translucent screen transmits light from a projector situated behind the screen. The picture may be indirectly projected on to the rear of the screen by a mirror, and in this case the projector is operated in front of, or at the side of, the screen. Rear projection produces brilliant pictures and is especially suitable for undarkened rooms. Since the picture can be viewed from both sides of the screen, this method of projection is very convenient for lecture demonstration and for the teacher who wishes to operate his own projector. The brilliance of the picture, however, depends on the material from which the screen is made, and may fall off rapidly at angles greater than $35°$.

When the picture is projected from the rear of the screen it is reversed left to right. This can be overcome by reversing slides in the slide carrier and by reversing silent films and filmstrips in the carrier or gate. The image may also be adjusted by a reversing mirror or prism and since sound films cannot be reversed in the gate they are reversed in this way. Some projectors are especially designed for rear projection and are then fitted with a special reversing prism to orientate the picture properly on the screen.

WALL SCREENS

The wall screen is usually attached to a spring roller housed in a metal or wooden box. The action of the roller is controlled by a ratchet-and-pawl mechanism which locks the roller when the screen is lowered. When the screen is pulled down a little past its rest position, the ratchet locking mechanism is freed and the screen rewinds itself. To clear wall obstructions, the screen may be suspended away from the wall on brackets. Roller screens without a dust box are available but are not recommended and all roller projection screens should be wound up when not in use.

DAYLIGHT SCREENS

A new type of rigid screen (*Figure 8.2*) measuring 102 cm square allows projection in normal daylight conditions. It is made of treated aluminium foil laminated to a spherically curved Fibreglass

Figure 8.2. Side view of the daylight projection screen showing its double curvature. Its spherically curved surface reflects directionally more light per unit area back to the audience (Courtesy Kodak (South Africa) (Pty) Ltd)

shell. The whole is mounted in a lightweight casing. This screen, designed to overcome the need for darkening small lecture rooms, has an adjustable mounting bracket for wall fixings, and is normally installed permanently in position.

PORTABLE SCREENS

Portable screens are a great convenience when the projector is to be used in different rooms. They roll into a narrow compact wooden box and are usually mounted on a spring roller. The screen can be erected in a few seconds and is braced with flexible folding spring-loaded stays which operate on the frame to keep the screen under tension. The carrying box serves as a base for the screen and applying pressure on the top rail lowers the screen immediately into its box. Similar screens can be mounted on a collapsible tripod.

372 OPTICAL PROJECTION

In this instance the screen is attached to a spring-loaded roller contained in a tubular case.

TYPES OF PROJECTORS

SLIDE PROJECTOR

The simple slide projector consists essentially of the basic components illustrated diagrammatically in *Figure 8.1*.

Modern projectors can project over long distances and are compact, multi-functional instruments (*Figure 8.3*). Some machines can project 5-cm and 6-cm slides and 35-mm filmstrips, and in some cases, when fitted with special attachments, may also be used for micro-projection.

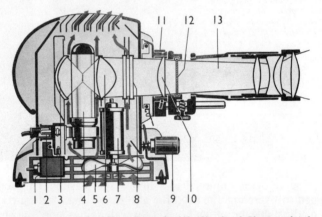

1 Changeover switch of blower motor for 110–120 volts 2 Plug-in socket for connecting cable to mains 3 Reflector 4 Pre-focus lamp 5 Blower 6 Aspherical condenser 7 Blower motor 8 Heat absorbing filter 9 Tilting device 10 Interchangeable field lens 11 Clamping screw to arrest the rotating slide stage 12 Slide carrier 13 Interchangeable projection lens

Figure 8.3. *Compact high-powered projector. This machine can be used for showing 5-cm and 6-cm (2-in and 2¼-in) slides and 35-mm filmstrips (Courtesy Ernst Leitz GmbH Wetzlar)*

With the aid of a small micro attachment the projector shown in *Figure 8.4* can be used for projecting general features of microscopic specimens, and if a micro attachment is fitted (*Figure 8.5*) three stages of magnification are possible. Vertical diascopic attachments are also available for the projection of wet or melting specimens which allow magnifications of up to $2400\times$.

Figure 8.4. Extended appearance of the projector with a charged slide magazine in position (Courtesy Ernst Leitz GmbH Wetzlar)

Figure 8.5. Slide projector fitted with attachments. The attachments give magnifications of up to 2400 × on a screen at a projection distance of up to 6 m (Courtesy Ernst Leitz GmbH Wetzlar)

A polarising attachment which is also available for use with the instrument is intended for the demonstration of double refraction and polarisation.

The magazine, now common to most modern projectors, permits slides to be loaded in order of projection and to be fed automatically or semi-automatically into the projector.

Some projectors also have interchangeable lenses, thus enabling the projector to be used in rooms of different sizes. Among the many other useful functions of these projectors is the prewarming of slides. The blower draws cool air into the projector housing and passes out warm air to prewarm the slides in the magazine, which prevents 'popping' during projection. When necessary the slides can also be changed in either direction by means of a built-in automatic interval timer. A remote-control mechanism may also be fitted which allows the lecturer himself, if he so desires, to focus the picture or to change the slides in a forward or reverse direction. This is done by means of an extension cable and a hand control. Yet another modern feature is the changing of slides by a synchronous cue control in conjunction with a tape recorder. When a recording is made the tape recorder is plugged into the projector. The tape passes through a special synchronising head and, by means of a press button, impulses are recorded on the tape at the required intervals. When the slides are subsequently shown the impulses activate the control mechanism of the projector so that automatic projection and the synchronisation of the slides with the recorded commentary are possible.

A pick-up unit which plugs into some projectors also allows slides to be changed from anywhere in the room by means of hand-operated ultrasonic whistles.

EPISCOPE

The episcope is used for the projection of the image of opaque objects on to the screen. The epidiascope, which is a combination projector, is capable of episcopic and diascopic projection. The principles of episcopic projection can, therefore, be suitably explained by describing this popular projector.

EPIDIASCOPE

The epidiascope is suitable for episcopic or diascopic projection and can therefore be used for the projection of opaque objects

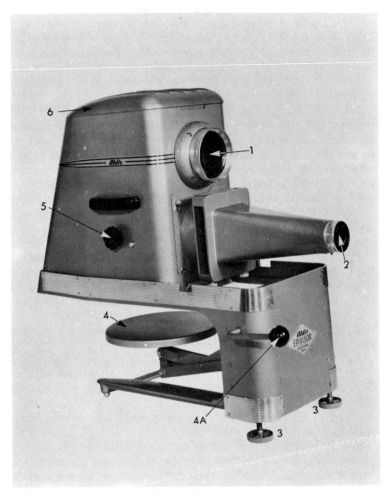

1 Aldis Anastigmat episcopic lens f/4 35-cm focus 2 200-mm or 250-mm Aldis projection lens 3 Tilting screws 4 Projection platform 4A Elevating handle for platform 5 Changeover knob 6 Top cover

Figure 8.6. Epidiascope. This instrument may be powered by a 500-W (convection-cooled) or 1000-W (blower-cooled) lamp. It takes a number of diascopic attachments which enable it to be used for micro-projection, lantern slides, and filmstrips. The changeover from episcopic to diascopic projection is by moving a mirror from in front of the lamp by means of an external knob (Courtesy Top Rank Television Rank Audio Visual Ltd)

and transparencies (*Figure 8.6*). The changeover from one optical system to the other is accomplished by simply changing the direction of the light from the projector lamp. This is done in some instruments by rotating the lamp itself so that the light is directed either through a transparency or on to mirrors which illuminate the opaque objects placed on the baseplate of the projector housing. Alternatively, the direction of the light is changed by moving the mirrors themselves, which has the same effect.

When the epidiascope is used for episcopic projection the light reflected from the opaque object is directed upwards on to another mirror surface set at an angle of 45°. The mirror, therefore, not only directs the image on to the screen but also reverses it, so that it appears on the screen correctly orientated. To obtain good pictures and to cover the larger objects used a lens of large diameter is needed. The projectionist himself can do much to improve the quality of the picture by restricting its size and that of the object, and also by using a good screen. The beaded screen gives good results if the seating arrangements permit it to be used.

Because the light reflected from an opaque object is less intense than that which passes directly from the light source through a transparency, episcopic projection requires a completely darkened room. In undarkened rooms, the pictures appear flat and uninteresting. Manufacturers have now overcome the relatively weak projection of the epidiascope by improving the cooling system and introducing projector lamps of up to 1000 W.

An efficient cooling system is an important feature of the epidiascope, since the fierce heat reflected on to frail material, such as biological specimens, can quickly ruin them. Even printed and photographic materials need protection, and to prevent these curling they are pressed on to the underside of a heat-resisting glass plate by raising the base platform.

The specimens or illustrations to be projected are placed on a platform situated at the bottom of the epidiascope chamber. The platform is usually raised and lowered by means of a lever to allow the introduction of material for projection. As the platform is lowered, light floods into the room and reduces the apparent brightness of the screen. Various methods of overcoming this fault have been tried. These include the provision of dark curtains suspended around the base of the projector which hang down to cover the wrists of the operator as he changes the material for projection. If the top of the projector stand is painted black this also helps absorb stray light. Sometimes a switch is fitted to dim the light while a change of specimen or illustration is made. Other instruments have diaphragm changers and the illustrations are loaded into sheaths before the projection begins and are ejected by the movement of a handle as they are used.

Some epidiascopes have a vacuum-operated platform which secures even small illustrations firmly in place. This type of platform avoids the necessity of mounting paper illustrations on black cardboard, or other firm backing material, as is often done. Even so, mounting illustrations required for regular use is still to be recommended, because mounting them on a card cut to the exact size of the platform greatly simplifies the orientation of the illustration on the platform.

If a projector has no special facilities for projecting a sequence of pictures, it is convenient to paste the illustrations on to a paper backing strip which is folded up in a concertina style. When the concertina is unfolded, the illustrations are presented in the correct order in which they are to be shown. Most epidiascopes have a rear viewing window made of special glass. This allows the operator to read printed matter and when necessary to write on it during projection. Another helpful device permits an illuminated arrow, or the silhouette of a pointer, to be directed to any chosen spot on the screen.

OVERHEAD PROJECTOR

The overhead projector has taken a long time to win a place in the hearts of lecturers but at last, owing to the ease with which it is able to present a wide range of visual media, it has attained great popularity.

This projector, formerly called a writing projector because its early use was limited to the projection of wax pencil markings, has now developed into a versatile and useful instrument. It can be used in lecture rooms under daylight conditions, thus enabling the students to take notes while lectures are in progress. It allows the lecturer to present illustrations and notes which are written or drawn and appear immediately on the projection screen. It can also be used as an episcope for the projection of slides, and for large transparencies such as photocopies and X-ray plates. Some models can project the enlarged images of pieces of apparatus and in this way scientific experiments may be observed while they are in progress. Portable projectors which fit into a case complete with carrying handle are now marketed.

Although the overhead projector may never entirely supersede the chalkboard it has already replaced it to a large extent. The chief reason for this is the amount of teaching time saved by the presentation of material prepared prior to the lecture. This is especially true of prepared drawings which would otherwise need elaborate blackboard presentation. Yet another advantage is gained by the lecturer in that he is able to face his class and thus maintain contact with them throughout his lecture.

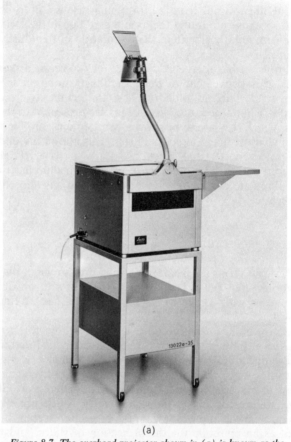

(a)

Figure 8.7. The overhead projector shown in (a) is known as the Diascriptor. It is fitted with a 1000-W lamp, has a writing area of 27 × 37 cm and can be used under daylight conditions. The same projector is shown diagrammatically in (b), opposite (Courtesy Ernst Leitz GmbH Wetzlar)

The optical system used for all overhead projectors consists essentially of a light source and a special condensing system necessary for large transparencies. The light is concentrated through the transparent surface, enters the projector objective situated directly above the writing surface, and is then reflected forwards by means of a mirror on to a screen. Different sizes of transparency stages may be used in the various projectors available and the user should therefore select a projector with a stage suitable to his requirements. Small stages 12.7 cm square are used for projecting scientific experiments and larger stages 20.3 × 25.4 cm for photocopy material. Even larger stages, 27 × 37 cm, may be used for handwritten material.

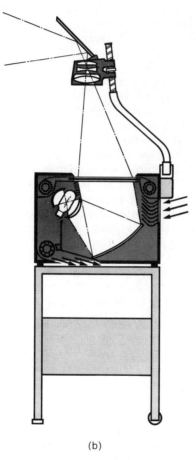

(b)

The projector shown in *Figure 8.7* has a hand rest situated immediately in front of an illuminated clear foil stage. This enables the lecturer to write or sketch quite normally on the foil by means of a grease pencil or Radiograph. The writing surface is available in rolls 50 m long, 38 cm wide, and 0.1 mm thick, also in special lengths of 10 m and 30 m. The rolls are wound on a spool which may be hand- or motor-operated. The writing surface can therefore be quickly changed as it is used up. The roll may also be moved backwards or forwards, and at any time during the lecture, material which has been previously shown may be returned to for further discussion. Single-sheet foils may also be used and these, or the rolls, may be prepared before a lecture, thus allowing complicated diagrams to be presented. Additional transparent originals such as basic circuits or site plans may be inserted under the writing foil at any time and the lecturer may make additional notes on them without damaging the original. At the conclusion of the lecture the rolls or single foils may be stored away for future use or may simply be erased with a soft cloth.

Because the projector may be used in daylight conditions, the screen should be opaque. It should be as large as possible and the audience should not be closer to it than two screen widths or further away than six screen widths. The tilted screen should be placed high up with its bottom edge not less than 1.5 m from the ground.

The position of the projector in relation to the screen is important and needs careful arrangement. It should be ascertained that good vision is possible from all parts of the room. The projector should be low down and supported at such a height that the lecturer may sit comfortably beside it without obstructing the projection or view of the audience in any way. Depending on local circumstances the overhead projector may be permanently positioned, and special tables are nowadays designed which are recessed to accommodate the projector. Alternatively, a mobile projector may be necessary, in which case it may be mounted on a trolley for use in various locations.

Preparation of transparencies

Heat process method

The heat process for the preparation of transparencies depends for its action on the degree of absorption and reflection of heat from the lines and background of the material being copied. The original drawing is rotated around a cylinder surrounding a tubular lamp which acts as the heat source. The speed with which the original passes around the cylinder can be varied, permitting various exposures.

To produce the transparency a special film is used in contact with the original. The originals may be pictures from magazines, pencil or indian ink drawings, typewritten material, or various other media. The film coating is affected by the heat absorbed by the lines of the diagram and the corresponding portions of the film surface become frosted. The affected parts of the film surface then show up as a black image when projected. By using coloured pencils on the transparency the lines of the image may be projected in colour.

Negative transparencies can also be produced this way, but in this case the drawing appears as bright lines on a dark background. There is no chemical processing involved and the transparencies can be produced in full daylight. The lines of the negative transparency can also be coloured but for this a felt-tip pen is used.

Diazo process

Transparencies suitable for the overhead projector may also be produced by the diazo process (see page 450) from translucent originals. The original drawing is made on tracing paper or tracing linen, placed on a sheet of sensitised diazo foil, and exposed to a strong u.v. light source. It is then processed in ammonia fumes. Although the foils are normally used over a black image to produce colour overlays, full-colour transparencies may be made by using diazo-chrome films.

Other processes

Spirit duplicator. If a clear sheet of acetate is used in place of copying paper on the spirit duplicator, a transparency suitable for projection is obtained. It is recommended that a number of copies be made first on paper until a good image is obtained. The clear image may then be produced on the acetate sheet.

Xerographic. Transparencies can be produced xerographically if the large xerox copying equipment consisting of camera and heat processor is available. With this equipment, too, the original image may be enlarged or reduced.

Commercial. Commercially produced transparencies covering a number of teaching subjects are now available and those complete with colour overlays are suitable for immediate use. Others, printed on translucent paper, may be modified to suit particular teaching purposes and finished off by the diazo or diffusion process.

Photographic method. To reproduce originals including full tone pictures photographically as 25.4×20.3 cm $(10 \times 8$ in$)$ transparencies, an ordinary small camera, e.g. 35 mm, may be used. An adjustable copying stand fitted with Photoflood lamps, such as is used for the production of lantern slides, is quite suitable for photographing the original. Finally, an enlarger is used to project the resultant negative directly on to a slow sensitive film to produce a transparency the correct size.

Reflex copying. Transparencies can also be produced by the reflex copying method (see page 446), which does not require a camera. Whereas this copying process is normally used for the production of paper copies from originals, transparencies can also be made in the same way by the diffusion transfer process. This process simply involves placing a sheet of sensitive paper in contact with the original and subjecting it to a light source. The photographic paper is developed to produce a paper negative copy. The negative paper is then used to produce an image on clear film. After development a copy of the original may be projected immediately.

FILMSTRIP PROJECTOR

The filmstrip projector is a miniature projector used to show still pictures photographically produced on 35-mm film. Prepared filmstrips covering a wide range of subjects in black and white or in colour are readily obtainable. The pictures are diascopically projected and the projector has an optical system which is basically identical with that of the optical lantern. The light beam is concentrated on the transparencies which are vertically or horizontally fed across the beam when the film is advanced. Coloured transparencies require more light than the black and white and when colour film is used the projector may have to be moved nearer the screen. The instrument must be well ventilated and for lamps above 250 W a heat filter and a fan should be fitted. Because the instrument and the filmstrips are easily portable and inexpensive these projectors are much in favour.

Filmstrip projectors vary in design and lamps of less than 100 W and up to 1000 W are used in them. Since the size of the transparency is small compared with other types of projectors, the best pictures are obtained by using high-wattage lamps. The projector must, however, be selected in accordance with the size of room and the type of duty for which it is intended.

For long-distance projection, a high-wattage projector is required. Such projectors are very suitable for educational purposes because the brilliant pictures they produce can be seen in a partially lit room. The high-powered projectors, too, can be relied on to give a reasonable performance when the other projection facilities are poor. Some instruments, equipped with low-voltage lamps, require a transformer which increases considerably the weight of the outfit. For use in rooms of reasonable dimensions, the 250-W projector is quite satisfactory but the use of projectors with lamps of wattage lower than 250 W is restricted to small rooms. Some projectors can operate from a car battery and are suitable for use when a mains electricity supply is unavailable.

Figure 8.8. Filmstrip and slide projector with 5×5-cm (2-in) slide carrier. An accessory also permits micro-projection using microslides or specimen tubes (Courtesy Rank Organisation, Taylor-Hobson Division)

Most miniature projectors are designed to accommodate 5×5-cm slides as well as filmstrip. The filmstrip holder can be removed and replaced by a slide carrier (*Figure 8.8*). Some instruments have a built-in slide carrier. Further attachments are obtainable which permit filmstrip projectors to be used for projecting micro slides. The standard of projection possible with miniature projectors of this nature cannot be expected to be as high as that with a specially constructed micro-projector. Nevertheless, the quality of the image is suitable for certain levels of instruction.

For the projection of single- or double-frame pictures, adjustable or self-contained masks may be supplied with the instrument and for the same reason the front condenser may also be interchangeable. An alternative means of adjusting the optical system is to move the transparency itself and for this purpose an adaptor is used which when fixed in position sets forward the film carrier and the lens. When lenses of different focal length are used, this may involve changing the condenser and this interchangeability is a feature of some filmstrip projectors.

Special types of projectors are now available which play a recorded commentary as an accompaniment to the film. The record is synchronised with the film, which is either manually advanced when a faint signal is heard in the recording or automatically advanced by electronic means. In some cases the film may also be advanced by remote control from any point in the room.

Filmstrip

The normal 35-mm nonflammable film with a cellulose acetate base is used for filmstrip and is made to a British Standard Specification. The length of filmstrips varies, and although they normally contain up to fifty pictures or frames some are longer. At the beginning and end of the film is a leader, marked START or FINISH as the case may be. The film is usually supplied in small metal containers with the film coiled inside with the emulsion side inwards.

For normal projection work two sizes of frame are used. These are the double-frame size 24×35 mm and the more commonly used 24×18-mm size. The filmstrip is perforated along each edge. The double-frame size has eight perforations to one side of the frame and the shorter sides of the pictures are adjacent. The single frame has four perforations on each side and the longer sides of the pictures are adjacent. The double frame is therefore passed through the projector horizontally and the single frame vertically. To allow this, the filmstrip holder can be rotated through 90° or 360°, at which angles the film holder clicks into position. The film is laced into the film advancing mechanism inverted and reversed. It should be held up and viewed so that the first frame is observed to be the right way up and the correct way round when looked at in the direction lamp to screen. The film is then inverted and inserted in the filmstrip holder. The movement of the film in the holder should be carefully tested to ensure that the film is properly engaged and advances freely.

Several simple methods are used to advance the film and these vary with the type of projector. They include the sprocket which is rotated by a knob or lever and engages in the film perforation. A take-up spool is often used which is similar in action to that in a roll film camera. To prevent the film from buckling, it is held between two glass plates. The plates open automatically as the film advances to allow it free passage, and as the film emerges it coils up owing to natural tension in the way it was originally wound. Some instruments have a film rewinding device. One is a slotted cylinder which the film enters to coil itself upon the inner walls. It does so in such a way that each successive coil winds itself within the preceding one, and having completely rewound itself is ready for further projection.

Obtaining filmstrips

Filmstrips may be hired or purchased at reasonable cost from various commercial firms and educational film libraries. They may also be made up to suit individual requirements from photographs, diagrams, or manuscripts submitted to the manufacturers. After a moderate manufacturing charge for the initial strip, subsequent copies can be produced at much lower cost. It is as well, before submitting material for making into filmstrip, to obtain from the manufacturers information concerning the size and quality of the material to be submitted and this will guarantee the best results. The size of submitted material should normally be in proportion to the frame size of the filmstrip picture.

Care of filmstrips

The filmstrip has no protective coating and is easily scratched; therefore the projector generally, and the glass guide plates in particular, must be kept free from dust. Great care should be exercised when handling or rewinding filmstrips, which should not be handled on the face. Finger marks can be cleaned off by laying the film on a flat surface and removing them with a clean soft rag. Filmstrip may have to be joined and the method used for cine film is employed. With cine film about 3 mm of the emulsion is scraped off with a knife at the ends to be joined and film cement is quickly and carefully applied. After a few seconds have passed the ends are brought together with the perforations in line. Splicers sold for the purpose simplify the operation and ensure that the perforations are correctly aligned. Although this method of joining is easily carried out at the ends of the filmstrip, it is difficult to employ it for making joins between the picture frames, as these are very closely spaced.

MICRO-PROJECTOR

Although the micro-projector cannot compete with the degree of definition and magnification obtained with a high-class microscope, or substitute entirely for intelligent microscopic exploration by the individual student, its many benefits are obvious. It is particularly advantageous to the teacher, who through this medium can interpret the important features of a slide or specimen to groups of students and explain the various points to be observed during their individual examination of the preparation.

The difficulty of sufficiently illuminating the preparation restricts micro-projection mainly to the projection of transparent objects. The amount of illumination required increases with high magnification.

Simple but effective micro-projection may be carried out by fitting certain attachments to the epidiascope, filmstrip projector, or slide projector. For better micro-projection, however, an ordinary compound microscope may be adapted, or a specially designed projector should be used.

MICRO-PROJECTION WITH THE COMPOUND MICROSCOPE

For more serious work demanding higher magnification and the examination of a small field, the compound microscope can be used for micro-projection purposes. The normal microscope optical system is used, whereby the light from the source is converged by a sub-stage condenser on to the preparation and the magnified image, produced by the microscope objective, is further magnified by the eyepiece. By means of a right-angled prism or reflecting mirror attached to the eyepiece, the image may be horizontally projected on to a screen or projected downwards on to a reflecting surface placed on the bench top.

Horizontal projection through the microscope eyepiece direct on to a wall or vertical screen is also possible with the aid of instruments such as that illustrated in *Figure 8.9*.

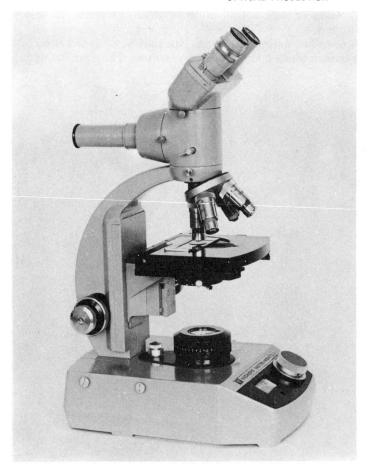

Figure 8.9. Illuminating base. By means of an insertable trip mirror unit the image can be deflected through the eyepiece of the horizontal tube on to a wall or vertical screen. The binocular and monocular bodies can be mounted juxtaposed to the assembly so that a vertical monocular tube is available for eye and camera attachments (Courtesy Vickers Instruments Ltd)

For screen projection, plain white or beaded screens are best, and although daylight projection at short distances is possible, a darkened room is desirable in most cases.

The adaption of an existing microscope for micro-projection, by manufacturing a light box to be used in conjunction with the microscope for illuminating purposes, is a relatively simple workshop matter. Commercially manufactured illuminating bases are, however, available and the stability of these instruments, and the advantages of their optical fitments, greatly facilitate the ease of projection.

A new illuminating base of this type shown in *Figure 8.10* consists of a base in which microscopes may be fitted. The illuminating base

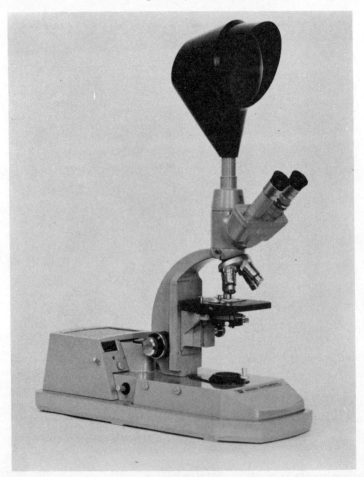

Figure 8.10. The research microscope shown in position within the illuminating base has a substage mirror located in its box base. The picture produced on the 15-cm diameter projection screen is brilliant enough for viewing in daylight (Courtesy Vickers Instruments Ltd)

light source is a high-intensity 100-W quartz iodine lamp and the lamp box houses a transformer, filament colour temperature recording meter, rheostat for adjusting the light intensity, condenser system, and iris diaphragm for controlling the light. By attaching a trip mirror photovisual unit the fully illuminated imaging beam may be directed either to a normal binocular eye end unit or into a vertical monocular tube with eyepiece to an attachable 15-cm diameter projection screen. By interchanging the monocular and

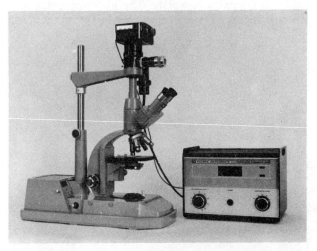

Figure 8.11. The illuminating base can be used for micro-photographic purposes using the latest photographic equipment. In this assembly the 35-mm roll film camera is fully automatic with electro-magnetic shutter and auto film wind, and is under the control of an automatic exposure device via a sensitive photoelectric cell fed from the beam-splitter of the photovisual head (Courtesy Vickers Instruments Ltd)

binocular heads, horizontal projection at any orientation can be achieved, while the binocular can be employed visually on tripping the mirror. The illuminating base also has a central socket to accommodate a camera support pillar and arm so that it can be used for micro-photographic purposes (*Figure 8.11*).

SPECIALLY CONSTRUCTED INSTRUMENTS FOR MICRO-PROJECTION

A projector described by Sumner[2] has a number of special features including a cooling trough for the projection of living specimens. The construction of the instrument is based on that of the optical bench and sturdy systems of this kind are necessary for good micro-projection. The removal of the microscope body permits the bar of the projector to be used as an optical bench for experimental purposes. A more recent fitting added to the instrument is a 5 × 5-cm slide carrier which accommodates two small pond life troughs up to 12.7 mm thick.

The arc is an excellent source for micro-projection and provides a point source of light which is necessary for small-field illumination. *Figure 8.12* shows a modern projector with arc lamp and microscope mounted on a common baseplate. The instrument has adjustable projectioning prisms to raise or lower the image on the screen.

Figure 8.12. *Micro-projector for the projection of screen images of microscopic specimens. The magnification can be changed rapidly by a single-knob control. The image is focused by the operation of the coarse and fine coaxial controls situated beneath the horizontal object stage. The stage itself also has a built-in mechanical stage (Courtesy Ernst Leitz GmbH Wetzlar)*

The arc lamp is accessible by pushing up the projector housing and the carbons may be separated by the manipulating knobs situated outside the housing. Once the current gap has been set the automatic feed with electromagnetic control ensures that no further attention is necessary and the arc can be seen through a sight glass. Either a.c. or d.c. current may be used for the arc lamp, but d.c. is best and gives a light intensity one-third greater. The lamp which burns at 55 V d.c. and has a load of 10 A is connected to the supply through a resistance or rectifier.

The micro-projector is capable of producing screen images of up to 2.5 m diameter and can project over distances of up to 23 m. The diameter of the screen image is determined by the particular eyepiece used. Eight different eyepieces are available to suit various projection distances. It is also possible to project general features without using an eyepiece and to change over to micro-projection by introducing an eyepiece.

Changes in magnification are effected by a single knob control. Four different magnifications of image are possible and the objectives are mounted in such a way that when a change in magnification is desired the specimen is introduced beneath the next objective in exactly the same position. The iris diaphragm of the arc lamp serves as an aperture stop for the two low-power objectives and a corresponding condenser is used with each objective.

A similar model can also be supplied which is fitted with a powerful air-cooled xenon burner. This projector is not only suitable for higher magnification but because of the increased light intensity permits projection with oil-immersion objectives. By means of special devices the brightness of the burner can be adjusted in steps from one-twentieth to full value. The instrument has a rotating and tilting projection prism and a mechanical stage for object scanning.

MICRO-PROJECTION FOR SPECIAL PURPOSES

Micro-projection is also used for the examination of fabrics, small machined parts, and other items, as a means of industrial control. The use of these industrial micro-projectors allows a regular inspection of parts without the necessity for the continuous use of a microscope. Many of the appliances for the gauging and inspection of production parts have been described in a handbook[3].

CINE FILM PROJECTOR

The cine film projector projects on to the screen a number of successive images each of which is momentarily retained on the retina of the human eye. The brain interprets this succession of images as a moving picture, but in order to create the illusion, the pictures must be projected at a rate of not less than sixteen per second. The pictures are photographically produced on film and a number of different film sizes are used. Most films are made of nonflammable cellulose acetate.

Sizes of films

The standard size of film is 35 mm but the 8 mm, 9.5 mm, and 16 mm are all commonly used and are known as sub-standard films. For educational purposes, by far the most important size is the 16 mm. The films are perforated so that part of the projector

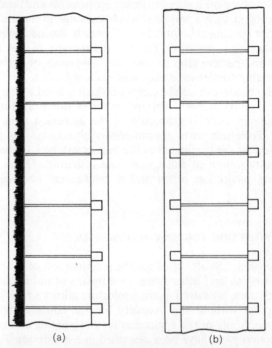

Figure 8.13. (a) 16-mm sound film with simple single-point, variable-area sound track. (b) 16-mm silent film. This film is used with those projectors which have claws and sprockets which engage in both sets of perforation

mechanism can engage in the perforations and carry the film forward. The actual positions of the perforations are illustrated in *Figure 8.13.*

The 16-mm film

The 16-mm film is perforated down the sides of the film, each perforation occurring between one frame and the next. Silent films carry the perforations on both edges. Sound films are perforated only on one edge and the other edge is used for an optical or magnetic sound track.

Sound films, projected at the rate of twenty-four pictures per second, are more widely used than silent films. Silent films are projected at sixteen pictures per second and silent or sound films can be used in modern projectors.

A reel of film is 122 m long, this being the unit length, and has 132 pictures to each metre. The film is wound on a spool, which in some projectors can accommodate up to 600 m. The spool from which the film is fed to the machine is called the feed spool and that on which the film is wound is known as the take-up spool.

Projector mechanism

Optical

The optical system of the film projector is contained in a smaller space but is otherwise similar to that of the optical lantern. Recent models may be fitted with variable-aperture plates, so designed that the picture may be shown in a wide-screen format; and by using a range of lenses the picture size can be varied to suit any screen.

The film spools are slipped on to spindles, themselves supported by arms which protrude from the projector body. The two spindles are identical so that the spools are interchangeable. The rear arm spindle is power-driven and takes up the film on the spool. To rewind the film without removing it from the machine, the front spindle may also be power-driven. To facilitate the ease of transportation of the projector, the spool arms either fold down or are detachable.

A hole passes through the centre of the spool. Spools for silent film usually have a square hole on one side and a round one on the other. When this type of spool is used, the square hole must go on to the spindle first so that it engages on that part of the shaft which has a square section and by this means the shaft rotates the spool. Sound spools have a square hole on both sides.

The spool is secured on the spindle by means of a ball catch or by a piece of metal which is hinged to the end of the shaft and folds down when the spool is in position. The front arm spindle may be extended to protrude on both sides of the arm, and a grooved pulley attached to the free portion of the spindle at the back of the arm accommodates a spring belt for rewinding or reverse drive purposes.

Gate

The film is fed from the front spool by means of a sprocket, the teeth of which fit into the film perforations and lead it to the projector gate which is made of hard, highly polished metal such as stainless steel. The projector gate can be opened down its length to allow the film to be inserted in it. One half of the gate is spring-loaded so that pressure is brought to bear on the other. The tension of the gate is important because the film must pass smoothly through it. The pressure of the gate is brought to bear on the edges of the film, but to avoid damage no pressure is applied to the actual picture area. Pressure plates at the gate hold the film flat and still during projection of pictures and overcome the inertia of the moving film.

Intermittent

By maintaining the film in a steady position the gate assists the film shift mechanism, otherwise known as the intermittent, to function properly. Various methods for moving the film forward, one frame at a time, have been tried. The most common form of shift mechanism consists of claws made of tungsten carbide alloy which engage in the perforations of the film and pull it down. The precise entry and withdrawal of the intermittent claws is important, since their clean action prevents damage to the film. Quite often double claws are used which ensure a smooth movement of the film, even though some of its perforations may have been previously damaged. The claws are actuated via gears driven by the projector motor and work in conjunction with a shutter which cuts off the light from the lamp at the precise moment when the film is moved down by the intermittent. The claw drive may be adjusted on some machines and can be used to correctly frame the projected picture. When the picture is framed in this way the initial centring of the picture is not disturbed.

Shutter

The rotating action of the shutter, which consists of from one to three blades according to the type of projector, cuts off the light beam and permits a smooth succession of still pictures to be projected on the screen. This prevents any observable flicker.

A take-up sprocket collects the film from the gate and moves it forward, and in the case of sound projectors the film then passes over the smoothing and scanning system before finally being rewound on the power-driven spool at the rear of the machine.

As the amount of film taken up on the spool increases, the rate of rotation of the spool must be adjusted to compensate for the increasing diameter and for the increase in the speed of winding which would normally result. To maintain a constant speed, various methods have been devised. In one case the tension of the driving belt is adjusted so that it automatically slips as the weight of the spool increases.

Film loops

Loops are left in the film above and below the gate and these must be of a size suitable to the type of machine. The bottom loop is standard in all machines. To assist the operator to adjust the size, the actual shape and size of the loop is usually indicated on the body of the machine. The function of the loops is vitally important, since they prevent any possibility of damage to the film due to the jerky movement of the intermittent which would otherwise work against the constant rate of the sprocket feed. The loops also ensure that the film remains stationary in the gate as each frame is projected. In the same way, the loops smooth the movement of the film as it passes over the sound head. Modern machines often incorporate automatic loop formers to ensure synchronisation between picture and sound.

Motor

The electric motor, housed in the projector chassis, drives the moving parts of the projector and operates the fan which cools the lamp and film. The motor should operate at its maximum speed and power and suitable reduction gears are used to drive the projector parts, such as the film drive and the intermittent. The fan, however, is not geared down and operates at the maximum motor speed to give the best cooling effects.

Electrical supply

Cine projectors operate on 110 V, which is lower than the mains voltage and therefore necessitates the use of a transformer for an a.c. supply. For d.c. and universal machines a resistance is used to lower the voltage. Care should be exercised to ensure that the connections from the mains to the input terminals of the transformer are correctly made. Because high-wattage lamps are used in most projectors, they should be operated only from power circuits and to ensure proper earthing a three-pin plug should be used to connect the instrument to the electrical supply.

Still picture projection

It is sometimes convenient to project still pictures and a clutch device is fitted to some projectors for this purpose. The clutch disconnects the film mechanism and for extra safety a heat filter simultaneously drops into place to protect the film. The filter cuts off some 80% of the light and reduces the brilliance of the picture to a very great extent.

Film reversal

Many projectors can be reversed by means of a switch so that those parts of the film which merit further discussion may be re-viewed.

ARRANGEMENTS FOR PROJECTION

SEATING

The audience must be seated in such a way that each person has an uninterrupted view of the screen. At least 0.5 m^2 should be allowed for each viewer. The seating should therefore be arranged so that the audience in any particular row can see over the heads of those seated immediately in front of them. Extreme viewing angles should be avoided. The front row of seats should be at least two picture widths away and the rearmost row not more than six times the width of the picture from the screen.

POSITION OF SCREEN

To prevent eyestrain, the screen must not be too high and the audience should preferably look down on it. For this reason, seating raised on steps is very suitable and this arrangement in lecture rooms assists the acoustic properties and allows students to see lecture demonstrations more clearly. When a level floor is unavoidable and the screen must be set above the heads of the audience, the bottom of the screen should be so arranged that it is only just above eye-level.

For the best results, the screen should be at right angles to the direction of projection, with the projector pointing to the centre of the screen. If upward or downward projection angles are necessary, such as when the projector is operated from a gallery, then the screen may have to be tilted or keystoning effects will be noticeable. Since the best position for the screen conflicts with the obvious blackboard position, the centrally placed screen may be inconvenient and this can be overcome by arranging the screen diagonally across one corner of the room. This gives a good angle of view to the audience and leaves the blackboard free for use. This screen position is very useful because slides can be projected in a partially darkened room and students can take down notes from the blackboard during the course of the lecture. The diagonal screen can also be used for projection from the lecture bench. It is advisable, however, to have a roll-up screen hung above the blackboard in addition to the diagonal screen and this can be used when an uninterrupted film show is given and the blackboard is not required.

POSITION OF PROJECTOR

The projector should be situated behind the audience and this avoids the distraction and inconvenience experienced when it is used from the body of the hall. The projector should be mounted at the correct height on a stable projection stand and it is best if a raised platform, on its own solid foundation, is provided at the rear of the hall for the exclusive use of the operator. The light beam should be well clear of the heads of the audience and any other obstruction. The operator should be protected from any inconvenience caused by the audience entering or leaving the room and the electrical leads to the projector should be arranged so as not to constitute a tripping hazard to passers-by. To avoid any possibility of the projector being pulled to the ground, leads taken along the floor should first be secured to the projector stand. Better still, a plug and socket join which is easily pulled apart may be incorporated in the projector lead.

POSITION OF LOUDSPEAKER

The loudspeaker should be positioned in front of the audience at a fairly high level so that the sound is not blanketed by the front rows of the audience. It should be tilted slightly downwards and directed towards the centre of the audience. The corner of the room is not the best position for the loudspeaker and it should be placed, as near as possible, in the position from which a lecturer would address the audience. If two loudspeakers are used they too should be similarly placed and should be reasonably close together.

LIGHTING

All light controls should be situated close to the operator and should be duplicated close to the lecture bench. A switch which controls indirect side lights is a decided advantage and the subdued lighting permits note-taking.

The angle of projection should be adjustable and a tilting device is usually fitted below the front of the projector. The machine should not, however, be tilted beyond the limit of the adjuster or damage to the lamp may result. Should further adjustment be necessary, the height of the projector should be altered by adjusting the height of the projector platform itself.

ACOUSTICS

The room should be acoustically good and it should be possible to exclude all noise from outside sources. If the acoustics are poor improvements may be effected by the use of curtains, or more permanently by attaching to the walls pegboard behind which is placed a layer of loosely packed felt or similar sound-absorbing material.

VENTILATION

Adequate ventilation is very important and due allowance should be made for the fact that window ventilation will be greatly reduced when the dark blinds are drawn. The room should always be well ventilated before it is used, but care should be taken not to lower the temperature too much.

DARKENING ARRANGEMENTS

It should be possible to darken the room completely. The normal method is to use opaque fabric blinds, guided by grooves in wooden or metal light-tight frames fitted inside the windows. Unless an automatic means of darkening the room is provided this process takes a considerable time and should always be done before the film show takes place.

CINE PROJECTOR PROCEDURE

Lacing the film

To lace the projector the film spool is placed on the forward arm of the machine. The film should be fed into the film gate so that the frames are upside down and reversed left to right. To ensure this is so, the operator should face the machine and lift up the free end of the film. If the film has been wound correctly the title or picture should appear the right way up and the right way round as it would appear on the screen. A lacing diagram is usually shown inside the projector door and this must be carefully followed. In projectors which accommodate sound and silent film, alternative lacing paths may be shown.

If the film has been wound the wrong way round and the spool has a square hole on one side and a round hole on the other, the winding cannot be corrected by reversing the spool on the shaft. In such cases the film may be wound off the spool and rewound on to it after reversing the spool.

Film spools should be inspected to see that they are correctly wound as soon as they are received. The spool is held so that the round hole is on the viewer's left and on lifting the end of the film the picture or title should appear right way up.

Checking and operating

A few tools, some essential repair material, and a spare projector lamp should be to hand during the projection. The machine must be carefully checked over beforehand.

Projection procedure

1. Switch on the mains and then the amplifier. While the amplifier warms up, the gate, sound head, and lens should be lightly cleaned. Only lens cleaning tissue or very soft rag should be used for the purpose.
2. Test the motor by switching it on. Switch on the projector lamp.
3. Focus roughly by observing the edges of the projected frame of light.
4. Test the sound. This is done by interrupting the exciter lamp beam with a piece of paper. The amplified sound of this movement will be heard in the speaker.
5. Switch off the lamp and then the motor. Leave the amplifier on.
6. Lace the projector. CHECK THE LOOP SIZES. Check the film tension.
7. Move the film forward by hand using the special advancing knob and make sure the film moves freely. Switch on the motor for a brief moment to see that the film takes up. Recheck the loops and the tension of the film.

    (Note that manual threading is not necessary with modern projectors. When the motor is switched on the film is trimmed with a built-in trimmer, and after setting a lever to the auto-thread position the film is inserted. It then automatically threads itself. An automatic loop restorer ensures that the loops remain at the correct size.)
8. With the volume control turned down switch on the motor. Switch on the lamp. (Note that some machines have three-stage switching—Stage 1, motor starting; Stage 2, motor at top speed and lamp heating; Stage 3, motor and lamp.) Delay switching on lamp until first title is in gate.
9. Adjust focus finely. Frame the picture on the screen.

As the film is projecting, a strict watch must be kept to see that the size of the loops remains constant and that the film is winding undamaged on to the take-up spool.

Switching off procedure. At the end of the film the switching off procedure is as follows:

1. Turn down volume control.
2. Switch off lamp.
3. Switch off motor when film is clear of mechanism.
4. Switch off at mains, remove plug and disconnect leads from the machine. Coil up the flex carefully.

Care of projector

All optical parts of the projector should be cleaned each time it is used. It should be more thoroughly cleaned and lubricated in accordance with the manufacturer's directions at least once a week, and should be serviced periodically by an expert mechanic.

RECORDING AND REPRODUCTION OF SOUND

Until quite recently, the recording of sound on film was done almost exclusively professionally. The introduction of magnetic striping has made this a relatively easy task which can be done by persons with no specialised knowledge.

TYPES OF SOUND TRACK

Two kinds of sound track are used on film: (a) optical and (b) magnetic stripe.

Optical sound tracks

Two optical methods are used for recording sound on film, the variable-area and the variable-density methods.

Variable-area method

The variable-area sound track is produced on film by using a microphone to pick up acoustic sound waves. The diaphragm or ribbon in the microphone vibrates, owing to the sound waves falling upon it, and causes small electrical impulses to be passed to an amplifier. From the amplifier the audiofrequency signals are passed to a galvanometer and the movement of an armature, situated between the poles of a permanent magnet in the galvanometer, is transmitted to a mirror. The light from an exciter lamp is directed on to the mirror and reflected through a narrow slit to record a sound pattern on the edge of a moving photographic film. The amplitude of the signals applied to the galvanometer are recorded as the width of the sound track, and their frequency as the distance between the peaks (*Figure 8.14*).

Figure 8.14. (a) Variable-density sound track. (b) Single-point variable-area sound track. The single-point is the simplest form of variable-area track. (c) Modern variable-area sound tracks. These have a different form from the single-point shown in (b) (Courtesy Rank Film Laboratories)

OPTICAL PROJECTION 403

(c)

Variable-density method

The variable-density track is produced in a similar way; it has a constant width and is made up of alternate bands of light and dark lines, the density of which varies.

SOUND HEAD

The sound head, incorporated in sound cine projectors, contains the means of transporting and guiding the film and the optical system for scanning the sound track. The film transport mechanism consists of the sound drum and two stabilising rollers, which together with the drum guide the film and keep its speed uniform. A constant film speed is necessary to prevent irritating variations in the pitch of the reproduced sound.

To prevent wear and scratches on the film, rotating sound drums are normally used. A balanced flywheel, usually attached to the shaft of the drum, also assists in smoothing out sound variations. It is important, too, if good sound reproduction is to be obtained, that the film is correctly tensioned around the drum. The width of the film is greater than that of the drum so that its edge, on which the sound track is printed, overhangs that of the drum and is scanned by a narrow and sharply defined line of light. The transmission of the light is thus regulated by the shape or by the density of the sound track. On 16-mm films the sound track is printed 26 frames ahead of the centre of the projected frame, and in this way the sound is synchronised with the picture being projected. The light which passes through the sound track falls on to a prism or a convex mirror and is reflected on to a photoelectric cell situated inside the sound head. This causes electrons to be emitted from the cathode of the cell which are collected, by the application of a positive potential, at the anode. The small electrical currents which flow from the photoelectric cell are then boosted by means of an amplifier and actuate the loudspeaker. The volume and tone of the sound emitted by the speaker are governed by controls on the projector. The volume should be such that the sound is comfortably heard by everyone in the audience. The tone should be suitable to the particular recording and the particular acoustic properties of the auditorium.

MAGNETIC STRIPE

The magnetic stripe is a relatively new means of recording sound on films and can be applied over the normal photographic emulsion. Existing films can also be treated in this way by processing firms. The magnetic stripe, or sound stripe as it is called, when applied locally to films with normal photographic emulsion, may be applied over the full width of the area normally occupied by the optical track. In this case it is known as 'full-stripe'. 'Half-stripes', half the width of the full-stripe, may be applied over existing optical variable-area or variable-density tracks, which leaves one half of the track available for reproduction. (This is not applicable to variable-area tracks of the single-point type which are, in any case, now little used.)

Quarter-stripes, also known as edge-stripes, can be applied to films with double perforations and are placed between the perforations and the film edge.

The sound stripe is scanned by a magnetic head which is separate from the optical scanning arrangements and actually comes into contact with the stripe on the moving film. In projectors made to project magnetic stripe film, the magnetic head usually plugs into a socket on the sound head. Separate heads are necessary for single- and double-perforated film, that is, full- or quarter-stripe, but the full track head is also suitable for half-stripe.

BASIC RECORDING COMPONENTS

For recording purposes certain basic components are necessary and the simple function of these is briefly described.

Microphone

The microphone transforms variations in sound pressure into corresponding small variations of electric current. The different types in use today may be divided into two main classes: (a) pressure-activated and (b) velocity or pressure gradient.

Pressure-activated type

The pressure-activated microphone consists basically of a diaphragm, which, owing to its movement effected by changes in pressure outside and inside the microphone, generates small electrical pulses.

Velocity type

The most commonly used velocity microphone is the ribbon type, in which a ribbon is suspended between the poles of a permanent magnet. The suspended ribbon detects sounds and by its movement generates small electric currents. This microphone is sensitive to sounds coming from in front or from behind but not to those from the side. Pressure-type microphones, on the other hand, respond to sound from all directions. The qualities applicable to various types of microphones are of great importance in sound recording, and often affect the choice for particular applications. Other qualities, of course, may be equally important, depending on the use to which the microphone is put. The many particular purposes microphones are used for are outside the scope of this book. It is worth noting, however, that when speaking into a microphone, a regular distance between the mouth and the microphone should be maintained. This keeps the level of the recorded speech constant. When a script is used, it should be held in such a position that there is no necessity to turn the head to consult it.

Magnetic recorders

Magnetic recorders record sound on wire or plastics base tape and need little skill to operate. The tape recorder is the one most used today.

Tape recorders

The tape used in tape recorders is treated with a ferromagnetic medium which is magnetised, usually longitudinally, as it travels over a small gap cut in a metal ring. When a series of amplified electrical impulses, which originate from a microphone, are applied to a coil wound around the body of the ring, magnetomotive forces are set up across the gap. These forces cause an invisible pattern to be recorded on the moving tape. By a reversal of this process, the magnetic recording can be transferred to the coil winding, and the electrical impulses are then amplified and passed to a loudspeaker. The recordings made on the tape may be immediately played back, or by the application of a permanent magnet to the moving tape, or by the use of a high-frequency head, may be erased. Some recorders have entirely separate heads for recording, replaying, and erasing purposes.

In addition to its many other functions, the tape recorder can be used in conjunction with the recording mechanisms of modern cine projectors, for the reproduction of sound effects, and commentaries on magnetic stripe on film. The sound recorded on the magnetic recorder is played back and rerecorded on the film sound stripe.

Because of the changes in length of the magnetic tape due to humidity, temperature, and tension, the tape cannot be driven at precise speeds on the tape recorder. This causes some difficulty when recording on film, because the speed at which sound projectors are driven is very constant. Precise synchronisation when tape recorders are used for recording on film stripe has, therefore, not been possible. Stewart[4] has pointed out, however, that a variety of methods of attaining synchronisation, depending on the requirements of the application, are now being used. Small discrepancies in synchronisation between the tape and the film may not be important, particularly if commentaries are prepared in sections and timed to fit into available periods. If the tape is used only for music or background effects, the difficulties associated with synchronisation do not arise. It is generally recommended that recordings should be transferred as soon as possible to the film stripe to minimise any dimensional changes. Tapes on which recordings have been made should not be allowed, in transit or storage, to contact one another or to come into contact with any magnetised material.

The speeds of tape recorders vary according to the type of machine, and by adjusting a switch some may be run at several different speeds. The standardised speeds are 30, 15, $7\frac{1}{2}$, $3\frac{3}{4}$, and $1\frac{7}{8}$ in per second (76, 38, 19, 9.5, and 4.75 cm per second). The higher speeds give better response, but for recording on film sound stripe $3\frac{3}{4}$ in per second is adequate for good reproduction. For the purpose of recording on film stripe, tape recorders should have an instantaneous stop and start action, and an accurate tape position indicator.

Amplifier

Cine projector amplifiers may derive their input from a photocell, magnetic head, microphone, or a number of other external devices. In modern projectors, any one or combination of the above sources may be involved and the amplifier may therefore be divided into sections. The function and the arrangements of these can be varied by a selector level. Each pre-amplifier section has its own volume control and the tone is also controlled. The outputs from the pre-amplifiers are mixed and pass to a power amplifier.

RECORDING ON MAGNETIC STRIPE

Projectors of the type illustrated in *Figure 8.15* may be used for recording sound on magnetic stripe on film, and recording aids,

Figure 8.15. Modern 16-mm sound projector suitable for magnetic recording onto film and playback of optical and magnetic sound tracks (Courtesy Rank Film Laboratories)

such as microphones, record players, magnetic recorders, and radio receivers, may be connected directly to them. By means of a recording head, the output signals from the sound source are recorded on the sound stripe as variations in the state of the stripe coating. Recording control is provided by the various controls on the projector itself and may be assisted by sound mixers and other projector accessories; sound from either optical or magnetic track can be played. Recordings of music or commentaries can be made on full, half, or quarter sound stripes and can be immediately replayed. Recordings can be erased and further ones made on the same stripe, or if a revision of the recorded material is necessary they may be partially erased. Because of the ease with which erasures can be effected and new recordings made, persons new to this method of recording can rehearse the procedure until perfection is attained. The versatility of these projectors may be illustrated by the following summary of their recording capabilities.

1. By placing half-stripe over existing optical track the latter can still be used, while the half-stripe is also available for extra recordings.
2. The sound from optical tracks can be reproduced on half-stripes.
3. By the use of the quarter track head, recordings may be superimposed on full-stripe recordings and the new and old recordings may be reproduced together. The superimposed recording may at any time be erased and replaced.
4. By using a special magnetic head, two separate tracks can be recorded on full-stripe.

For best reproduction quality, the full-stripe should whenever possible be used and the machine run at sound speed when recordings are made. The projector is also provided with a neon lamp recording level indicator which indicates that the recording force has reached a safe maximum. The safe maximum should not be exceeded, for although it is advantageous to record at as high a level as possible, too high a level causes distortion due to overloading. Level indicators show only the peak level of the recorded sound. The relative loudness of music and other background effects should be judged aurally by rehearsals or may alternatively be monitored by headphones when recordings are made.

LANTERN SLIDES

Slides—providing they are good slides—assist audience understanding of the lecture subject matter. By good slides I mean carefully prepared slides which can be read clearly by the furthermost member of the audience, which help to eliminate lengthy verbal descriptions, and which stimulate audience interest.

Although this description is concerned with monochrome slides, mention must be made of colour slides, which are ideally suited for natural science (particularly botany and zoology) and medical lectures, and can also enliven other subjects.

Colour can be an essential part of the object photographed, particularly with plants, animals, medical specimens, and stained sections of these. As colour film is usually processed by the manufacturer, much of the labour of slide making is eliminated.

Monochrome slides, however, are more likely to be required for lectures, and are frequently requested at very short notice, so we must consider how these may be made using relatively simple equipment.

The standard 5-cm (2-in) square slide evolved as a consequence of the popularity, efficiency, and versatility of the 35-mm camera has, except in a diminishing number of diehard institutions, almost completely replaced slides of other sizes.

The 5-cm square slide can be made either on 5-cm square glass lantern plates or on 35-mm film. Both of these are considerably less expensive than the obsolescent 8.3-cm ($3\frac{1}{4}$-in) square slide, and the weight is only a fraction of that of the old slide—a point which will be appreciated by anyone who has had to transport a hundred or so of the 8.3-cm square slides to a distant lecture or conference.

The modern slide projector is also a compact, fairly portable instrument. Indeed, many lecturers prefer to take their projector with them to strange lecture halls. This can be unfair to the projectionist, and may lead to inefficient projection unless the lecturer is prepared to act as projectionist.

Photographing the original

Apparatus

The original should be photographed on a slow, fine-grain panchromatic film suitable for continuous tone. Ilford Pan F and Kodak Panatomic X are suitable for this, and are readily available in 35-mm cassettes, or in bulk rolls up to 30 m in length which may be loaded into one's own cassettes as required. Using bulk rolls in this way saves considerably in cost. The position giving the best or most informative view of the apparatus should be chosen. All extraneous items in the vicinity of the apparatus to be photographed should be removed. A plain background can be placed behind the apparatus to eliminate other equipment, furniture, etc., which may intrude on the area to be photographed.

As the films suggested above are comparatively slow, the camera should be firmly supported on a tripod or other stand, and additional lighting (tungsten lamps or electronic flash) should be provided, as the normal lighting in laboratories is usually low in intensity and seldom in a suitable position for photographing the subject.

WARNING: Expendable flash bulbs can be dangerous in an explosive atmosphere.

When arranging the lighting, excessive flare or reflections from glass or highly polished metal surfaces should be avoided as far as possible. A polarising filter over the lens often helps in this respect.

The camera lens should be stopped down if the greatest possible depth of field is required. This leads to comparatively long exposures, hence the need for a camera support.

Flat objects

These include line drawings, text, photographic prints, herbarium sheets, etc. They are most easily photographed by placing them horizontally, with the camera above them on a firm support. The illustration should be illuminated evenly, and the film plane must be parallel to the object plane.

Photographic prints and herbarium sheets may be copied on slow, continuous-tone film, such as those suggested for apparatus.

Black-and-white line drawings are most successfully copied on an extremely slow, high-contrast, virtually grainless film such as Kodak Microfile or Ilford Micro-Neg Pan. Since these extremely slow films have very little latitude, exposure is critical.

A permanent copying set-up is a great advantage here, as once the correct exposure has been obtained by means of a test strip it can be repeated whenever necessary.

Stains on the original, for example yellow staining on old illustrations or on old herbarium sheets, can be hidden by the use of a suitable photographic colour filter over the lens during exposure. The coloured lines on graph paper can also be subdued or even eliminated by the same method.

Preparing the line illustration

The original should be in bold black on white. Indian ink on white card gives the best result, providing the card surface is suitable for ink drawing. Bristol board is ideal, but rather expensive. Whatever material is used it is essential that the maximum black be obtained and that the background material be really white.

The data on the slide should be simplified as much as possible. The audience cannot be expected to absorb complex detail in the short time a slide is projected.

Letters and lines should be bold, so that they may be seen clearly by the furthermost members of the audience. Delicate, fine line, which is attractive in a book illustration, is often unsuitable for projection in a large hall.

A broad lettering nib or a felt-tip pen should be used for the originals or, better, stencils or rub-on transfer letters. These produce uniform, black letters, which are much neater than the average freehand script.

If it is desired to add caption or key-letters to a photographic print, these can be made on suitable small pieces of card and placed on the print before copying. A clean glass cover will keep the cards flat on the print. The photograph is therefore preserved undamaged and both photograph and key-letters can be re-used for other purposes.

Processing the film

The first principle of producing a good slide is to take a good photograph. Assuming one has a good photograph in the exposed but undeveloped film, there are two methods of producing slides.

Reversal processing

This method produces a positive image ready for projection. There are numerous formulae and processes for producing positive images, and although these are similar it is advisable to seek advice from the manufacturer of the particular film you are using.

Basically, however, the reversal process goes as follows:

1. Develop the film and wash in water.
2. Bleach in a dilute potassium dichromate–sulphuric acid solution and wash.
3. Clear in a sodium sulphate solution.
4. Briefly expose to light to fog the film.
5. Redevelop the film and wash.
6. Fix and harden.
7. Finally, wash in running water.

The total time involved for the above technique is from 45 minutes to one hour, depending on the film and process used.

In the case of continuous-tone transparencies, the exposure to light (4) can be omitted if a fogging agent such as hydrazine is added to the second developer.

Negative–positive process

This method produces a negative from which a positive is made later. It has some advantages over the reversal system, the most important of which is that the negative can be kept for future use to make paper prints for publication, extra slides for other lectures, or a slide to replace the one lost by your lecturer in some remote lecture room.

The negative–positive process goes:

1. Develop in a suitable developer, rinsing in a stopbath.
2. Fix and harden.
3. Wash in running water.

The time taken in this procedure is rather less than 30 minutes.

When the film is dry, a positive is made on 'safety positive film' by contact printing or projection. This must also be developed, fixed, and washed, then dried before mounting. Safety positive film can be handled in a suitable safelight and dish-developed, whereas the negative film has to be processed in total darkness.

In an emergency, the negative of a *line drawing* can be used as a slide, giving white lines on a black background. If this type of film is used, it should be remembered that a normal pointer is virtually useless. An illuminated pointer stick, or a pointer torch of the type which projects a small bright arrow image on to the screen is much more efficient.

Lantern plates

These consist of thin glass plates, 5 cm square, coated with photosensitive emulsion. They may be used in place of the safety positive film, particularly for printing by projection.

After processing, the emulsion side of the lantern plate is protected by a 5-cm square cover glass and the two bound together with binding tape.

Film positives, once washed and dried, should be mounted in suitable frames. The frames, or mounts, can be obtained in many forms. They can consist of 5-cm square cover glasses between which, suitably masked, the positive transparency is sandwiched. The edges are then bound with black self-adhesive tape. Others consist of plastics or aluminium frames, the halves of which clip together to hold the cover glasses and transparency in place; if these are used they should be selected carefully, as some makes are prone to spring open if roughly handled or overheated in the projector.

In hot, humid climates, plastics frames without glass may be advisable, as in such climatic conditions there is a danger of fungoid growths on film enclosed in glass.

The finished slide should now be labelled. At the same time it is 'spotted' so as to enable the projectionist to insert the slide correctly into the projector. A spot is placed at the bottom left of the mount when viewed in the hand. The slide is inserted into the projector with the spot at the top right corner of the mount, facing the lamp.

Projection

Check everything before showing your slides—last-minute disturbances do not contribute to efficient projection.

Check that the slides are in the correct order, and correctly positioned for projection so that the projectionist does not have to re-position them.

Check that the projector is at the correct distance and in focus.

If synchronised sound is used, position the speaker at the front raised on a table, not on the floor.

Ensure that there is adequate ventilation in the lecture hall, to avoid drowsiness.

Arrange a signal to warn the projectionist when to change slides.

Slide storage

Many types of storage boxes for slides are available. Some are in the form of drawers subdivided to hold groups of slides, others are boxes fitted with grooves to space the slides. Another type consists of a tray with a transparent bottom. The slides lie flat in the tray and can be illuminated from below for sorting.

If set lectures are to be repeated, the slides may be stored in the magazine used for the automatic type of projector.

Slides should be classified for easy sorting and kept in a dustproof container. In some areas of high humidity, a suitable desiccant should be kept in the container. The small packets of silica gel supplied with new microscopes and other instruments are ideal. They should be dried occasionally by warming in an oven.

REFERENCES

1 Hartung, E. J., *The Screen Projection of Chemical Experiments*, Cambridge University Press, London (1953)
2 Sumner, W. L., *Visual Methods in Education*, 2nd edn, Blackwell, Oxford (1956)
3 Hilger and Watts, *Hilger Handbook of Projectors for Gauging and Inspection*, Hilger & Watts Ltd, London (n.d.)
4 Stewart, W. E., *Magnetic Recording Techniques*, McGraw-Hill, New York (1958)

9
Laboratory records and technical information

Efficient laboratory organisation and administration entail the keeping of accurate records and the compilation of other kinds of information. For this purpose a definite filing system is required, the need for which has been adequately stressed by Lampitt[1].

FILING SYSTEMS

It has been said that no two filing systems are the same. This is certainly true of scientific departments, where the method of filing and the nature of the filed material vary considerably from one department to another. In the following pages the fundamentals of some filing systems are explained, but the numerous finer points necessary for the ultimate perfection of these systems are outside the scope of this book. These are more fully covered by the other excellent publications mentioned in the bibliography.

Filing may be defined as the systematic keeping of records or material in such a way that they may be quickly located when required. Good filing systems should therefore be (a) simple, (b) suitable, (c) accessible, and (d) adaptable.

MATERIALS REQUIRED FOR FILING

The extent of the equipment and accessories used for filing purposes is determined by the requirements of the system, the amount of material to be filed, and the amount of money available. Ingenuity may, to a large extent, overcome lack of money and effective filing arrangements have been set up with nothing more pretentious than wooden boxes and home-made guides. The value of any collection of filed material is judged in terms of its contents and not by the elegance of its housing. This point is stressed, because in further discussion references have been made regarding equipment which may be beyond the financial resources of many science laboratories. Nevertheless, the comfort associated with modern filing equipment and the ideas incorporated in it should not be overlooked.

PRIMARY INDEX GUIDES

To indicate the main divisions within a system, primary guides may be used. In alphabetical arrangements, for instance, they are used to indicate where the material filed under one letter of the alphabet begins and ends. In order to define these divisions more clearly, the guides have tabs which protrude above the tops of the folders and on these various captions are printed. The captions vary according to the system used. The tabs appear on the guides in staggered positions so that any one tab is not masked by another. The filed material is normally kept in folders and up to ten of these may be filed behind each guide. When large numbers of folders are used, the number filed behind each primary guide will be in excess of ten and secondary guides then become necessary.

SECONDARY GUIDES

Secondary guides are used to subdivide the folders within the range of the primary guides, and in this case their captions indicate the extent of the subdivisions. In alphabetical arrangements, for instance, if forty folders are used between the primary guide *A* and the primary guide *B*, then three secondary guides would be required for proper subdivision.

FOLDERS

The filed material is kept together in groups in individual folders. These are made of stout paper and are available in a variety of weights, grades, and styles to suit the material which they are required to hold. The better the grade of paper, the longer the life of the folder. For this reason, the amount of handling the folders will receive must be considered when a choice is made. The life of the folder is reduced, and the efficiency of the filing arrangement is impaired if the folders are overfilled, but certain types are designed to allow a reasonable amount of expansion. The folders themselves may also bear captions and in some cases staggered tabs are used on them. Some tabs are made of celluloid and slips of paper bearing the caption may be slid into them. Special rolls of perforated paper in various colours are available for this purpose.

MISCELLANEOUS FOLDERS

The individual folders are brought into use when sufficient related material has accumulated to warrant a file of its own. Meanwhile, the material is kept in a miscellaneous folder which is normally filed at the back of the group of folders which it serves. The miscellaneous folder bears the same caption as that of the section guide. The material kept in it should be in alphabetical order and when several pieces of related material are kept in the folder they are arranged in date order.

FILING EQUIPMENT

FILING CABINETS

A large variety of filing equipment is available and is designed to cope with material ranging in variety and size from microfilm to large drawings (*Figure 9.1*).

The well-known vertical filing cabinet may be purchased with from one to five drawers. Since the weight contained in the drawers may be considerable, the strength of the slides, and the ease with which these can be manipulated, are the important features in this type of cabinet. Inside the drawers some easily adjustable means of supporting and tensioning the folders should be provided.

The suspended method of filing is also becoming very popular. In this arrangement, paper pockets are connected together in concertina style by means of metal channel pieces. The ends of the metal channels are supported on side rails, as illustrated in *Figure 9.2*. A roll of tear-off paper strips is used for reference titles which, when typed, are slipped into transparent card holders fixed on the tops of the channels. The reference tabs are clearly visible and cannot be so easily damaged as the protruding tabs common to other filing arrangements. For smaller sizes of materials, such as

Figure 9.1. Plan file (Courtesy Vickers Roneo Ltd)

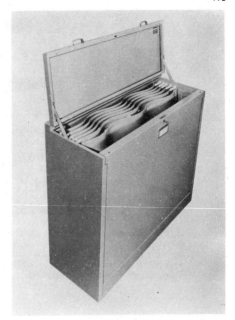

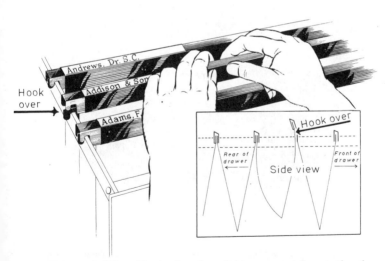

Figure 9.2. Suspended filing. Showing how the pockets are connected one to the other. The front part of the channel of the second pocket is hooked over the suspension bar holding the back of the first pocket until they are all linked up (Courtesy Vickers Roneo Ltd)

Figure 9.3. Lateral filing. Kraft pockets are hung from suspension frames which can be attached to metal or timber supporters. Steel units are also used which may have open fronts or may be fitted with blinds or steel curtain doors. The suspended pockets are linked together in a concertina style (Courtesy Vickers Roneo Ltd)

Figure 9.4. Visible filing. This revolving unit can carry as many as 56 000 references. The items of information are typed on separate strips and arranged in the desired order on the panels. Each strip can be protected by a transparent holder (Courtesy Vickers Roneo Ltd)

photographs, similar filing arrangements may be employed and the cabinet drawers may be divided to provide half-size or other sizes of pockets as required.

The lateral arrangement for filing is now also widely used. In this case, the folders are inserted into the sides of suspended pockets and it is claimed that the system saves much space (*Figure 9.3*).

VISIBLE FILING EQUIPMENT

If a set of records must be constantly referred to, and there is no necessity to remove the record itself from the file, visible filing methods may be employed. The information is normally recorded on cards. A number of different means of visible recording may be used and an example of these is the revolving unit shown in *Figure 9.4*.

Visible card cabinets occupy little space and are efficient and accessible (*Figure 9.5*). The cards and the cabinets which contain

Figure 9.5. Visible card records (Courtesy Vickers Roneo Ltd)

them vary in size in accordance with the needs of the filing system and the information to be recorded. The cards themselves may be ruled and printed in accordance with the user's requirements and points of particular interest may be pinpointed on them by using small coloured signals.

FILING ARRANGEMENTS

Two basic filing arrangements are used, (a) alphabetical and (b) numerical.

ALPHABETICAL

The basis of all filing systems is the alphabetical arrangement. It is necessary to vary the manner of the application of the alphabetical arrangement in accordance with the nature of the items filed and certain rules must be applied.

Alphabetical arrangement of names

Individuals

In dealing with correspondence, student records, and other matters, it is necessary to file according to the name of the individual. Names of individuals are arranged in alphabetical order in accordance with the exact sequence of all the letters which make up the surname. This works very well until two similar surnames are encountered and the system must then be extended to overcome this difficulty. All names are therefore divided into units, the surname being unit 1. Example:

Unit 1	Unit 2	Unit 3
Smith	Thomas	Harold

When sorting names into alphabetical order for purposes of filing, those surnames which are identical, that is to say those which are identical in unit 1, are considered for unit 2. If both units 1 and 2 are the same then unit 3 is considered. Unless very large numbers of names are involved, units 1, 2, and 3 are unlikely to be identical but should this happen then some additional information concerning the individuals must be added to the records to differentiate between them. Many other minor difficulties arise and a number of rules are applied to overcome them. Two of these rules are:

1. Names with prefixes are considered as though the prefix were part of the name. Example:
Angus John MacDougal—MacDougal Angus John.

2. When surnames are used on their own they precede surnames which are followed by Christian names. Similarly, the stem of a name such as Burn precedes other names which have an addition to the stem, such as Burne or Burnett. Example:

Unit 1	Unit 2	Unit 3
Burn		
Burn	A	
Burn	Arthur	P
Burn	B	
Burn	B	C
Burn	B	Charles
Burn	Bernard	
Burn	Bernard	H
Burn	Bernard	Harry
Burne	A	
Burne	A	H
Burnett	S	
Burnett	S	P

Company names

Although the names of companies present somewhat more difficulty, the method of mental division of the name is applied in the same way.

Unit 1	Unit 2	Unit 3
Scientific	Supply	Company

If names of individuals appear in the company name they are transposed—F. H. Smythe & Co. would be divided up as follows:

Unit 1	Unit 2	Unit 3	Unit 4
Smythe	F	H	& Co.

For filing purposes the words A, AN, and THE are ignored.

NUMERICAL

Numerical filing arrangements are particularly suitable for continuous projects and active folders are added as the project progresses. A card index is always used in this kind of filing arrangement. The cards are filed in alphabetical order and each one is related to a particular folder. On the card is written the number of the folder and the essential information concerning its contents. The folders are numbered consecutively and, as for other systems, new folders are begun when sufficient material concerning any particular matter justifies it.

The folders are kept behind guides in numerical order. The primary guides are numbered in hundreds or some other convenient unit and supplementary guides are used at intervals of ten folders. As the folders become full, or when extra ones are used to supplement the original, a new folder is started. The new folder is given the same number as the original folder, but this is followed by a dash and another figure or by a letter. Example: The folder following number 300 becomes 300–1 and the subsequent folders 300–2, 300–3, or 300*a*, 300*b*, 300*c*, and so on. This way, when figures are used an infinite number of subdivisions may be made but if letters are used, as they might be in smaller filing arrangements, the number of subdivisions is limited to twenty-six. The new folders are recorded on the appropriate card in the index. This way, any new matter which is filed and is closely related to that in the original folder will be found as a subdivision in the folders adjacent to it. Example: If the material in the folder numbered 300 were concerned with high-vacuum pumps, the folders following it might also be concerned with pumps of the same class but of different types, makes, or performances.

Miscellaneous folders are also used with the numerical system. The index cards which refer to the material stored in these folders should be clearly marked in some way to indicate this. The cards may for instance be marked at the top in lead pencil with the letter M.

DEWEY DECIMAL CLASSIFICATION

The Dewey Decimal Classification is well known for its application in libraries. The system, used for classified filing arrangements, limits the number of main subject headings to ten. The ten main groups are numbered 000, 100, 200, up to 900 inclusive. Group 000 is used for material too general to be classified under any other of the nine remaining groups. All the main subject headings may again be divided into nine more groups, for example, 510, 520–600, and further subdivision is effected by making yet another nine groups—520, 521, 522–530. The subdivision may be further extended by the use of decimal points—520.1, 520.2, 520.3–520.9—and further numbers following the decimal point may be added. When this system is used all the material must be carefully allocated.

SUBJECT FILING

Subject files may be alphabetically or numerically arranged, or may be classified by the Dewey decimal system. The subject arrangement is of great importance to technicians, since this is the method used in filing the technical material with which he is primarily concerned. The chief difficulty is to allocate a subject heading to the material which accurately describes it so that it can be easily found when required. The difficulty of accurately identifying the subject has been dealt with by Vickery[2]. Extensive cross-referencing in the card index is necessary and when a choice of main headings is made these should be kept as short as possible. In subject files the cross-reference may be the '*see*' reference, used where nothing is listed under the heading given. Example:

        Acid muriatic   *see* Acid hydrochloric

The '*see also*' reference is used where material allied to that sought might prove useful. Example:

        Ovens   *see also* Thermostats. Elements, Heating

Alphabetical system

Dictionary method

In this method of subject filing, the successive folders are kept in alphabetical order. The material in adjacent folders may therefore be entirely unrelated. Example:

        Thermometers, Thimbles, Traps.

Encyclopaedic method

If the encyclopaedic method is used, the subjects are arranged in classified order. The main headings are arranged in alphabetical sequence, as are the sub-headings which are related to and follow the main heading. Example:

        Ovens electric,
                Drying
                Miniature
                Thermostatically controlled
                Water

GEOGRAPHICAL FILING

Material is sometimes filed in accordance with its geographical association, and in scientific establishments this may be applicable when information has to be collated on a geographical basis. In this filing arrangement, the primary guides indicate the main division and the secondary guides the smaller divisions of the geographic unit selected. The folders between guides may, if required, also be identified alphabetically. Card indexes may also be used with geographical filing methods.

MEMORY FILES

In administering laboratories many things, such as the regular maintenance of equipment, preparative work for lectures, and matters which arise out of correspondence, require attention sometime in the future. The memory cannot be relied upon when numerous matters are concerned and a memory file may be used to ensure that attention is given to these. A number of systems may be adopted for this purpose, but the one which commends itself is the use of a card file. This simple system merely entails noting the details of the matters which require future attention on 5×3-in (127×76-mm) cards. The cards are then filed behind guide cards each of which is marked with the name of one month of the year. A further set of cards numbered 1 to 31 is placed behind the current month guide. One of these date guides and the information cards behind it are removed each day and any necessary action is taken. The date guide is then placed behind the following month guide so that the cards are kept in a divided state for daily reference.

FILING CONTROL

FILING THE MATERIAL

When any material is filed it should be marked in some way to indicate the division of the filing system in which it should be placed. The usual way of doing this is simply to underline some outstanding feature of the material such as the name, subject, signature, or location. This of course depends upon the nature of the filing system in use.

The material should also be marked in some way if cross-referencing is necessary. Cross-references should always be made when any future difficulty in finding the material might be experienced. If several headings are involved, the matter should be filed under what is considered to be the most important of these and should be cross-referenced by the subjects of lesser importance. In technical articles, for instance, a number of subjects may be involved. For correspondence or other filing arrangements for which a card index may not be necessary, the material may be cross-referenced by placing distinctively coloured sheets of paper in selected individual folders in which the original document might be sought. Information is given on these papers by which the original document may be located.

DISPOSAL OF OLD MATERIAL

Essential records should be kept permanently in the files but other material which becomes redundant with the passing of time should be periodically destroyed. This keeps the filing arrangements up to date and active. The materials no longer active enough to remain in the files, but for record purposes having a further useful life, should be removed from the active files and carefully stored.

RETURN OF BORROWED RECORDS

Folders may be occasionally borrowed, and although a strict charge-out system as practised in large organisations may be unnecessary, some method of indicating that the folder is on loan is a wise precaution. One method of doing this is to put a card, which has the caption OUT on its tab, in place of the folder which has been removed. The face of the card should be ruled and the name of the borrower, the nature of the material, date, file number, and any other relevant information, should be recorded on it. When folders are lost this should also be indicated by placing a spare folder with the caption LOST on its tab in place of the missing file.

FILING SPECIAL MATERIALS

TRADE CATALOGUES

Most firms are only too pleased to include technicians on their mailing lists and much useful information is to be gained from their trade catalogues. The larger ones contain so much information of a diverse nature that they should be treated like books and stored on bookshelves for reference purposes. The catalogues should be listed in a card index and reference made to their shelf position. Trade catalogues generally, as pointed out by Collison[3], would be ideally indexed by specific subjects, in either alphabetical or classified order. Unfortunately, the smallest of these, including pamphlets and brochures, cover such a wide variety of products that even the broadest subject headings are useless. The catalogues are therefore best arranged under the names of the manufacturers. The author has found, however, that since a duplicate of most of the smaller catalogues comes to hand, that it is possible, when space allows, to arrange additional files by subject. The subject of the greatest technical interest to one's own department may be selected from these catalogues or leaflets as the main subject heading. An abbreviated index of the subjects covered by the filed pamphlets can be allocated to the spine of the box files themselves, if these are used for filing. This affords a rapid method of finding information concerning specific items of equipment at short notice.

Catalogue or box files are the most suitable means of storing small catalogues, which may be alphabetically separated in the files under manufacturers' names by means of guides made to suit the size of the file. Alternatively, small catalogues and pamphlets may be kept in steel filing cabinets provided strong folders and stout guides are used. An alphabetical card index of manufacturers' names should also be kept, so that folders or files may be quickly located.

ILLUSTRATIVE MATERIAL

In many laboratories, especially in teaching institutions, a collection of illustrative material, such as photographs and drawings, may be required for display or reference purposes. The usefulness of such collections in connection with the epidiascope, needs no elaboration. To build up a worthwhile collection, a high degree of selectivity in respect of the quality of the illustrations is necessary. The size of the illustrations may also be important if they are to be used with an epidiascope. The scope and nature of the subjects embraced in the illustrations are dictated by the interests of the department concerned, and in teaching institutions by the requirements of the lecturing staff.

Since illustrative material is concerned with subjects, a subject or classified filing system should be employed. For most laboratory purposes the subject filing system is adequate and suitable. A card index in which extensive cross-references are made is also necessary. The material may be filed loose or may be affixed at its four corners to stiff paper mounts with gum. Illustrations filed loose should be contained in folders of such a pattern that the contents cannot easily fall out. Although loose filing saves space and expense, mounting the material is undoubtedly the best. In this way the life of the illustration is prolonged, filing is made easier, and if the mounts are a suitable size the handling and orientation of material used in the epidiascope is assisted. To enable the illustrations to be found quickly in the file, the mounts should all be marked in the same position on the top edge, with the subject heading.

Certain illustrations of special importance may appear in books or may be bound in other ways so that illustrations cannot easily be taken out. These should also be referred to in the card index and may be identified by using coloured cards. The general information given on the index card may be supplemented by a short account of the illustration written on its lower half. Such information proves useful when illustrations suitable for a particular purpose are sought. Further detailed information concerning the illustration may also be fixed to the back of the mount itself.

The arrangements for filing illustrative material, as any other, are largely governed by the amount of money which can be spared for equipment. Box files or catalogue files are suitable, and if these are used information regarding their contents can be affixed to the spines. Vertical cabinet files, however, are neater, permanent, and more efficient.

The major problem, that of storing display charts, pictures, maps, and similar large items, still remains. The large plan or blueprint files especially made for the purpose are expensive. If used they should have large drawers which should also be shallow to restrict the number of diagrams stored in each drawer. This makes location easier and less damage is likely to be caused to the drawer contents. A simple and inexpensive storage arrangement consists of suspending the diagrams vertically from suspension rods or wires in the manner of an open wardrobe. A curtain should surround the frame to keep out dust. An alternative method is to store the material in large portfolios.

TECHNICAL INFORMATION

Technical information required for future reference may be indexed and filed in much the same way as has been outlined for illustrative material. Indeed, if the filed material is not extensive, the two classes may share a common arrangement. The material, such as extracts or cuttings from various publications, is usually filed by subject and a card index is required.

The arrangements for filing depend on the amount of material which will be filed and the facilities available. Scrapbooks, box files, loose-leaf files, or vertical filing cabinets represent some of the arrangements used.

Vertical filing is the best method because the material is more easily placed into divisions and subdivisions and speedy access to the information is afforded. The original cuttings or articles or copies of these, are kept in folders in the usual way and the main subject heading is recorded on the folder. The cuttings are more easily handled and filed if they are mounted on paper of a size suitable for the folder. Loose material should be kept in envelopes or folders with closed sides. If folders and envelopes are used in the same file, they should be the same size. When any one article consists of several papers, these should be securely fastened together. If box files are used, the main details of the contents should be outlined on the spine of the box and any further details given inside the cover.

With mounted material, the subject and full reference details concerning the source of the information should be written on the top of the mount. If the material is unmounted, the information should be written on the material itself. Old material which becomes outdated or corrected by later information should be discarded.

LANTERN SLIDES

Lantern slides are easily damaged and the most satisfactory way of keeping them is in special boxes or cabinets, which are manufactured for this purpose. The slides are stored vertically and are held in position in slots. For small collections slide boxes may be used, each of which may hold up to 100 slides. Larger collections may be compactly housed in wooden or metal cabinets. In some cabinets provision is made for the individual indexing of each slide.

So that slides may be found quickly when required they should, like other illustrative material, be systematically filed. The method of filing depends on the size of the collection and in some libraries, where large collections are kept, the Dewey decimal system is used. Small collections, on the other hand, which may be kept in a slide box, do not require an elaborate filing system and a list of slides stuck on the underside of the lid of the box is usually sufficient. Each slide should be given a number, which should be placed by the side of its title. The number should be repeated at the slot in which the particular slide is placed.

Assuming that most collections handled by laboratory staff are of a moderate size, the best method of indexing them is to use a card index system in which the slides are entered by subject. The cards should be cross-referenced where necessary. The classification or location number and description of the slides should be entered on the appropriate cards. The slides should be arranged in a cabinet either numerically by classification number or alphabetically by subject. Titles and classification or location numbers may be written on the masking or on the slide mount. Although this system permits individual slides to be methodically selected, it involves a considerable loss of time when a set of slides for a particular lecture has to be found. It is advantageous, therefore, particularly in teaching establishments, to keep the slides in sets. One solution to the problem has been suggested by Rothchild and Wright[4], who advocate the use of coloured masking tape to assist filing.

STORING AND FILING FILMSTRIPS

Filmstrips may be filed and indexed in a similar way to lantern slides and should be kept in small tin, plastics, or paper containers, in cabinet drawers. The titles and classification or location numbers should be written on labels affixed to the top of the containers. The title and a short description of the filmstrip should be entered on the index card. Other recorded information should include the number of frames and, in the case of commercially produced filmstrips, the name of the producing company and date of release. If the filmstrip is accompanied by a disc recording, the details of the recording, such as the number of sides, diameter, speed, and playing time, should also be noted. These details should be entered following the details of the filmstrip itself. In the case of recordings made on tape, the number and size of reels, speed, width of tape, and playing time should be given.

FILMS

Commercially manufactured films are so readily available from film libraries that it is usually unnecessary, for individual departments in scientific establishments, to keep them in large numbers. Instructional films which may have to be repeatedly shown, however, and those made by institutions for their own experimental or record purposes, may constitute a collection of considerable size. Irrespective of the size of the collection, a suitable system of indexing and filing should be adopted.

Because film libraries cater largely for teaching establishments, they generally group their films by level of instruction and by subject. A Dewey decimal system of subject classification is commonly used and films are alphabetically arranged in the catalogue index by title and subject matter. A classification number given at the side of the title acts as a guide to the classified body of the catalogue. Methods of indexing and filing in laboratories may be much simpler, but this of course depends on the amount of material involved and the special requirements of the department.

For filing films a card index system is again the simplest method and may, where applicable, be combined with one used for filmstrip records. If the card index is combined in this way the titles of the films should be followed by the words MOTION PICTURE and filmstrip titles by FILMSTRIP. Serial numbers assigned to films are often preceded by F.S. or M.P. for identification purposes. Films are identified by title and not by author or in other ways and should, therefore, be entered by title on the index cards. If a film has no title, a descriptive one should be chosen and written in brackets, on the card. In the case of commercially produced films, it is customary to enter the name of the producing company after the title. The accession number and the classification or location number should be entered at the top of the card. Other useful information which should be recorded includes the date of receipt, length (or running time), number of reels, sound or silent, colour, and width of film in mm. It should be stated whether the film is inflammable or nonflammable. Notes should be made on the record cards of descriptive material, such as lecture notes, which may accompany films. It is generally sufficient if the above information is followed by a short summary and this may be written on the lower half of the card or on its reverse face. In some institutions, detailed information concerning films is required. The position on the film of important experimental details and techniques may have to be recorded, in which case, in addition to the reel number, the film footage at which the point of interest occurs should be given. Such detailed film analyses involve a more complicated indexing arrangement.

Storage of films

Before they are filed all films should be marked with a number corresponding to that allocated in the card file. The number should appear on the film leader, the spool, and on the container. Films should be kept in fireproof cabinets and may be stored horizontally one above the other or vertically in holders, on shelves of sufficient depth (*Figure 9.6*). Drawers are also suitable, provided these are shallow and that the films are arranged in single layers. Whatever

Figure 9.6. Film and filmstrip combination storage cabinet. This cabinet can accommodate 700 filmstrip containers and 180 122-m reels of film. Its overall size is 183 cm high × 120 cm wide × 28 cm deep (Courtesy Naumade Products Corporation)

method is used, the titles should be visible, and if the films are horizontally or vertically stored should appear on the edge of the container. Similar storage is suitable for tape recordings for use with filmstrips. All films which have been used and returned should be carefully checked to see that they are undamaged.

Film, whether it be in the form of filmstrip, slides, or motion pictures, should be stored under suitable conditions. It should be kept cool and dry and the relative humidity should not exceed 25–50%. Temperatures of 21 °C or less are best. Colour film should be kept in the dark, and since it is affected by acidic gases such as sulphur dioxide and hydrogen sulphide, should be kept well away from laboratory and other harmful atmospheres. In very damp climates, it may be necessary to use silica gel as a desiccant to keep the film dry.

DOCUMENT COPYING PROCESSES

A means of copying and reproducing documents is an essential need in most scientific establishments. The copies may be needed for announcements, for recording of information, for the extraction of information from literature on short loan to the institution and for many other purposes.

The copying and reproduction of the documents or drawings may be satisfactorily executed in various ways. To satisfy the particular needs of any establishment a method should be chosen which is practically and financially suitable.

Copying and reproduction methods, such as typewriter carbon copies, rubber stamps, and cut-out stencils, are suitable for some purposes and are so commonly used as to need no explanation. Other methods, which are not quite so well known, merit attention and their description will serve to introduce more modern means of document copying.

GELATINE DUPLICATORS

The number of copies which can be produced with gelatine duplicators is limited, but the process has the merit of being simple and inexpensive. Gelatine is melted, poured into shallow trays, and allowed to cool. A paper master is prepared by writing or drawing with special inks or pencils or by typing on special carbons. Coloured materials may be used and this enables multi-coloured copies to be produced. If the master copy is typed the carbon paper is inserted in the machine facing the master. Another piece of paper is placed in front of the carbon to protect it and to provide a copy for checking purposes.

The master copy is stroked face down on to the gelatine and is allowed to remain in contact with it for several minutes until the dye is absorbed. The master is then removed and copies are made on absorbent duplicating paper by stroking each sheet on to the gelatine and immediately peeling it off. The number of copies which can be obtained depends on the dye medium used in the preparation of the master. Approximately twenty-five copies are obtainable from pencil masters and seventy-five copies from ink masters.

When the run is completed, the surface of the gelatine is removed by washing with warm water and it is then ready for further use. Made-up gelatine pads may be purchased in single sheets or roll form. The latter take up to fifty master sheets in succession.

STENCIL DUPLICATORS

The stencil used with stencil duplicating machines consists of a sheet of tough tissue paper coated with wax. The wax is cut by the action of a typewriter's typeface. This allows the ink to penetrate the stencil and to be absorbed by the copies as they are produced on the duplicating machine. In addition to the waxed paper sheet, the stencil has an intermediate protective tissue and a backing sheet. The protective tissue, which is used to prevent the waxed sheet and the backing sheet from sticking together, is removed before the stencil is cut. In its place a carbon paper is inserted so that the face of the carbon is in contact with the back of the stencil. This makes the stencil legible for checking purposes. Two carbons may be used back to back and in this way a perfect reference copy of the master sheet is obtained on the backing sheet. When the stencil is typed the typewriter ribbon is put out of action to ensure a clear-cut stencil. Typing mistakes can be corrected by gently inserting a pencil to separate the carbon and the stencil and applying correction fluid over the mistake on the front face of the stencil. The obliterated portion may be typed over.

Stencils may also be prepared by writing on them with a stylus or by marking them with other suitable drawing instruments. For drawing or writing purposes the stencil is placed on a suitable flat plate. For use with the stylus a plate with a roughened surface is recommended. Plain glass or translucent plates, which may be illuminated from below, are widely used, and these facilitate the transference of matter from books and other literature. Letter and symbol guides are available which are particularly useful for science subjects.

Stencils may also be cut on printer's type and manufacturers offer a special service in this connection. Another method of preparing stencils is by means of a brush or felt pen. A special ink is used which eats away the stencil surface. This method is particularly useful for the speedy preparation of notices and bulletins.

Photographic stencils

A photographic contact process may also be employed for the production of stencils. The original is prepared in lightproof ink on a transparent medium, placed in direct contact with the stencil, and exposed to the light from a Photoflood bulb. The light does not affect the surface of the stencil where it is protected by the outline of the original, but the remainder of the surface becomes hardened. When the stencil is washed the unhardened surface is removed, leaving the outline of the original on the stencil.

A similar photographic process allows originals, including halftone or original photographs, to be photographed directly on to a stencil. Enlargements and reductions of the original may be made. After processing the stencil copies are produced in the usual way.

Electronic stencils

Stencils may be made by an electronic process. The copying machine consists of two cylinders which rotate side by side. As the original, which is attached to one of the cylinders, is scanned by a photoelectric cell, the copy is reproduced on a stencil on the other cylinder. Original photographs can be reproduced in this way.

STENCIL DUPLICATING MACHINES

Stencil duplicating machines may have a single or a double cylinder. In the double-cylinder type, the ink is distributed to the cylinders by means of a roller. The ink penetrates a silk sheet which is in contact with the outside of the cylinders and then passes through the stencil to contact the copy paper.

In single-cylinder machines, the cylinder itself holds the ink supply (*Figure 9.7*). The ink passes through a perforated portion of the cylinder on to a pad which distributes it evenly to the stencil. The ink then penetrates the stencil, where it has been cut by the type face or stylus, and produces a copy on the copy paper. Various colours of inks may be used and some machines are so constructed that the cylinder itself can be easily changed for work involving another colour. Multi-coloured copies may be obtained by placing a protective cover over the black ink pad and attaching a clean pad above it. Coloured inks are worked into the clean pad with a brush in positions which correspond with appropriate areas on the stencil.

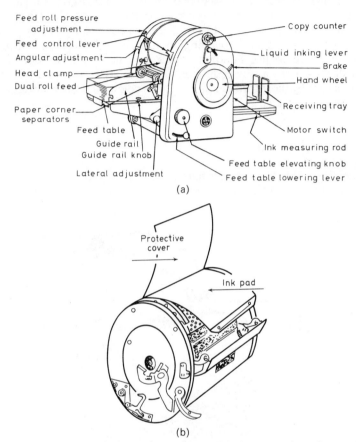

Figure 9.7. (a) Mimeograph. In this type of stencil duplicating machine, the stencil is placed on the outside of the cylinder and the paper is fed between the cylinder and the impression roller. The roller rises automatically and causes the paper to touch the stencil. (b) Mimeograph cylinder. The cylinder holds a supply of ink which flows through the small perforated holes of the cylinder diaphragm. A pad is attached over the perforated diaphragm and distributes the ink to the stencil with which it is in contact (Courtesy A. B. Dick Co., Chicago)

This method is suitable only when a distance of 2.5 cm can be left between the areas in which the colours are to be used. Another method consists of passing the copies several times through the machine and using a separate stencil for each run.

SPIRIT DUPLICATORS

The action of the spirit duplicator depends upon the removal of an aniline dye from a transfer sheet on to the glazed side of a paper master sheet. A spirit solvent is used to remove some of the dye from the master and to transfer it to the sheets of copy paper as they pass through the machine. Depending upon the model of machine, from two hundred to five hundred copies may be obtained from one master. As many as seven different colours may be incorporated in the master by the introduction of various coloured transfer sheets as the master is made.

Producing the copies

The glossy side of the paper master sheet is placed in contact with the coated side of the transfer sheet. When masters are typewritten,

Figure 9.8. Preparation of a typewritten master sheet for use with the spirit duplicator (Courtesy Block and Anderson Ltd)

a paper backing sheet is also used (*Figure 9.8*). The master is typed through the typewriter ribbon and the positive side of the master is used for checking purposes.

Masters may be prepared in a similar fashion by writing or drawing on the paper with a hard pencil, stylus, or ball-point pen.

A backing paper in this case is not required, but a writing platen with a smooth hard surface is placed behind the transfer sheet.

The negative impression formed on the back of the master is used for reproducing the copies and the master is attached to the drum of a machine by means of a clip (*Figure 9.9*). The machine

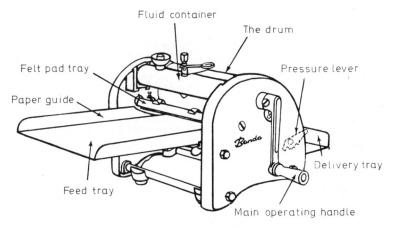

Figure 9.9. Spirit duplicator (Courtesy Block and Anderson Ltd)

may be hand-operated and the paper is fed into it manually or automatically. Electric machines have automatic feeds. If a full run is not completed the master sheet may be removed and stored for future use.

As the blank copy sheets are fed into the machine they come into contact with a felt pad or roller which is damped by spirit fed from a small tank. Small machines take paper up to foolscap size but larger machines are available.

By using a special negative paper to photocopy the original it is now possible to combine spirit duplicating with the photographic transfer processes which are described later in the chapter. After development the unexposed parts of the negative carry a laterally reversed colour image of the original. The negative can now be used on the spirit duplicator as a transfer original and up to 100 copies may be produced.

XEROGRAPHIC PROCESS

Xerography is a comparatively recent method of reproducing images on to ordinary paper or any other convenient surface from a paper or film original. It is a dry electrostatic process which does not need a darkroom or darkroom materials and is used extensively in automatic copiers and printers (*Figure 9.10*).

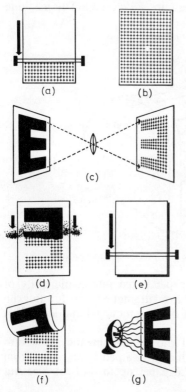

Figure 9.10. How xerography works. (a) A specially coated plate is electrically charged by passing it under wire. (b) The surface of the plate becomes coated with a positive electrostatic charge. (c) Copy (E) is projected by a camera and the positive charge on the plate remains only in the area of the projected image. (d) Negatively charged powder adheres to the positively charged image area. (e) A sheet of paper is placed over the plate and is given a positive charge. (f) The powder attracted from the plate by the charged paper forms a direct positive image. (g) After the print is heated for a few seconds the powder fuses and a permanent print is formed (Courtesy Rank Xerox Ltd)

The process

The process depends on the photoconductive properties of selenium; that is, selenium conducts electricity only when exposed to light. In its simplest form, an aluminium plate coated with selenium is charged electrostatically with the light excluded. It is then exposed to the document to be copied in a camera in the normal way. The charge is dissipated by the light falling on the plate, thus leaving an electrostatic image. A special ink powder is then cascaded over the plate and adheres to the image by electrostatic attraction. The sheet of paper or any other material, such as offset master stock, is brought into contact with the plate, and the ink powder is transferred from the plate to the paper. The ink powder, being thermoplastic, is then fused into the paper by heat or chemical means.

The most common application of xerography is in the automatic copying machine, a popular type of which is the desk copier. Any number of copies from one to fifteen may be obtained by setting the controls on the machine and if the knob is set to M (multiple) copies are delivered continuously at the rate of nine per minute. Originals up to a maximum size of 21.6×51.1 cm and a minimum size of 14×14 cm may be reproduced. Even smaller originals may be handled if they are first enclosed in a transparent document carrier.

A large free-standing machine is more versatile and reproduces from originals up to 22.9×35.6 cm in size. Depending on the length of the copy paper, speeds of from fourteen to sixteen prints per minute may be obtained. This machine also reproduces from bound books.

Although both the copiers mentioned can print on master stock, another machine, known as the Mastermaker, is designed especially for this purpose and can produce paper or metal masters, transparencies, and dyeline translucencies. From paper or metal masters numerous copies can be produced quickly by the use of offset lithographic machines.

Yet another range of xerographic printers is available for enlarging and printing images stored on microfilm. This may be in roll form, or unitised, that is, single frames of microfilm mounted in standard punched cards. Large continuous-output printers can enlarge from microfilm and print on to a roll of ordinary paper.

Other uses of xerography, e.g. to produce from industrial X-rays, manufacture printed circuits, and transmit information over long distances, are being tested.

OFFSET LITHOGRAPHIC PROCESS

The offset lithographic process is one of the most efficient methods of duplicating and printing. It was formerly only used professionally, but the introduction of efficient compact machines made this inexpensive method of duplicating and printing available to educational and other establishments.

Briefly, the process consists of the preparation of a flexible plate made of paper, zinc, or aluminium. Paper plates are less expensive but have a shorter life and do not give quite the same excellent reproduction as do metal plates.

Typewritten master plates are easily prepared or may be drawn with special reproducing inks and pencils. Errors may be removed with an eraser and the corrections can be written over. Other methods of preparation include proofing the master plate directly from letterpress and photographic reproduction by exposing a negative on to a sensitised master. Black-and-white line or halftone originals may be copied in this way. The masters are easily filed and stored for future use.

The method of reproduction depends basically on the lack of affinity which the greasy printing ink and water have for each other. When a mixture of ink and water is applied to the master plate,

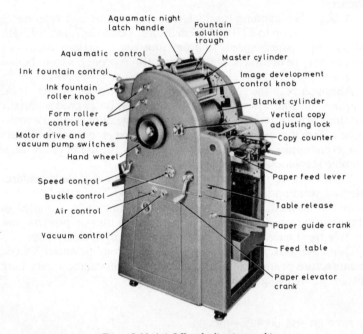

Figure 9.11. (a) Offset duplicating machine.

the ink adheres to the printed portion of the master but is repelled by all other parts of its surface. The ink is transferred from the master to a rubber roller and from the roller to the copy paper. Three main rollers are involved and these are known as the master cylinder, the blanket cylinder, and the impression cylinder. The master is attached to the master cylinder, and ink is transferred to it by an inking roller from an ink reservoir. The correct amount of ink is then transferred to the rubber-faced blanket cylinder. When the copy paper is passed between the blanket cylinder and the impression cylinder, the image is transferred from the rubber blanket to the copy paper (*Figure 9.11*).

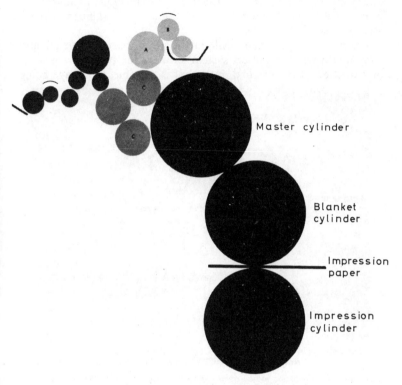

Figure 9.11. (b) Principle of the offset process. A grease-receptive image is placed on a master by means of typewriter, carbon paper, pencil, or crayon, or through a photographic process. When the matter is placed on the duplicator, a moistening agent (first etch and then fountain solution) and ink are applied. The moistening agent is repelled by the grease-receptive image but is accepted by the non-image area of the master. The ink adheres only to the grease-receptive image and is repelled from the moistened non-image area of the master.
Ink covers all twelve rollers in the ink–water system, including the roller in the Aquamatic fountain. Roller A, covered with ink, picks up fountain solution from ductor roller B and transfers it to the rest of the ink rollers. Ink from rollers C, now covered with a combination of ink and water, contact the master to continually supply it with ink and water (Courtesy A. B. Dick, Chicago)

The speed of printing is exceptionally fast and, depending on the type of image, up to 50 000 copies may be obtained. The paper is fed to the machine by a friction bar or by suction plates and may be manually or automatically operated.

CAMERA COPYING

Any kind of document, including large drawings, may be reproduced on paper by using a special copying machine as illustrated in *Figure 9.12*. Such machines are semi-automatic and can be operated by unskilled persons.

The machine consists basically of an adjustable camera back and copy board with an optical system between the two. Enlargement or reduction of the original document is possible, the image being

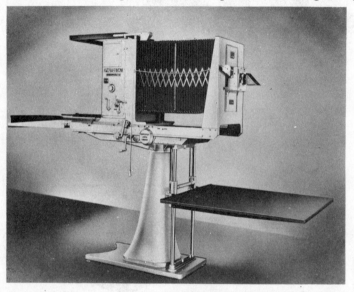

Figure 9.12. Photostat photographic copying machine (Courtesy Kodak (South Africa) (Pty) Ltd)

produced on paper supplied in spools up to 45.7 cm wide and 106 m in length. Exposed sheets may be automatically cut from the roll and processed in trays on the machine.

A similar machine, the Statfile recorder, makes rapid photographic copies of drawings up to 101×152 cm in size. This can be done on half-plate (12×16.5 cm) film negative but a more recent model uses 70-mm roll film. This film, which is 30 m long, accommodates up to 300 drawings each reduced to a 7×10.2-cm negative size. Up to 18 000 negatives may be stored in a container measuring only $76.2 \times 43.2 \times 50.8$ cm. The negatives can be enlarged on special viewers and the 70-mm recorder can itself be used to produce copies of the original drawings by enlarging the negatives back to their original or any other intermediate size.

Microcopying

The prime objects of microcopying are to save storage space and to provide extra security in the storage of document copies. As a true photographic image is obtained, it is unnecessary in most cases to keep the original document.

The microfilming of documents may be carried out on automatic, semi-automatic, or manually operated machines, and the images of the document are recorded on 16-mm or 35-mm film of up to 30 m in length. After development and processing the film is then read on a microfilm reader.

16-mm microfilming is normally done with automatic table recorders capable of photographing loose-leaf documents up to

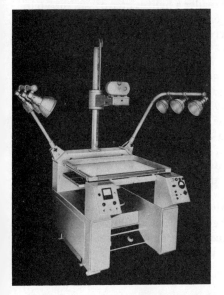 

Figure 9.13. (left) Semi-automatic 35-mm film recorder suitable for microfilming drawings and books. When books are recorded a resilient support or book cradle is used which compensates for the fluctuating thickness of the two halves of the open book. (right) Microfilm reader. The illustration shows a feeder reel in position for advancing roll film. Filmstrips and sheet film are advanced in special carriers (Courtesy VEB Carl Zeiss, Jena)

30.5×35.6 cm in size at a rate of up to 200 a minute and 3000 letter-size documents can be accommodated on 30 m of film. The same length of film can take much larger numbers of smaller documents which may be copied at even greater speeds. Some table recorders, in addition to the automatic production of microfilm copies, may also, by lowering a viewing screen, be used as microfilm readers.

For microfilming drawings and books of various shapes and thicknesses, larger machines using non-perforated 35-mm film are used and consist of a camera supported on a stand above a copying table (*Figure 9.13a*).

After copying, the exposed film is developed in a separate developing apparatus in which the film is run through a series of light-tight vessels containing the processing solutions. This operation may thus be done in ordinary room illumination.

To produce positive film copies from negatives, a constant printing method is used, the film being passed rapidly through a copying instrument at a constant speed.

For reading purposes micro readers are used. These may be capable of taking filmstrips and roll or sheet film. As the film is manually advanced through the reader a light source passes through the film to produce an optically enlarged and easily read image on a frosted glass or white plastics screen (*Figure 9.13b*).

NON-OPTICAL PHOTOCOPYING PROCESSES

The duplication of material may also be successfully accomplished by non-optical processes. These are of particular importance to establishments not equipped to carry out photographic processes and offer an attractive method of reproducing handwritten, printed, typewritten, or drawn originals at a reasonably low cost. A number of different types of machines are available for this purpose, but generally speaking these employ only two basic methods, known as the reflex and the direct positive methods. For success in both of these an intimate contact between the original and a paper coated with sensitised emulsion is necessary. Many different types and weights of document copying paper are available.

Reflex process

The reflex process involves the use of a slow document copying paper and can therefore be carried out in subdued light. The principle of the process is illustrated in *Figure 9.14*. A sheet of sensitised paper is placed in contact with the original and the weight of a heavy sheet of plate glass, placed on the top of these, ensures close contact between them. The light which passes through the glass plate also passes through the back of the photographic paper and evenly fogs it. The paper is also further exposed by the light reflected back from the original. Since more light is reflected from

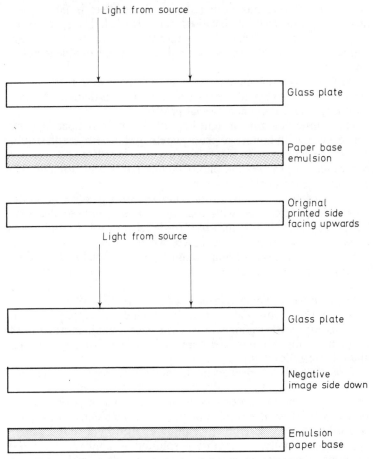

Figure 9.14. The reflex process (top) and the transmission printing process

the light portion of the original than from the darker portions, a laterally reversed image is produced on the paper when it is developed. The negative is then washed and fixed in the normal fashion. The negative, after drying, is ready for the production of any number of positive copies by normal transmission printing.

Direct positive process

By the introduction of a high-contrast paper capable of producing a direct positive image from a positive original, the time taken for document copying has been reduced. Since there is no negative made, only one sheet of sensitised paper is used. The speed of the paper is slower than that of the paper used for reflex printing and a greater intensity of light is required to give a reasonable exposure time.

With the advancements made in photocopying methods the arrangements as shown in *Figure 9.14* are now little used. Modern photocopying is done in a special printer.

The printer is simply a light box fitted with a number of lamps which evenly illuminate a glass platen. A lid with a soft rubber pad attached to its underside is lowered on to the platen and pressure is brought to bear on the photocopying paper and original. This ensures good contact between them.

To facilitate the change from reflex to direct positive processes, most printers are fitted with combinations of lamps so that extra lamps may be switched into use for direct positive printing. Most units are fitted with electric timers which allow exposures to be preset. In some cases vacuum pumps are also fitted in the printer to ensure good contact between the original and the sensitised paper.

Direct positive paper can be used in normal, artificial, or subdued lighting. This colour-sensitive high-contrast paper allows yellowed or faded originals to be used since a weak contrast in the original is improved in the copy. Direct positive paper is also suitable for reflex printing.

Wet methods of processing in document copying have been largely superseded by semi-dry processes. A porous block is used in one such method to absorb excess moisture. The exposed paper is placed with its emulsion side uppermost on the block and the paper is brushed with a developing solution applied by a rubber sponge applicator. Finally, a stabilising solution is applied. Automatic methods of processing copies are also available in which the copy passes through the developer and emerges dry from the machine.

Photographic transfer process

In the photographic transfer process the original document, which may be single-sided translucent, transparent, or opaque double-sided, is placed in contact with the photographic paper in a printer. Adjusting clamps allow the level of the pressure platen of the printer to be raised for the insertion of books or other bound literature (*Figure 9.15*).

After exposure the negative produced is placed in contact with a sheet of transfer paper and passed through a louvre in the side of the unit which contains a developing solution. The papers are propelled through the electrically operated unit by rollers and emerge together in a damp condition. They are left in contact for a few moments, and when separated a black-and-white copy of the original is

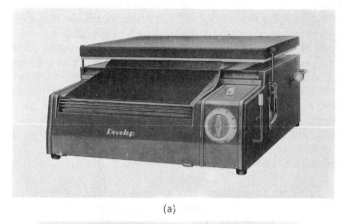

(a)

(b)

Figure 9.15. (a) Photocopier. The adjustable-height lid allows loose sheets or pages in books of any thickness to be copied. (b) The lid of this machine is hinged at one side and pressure is applied by hand. Books can be copied without loss of inner margin and the lid is removable to allow originals larger than the copying surface to be reproduced in parts (Courtesy Develop Kommanditgesellschaft DR. Eisbein & Co.)

obtained. When multiple copies of the original are required the above process is inconvenient and slow but since translucent copies can also be made these are suitable for reproduction by the diazo process.

A similar contact printing process has now been introduced in which a matrix paper is used. The matrix is placed in contact with the original in a copying machine and exposed to light. It is then floated in an activating solution and after twenty seconds is withdrawn and simultaneously brought into contact with a sheet of copy paper. The image is transferred as the sheets pass through a squeeze roller which removes excess moisture and the copy sheet is then separated from the matrix. The matrix may be reimmersed in the activator and the process repeated to give up to six copies. If larger numbers of copies are required, a translucent master may be made from two-sided or opaque originals which can be used for producing copies by other methods. Offset masters may also be produced directly by this process by bringing the matrix, after activation, into direct contact with a paper offset master. The master is then stripped away, the image etched if necessary, and then hardened with a special hardening solution. The plate is then ready for the offset duplicator.

Diazo process

The diazo process is used for the reproduction of a translucent original. The sensitised paper which contains a dye is placed in

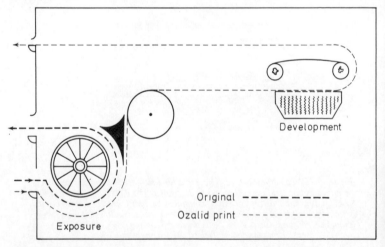

Figure 9.16. Basic principles of the diazo process (Courtesy Antara, New York)

contact with the original and passed into a machine in which the sheets are conveyed by a belt around a mercury lamp (*Figure 9.16*). The ultraviolet source activates the dyes on those parts of the paper not masked by the matter on the original. The paper is 'developed' in the machine as it passes across a container giving off ammonia vapour. Liquid developers are sometimes employed. The whole process may be conducted under normal lighting conditions and different colours and weights of sensitised copy paper, transparent foils, and linen tracing paper can be used.

REFERENCES

1 Lampitt, L. H., *Laboratory Organization*, Royal Institute of Chemistry, London (1935)
2 Vickery, B. C., *Classification and Indexing in Science*, 2nd edn, Butterworths, London (1959)
3 Collison, R. L., *Cataloguing Arrangements and Filing of Special Materials in Special Libraries*, Aslib, London (1950)
4 Rothchild, N., and Wright, G. B., *Mounting, Projecting and Storing Slides*, Universal Photo Books, New York (1956)

10
Organisation of demonstrations and exhibitions

Scientific demonstrations are an excellent means of introducing the elements of science to the public and scientific facts to students. They serve, too, to draw the attention of scientists to new methods and techniques and acquaint them with the work being done in fields of science allied to their own. The various types of demonstrations and exhibitions may be classified as follows:

1. Permanent and semi-permanent exhibitions.
2. Lecture and laboratory demonstrations.
3. Open-day exhibitions.
4. Exhibitions arranged by scientific societies.

PERMANENT AND SEMI-PERMANENT EXHIBITIONS

In teaching institutions specimens and other small items for display are often stored in boxes and cupboards, and from time to time are brought out to illustrate a lecture. This kind of exhibition undoubtedly serves a useful purpose, but because the specimens or models are sometimes poorly exhibited and may be seen only for short periods at a distance, the interest shown in them by students cannot always be described as enthusiastic.

PERMANENT EXHIBITIONS

Interesting specimens, samples, or models may be shown to more advantage, and can be better inspected if they are exhibited as a permanent display. They should be set out in glass showcases in museum style and in many teaching establishments special rooms are used for this purpose.

The showcases must be roomy and constructed in such a way that they can be viewed from all sides. To allow individual items to be removed for lecture demonstration the contents of the cases must not be congested. Specimen rooms of this kind are particularly valuable in biological laboratories.

Although such museums serve an excellent purpose, it may be said that they tend to take on a lifeless appearance and are not frequently visited by students. Also, because of the mass of specimens and the size of the collection, there is usually so much to visually digest at once that little of what is seen is retained by the visitor. This may explain the popularity of the special exhibitions which are arranged from time to time by the national museums and at which 'live' exhibits are shown. These exhibits create a great deal of interest and attract many visitors.

SEMI-PERMANENT DISPLAYS

For teaching purposes, semi-permanent exhibits presented in an instructive way have more effect on students than do those in the teaching museum. Considerable thought must be given to the presentation and style of the exhibits and they should illustrate clearly the principles of the subject with which they are concerned. They should also be provocative and sometimes problematical. This type of display induces the student to think about the exhibit and arouses his interest and curiosity. It also provides the necessary encouragement to seek further knowledge of the subject illustrated.

ARRANGEMENT OF EXHIBITS

Exhibits connected with science subjects require good arrangement and proper presentation. Good presentation requires efficient preparation and it is essential first to plan the exhibit on paper. The arrangement and size of the exhibit should be related to the size of the lettering and descriptive material.

The exhibition should be arranged in a colourful style and exhibits which are themselves colourless should be enlivened by colourful surrounds. On the other hand, others may be better exhibited using a quiet background. In some cases exhibits should be mixed so that one may enhance the other but the mistake of crowding them should be avoided.

DISPLAY OF EXHIBITS

The purpose of the display is that it should be seen by as many people as possible. Bacon[1] has described how wall cases may be easily made up to suit the size of particular exhibits and hung on the walls of the laboratory or in corridors. Since visual teaching is so important, however, it is worth while providing permanent facilities for this method. *Figure 10.1* shows glass-fronted wall cabinets 13 cm deep, which were set into the walls of a chemistry department, at

Figure 10.1. Glass-fronted wall display cabinet

little cost, during a building renovation. These shallow cabinets were designed by the author and consist of hardwood frames backed with pegboard and fitted with adjustable and removable glass shelves. The cabinets have sliding glass fronts. Strip lights were placed above them to provide illumination and the cabinets were positioned in a corridor along which students had to pass to reach the chemical laboratories.

Corridor exhibits may be arranged either by students as part of their training or by members of the staff. They may be of a general nature, or directly related to the subject matter of current lectures. To sustain student interest the contents of the cabinets should be changed frequently, and occasionally local industries and manufacturers should be approached to subscribe items of interest for exhibition. There is usually no lack of co-operation in this respect and the generosity of the donor may be suitably acknowledged by a card placed with the exhibit. If the displayed items are fully enclosed, they do not deteriorate, nor are they covered with dust as is usual when models are displayed. When removed they are in good condition for storing and for display at a later date to future students. The general effect of the cabinets, particularly if the exhibits are colourful, enlivens the corridors and makes them more cheerful and interesting. Such displays, too, are a source of interest to visitors and help to illustrate the methods of teaching employed by the department.

LECTURE DEMONSTRATION

The rapid advances in chemical theory do not appear to be accompanied by similar advances in lecture demonstration. Lecture experiments, as performed many years ago, are still with us today. An improvement in demonstration techniques has been more obvious in recent years, but there still remains a tendency to carry out lecture experiments in an old-fashioned way.

Lecture demonstrations should be up to date in the manner of presentation and investigative in character, and must convey facts to the student in a convincing fashion.

Lecture experiments are an essential part of teaching and do more than any other method of instruction to produce an enthusiastic response to the subject from the student. In the student's formative years, when he is as yet unacquainted with the names and manipulation of the various pieces of apparatus, the fact that demonstrations can be of value cannot be disputed. Provided the nature of the apparatus and the grade of experiments are in keeping with the seniority of the class to which they are demonstrated, then all undergraduate students can benefit from lecture demonstration. It has been said that students attending lecture demonstrations do better in written tests than do those who themselves perform the experiments in the laboratory.

PURPOSE OF LECTURE DEMONSTRATION

Before considering the best practical way of presenting the lecture demonstration, it is helpful to dwell upon the purpose behind it. This may be summarised as follows:

1. Apparatus is introduced to the student in such a way that he is able to understand its uses and limitations.
2. The student is encouraged to adopt by imitation the correct methods of use of the apparatus.
3. Time is saved and the number of teaching staff required to provide individual laboratory instruction is reduced.
4. The student is shown experiments which because of the danger, cost, or complexity involved he could not perform in the laboratory.

PREPARATION FOR DEMONSTRATIONS

The preparative work necessary for lecture demonstrations absorbs a considerable amount of time and in past years has interfered with the normal duties and research work of the lecturer. This was possibly the main reason for the decline of lecture demonstration. This has been overcome by the employment in almost every advanced institution, and in many schools, of trained laboratory technicians well versed in laboratory arts. It is to these technicians that the work of preparation may now be safely entrusted. The amount of time required for the preparation, however, remains unaltered and sufficient notice should always be given to the technician to allow him to prepare efficiently for the lecture to his own, and to the lecturer's, satisfaction.

PRINCIPLES OF DEMONSTRATION

The object of any experiment should be explained before its commencement. It should illustrate as clearly and simply as possible the principle it is meant to convey. This does not mean advanced pieces of apparatus cannot be incorporated in the experiment. For example, a student may not understand the inner workings of an oscilloscope yet may well be able to interpret what he sees upon its screen.

A demonstration will be more likely to be remembered if the experiments performed are few in number. An agile, efficient, and enthusiastic manipulation and an air of personal enjoyment are required to bring out the best response from the class.

Spectacular experiments are desirable, but these should not be performed regardless of others which, although less spectacular, present the facts more clearly. Nor should the demonstration take on the aspect of a conjuring act since it is meant to reveal as many facts as possible.

The timing of demonstrations is important and adequate instruction should be obtained beforehand. A good lecture demonstrator can start the experiment at the exact moment when the lecturer requires it, while a badly timed demonstration, or the failure of an inadequately tried out experiment, can distract the audience and spoil a whole lecture.

It is important that experiments be conducted so that the occupants of the front seats are in no danger. Charts with clear drawings of the experimental set-up will, if displayed, save the time which would otherwise be spent on blackboard artistry.

METHOD OF DEMONSTRATION

All apparatus should be large so as to be clearly visible to everyone in the lecture theatre. The experiments should be arranged on the bench in the sequence in which they will be demonstrated. The centre of the bench should be left clear for a working space and to allow the class to see the experiment clearly. The apparatus is more clearly seen from the student's point of view if the lecturer keeps his hands out of the way as much as possible. Overloading the bench serves only to create confusion in the minds of the audience and excess apparatus should be removed as the demonstration proceeds. A few spare parts should be provided to replace things likely to suffer damage during the experiments. Orderly arrangements do much to ensure the success of the demonstration and encourage the student to employ tidy methods in the laboratory.

AIDS TO DEMONSTRATION

Aids such as large models, charts, diagrams, specimens, visible indicators, and any others which help clarify the demonstration should be employed. Should it be necessary to resort to the blackboard for diagrammatic explanation, coloured chalk should be used.

Projection equipment for all sizes of film and slides and other visual aids, such as the epidiascope and micro-projector, should be available.

The film projector which provides moving pictures as opposed to the static ones shown by other projectors has added advantages as an instructional medium. Perhaps the most important advantage is that students may be shown demonstration experiments and equipment which are normally beyond the purchasing power of some teaching establishments. Films also allow experiments to be repeatedly shown and may be used to demonstrate slow or fast processes which are normally beyond the perception of the human eye. The recording of the film dialogue permits also accurate and timely observations to be made which accentuate the essential features of experiments to the students. Complicated operations are also better demonstrated and applied scientific methods, including the extraction of raw materials in faraway places, may be brought home to the student in the lecture theatre.

The films shown should be short ones as these maintain the interest of the student without cutting too deeply into his practical period time. All films should be previewed by the lecturer as many sponsored films are over-endowed with advertising and may waste the students' time.

The permanent impression conveyed by films is demonstrated by experiments which have been carried out on students. It has been reported that the performance, in conducted laboratory tests, of students taught by film has surpassed that of others who themselves did laboratory experiments in the normal way.

Undoubtedly the demonstration of analytical procedures and the correct use of laboratory equipment by means of film has proved to be an important educational aid. It also presents an economical teaching method because experiments may be repeated without time-consuming preparation and recurrent expense. The success of film teaching has prompted the preparation of complete courses on film, some complete in themselves and shown in conjunction with teaching manuals, others supplementary to normal teaching.

An overhead projector enables the lecturer to project notes and sketches directly on to a screen as he writes them on a pad. He can face his audience at the same time and has no need to darken the room. Modern projectors allow numerous changes of slides to be made, without the assistance of an operator, by remote control.

Nowadays, too, many instruments specially designed for lecture use are available. Voltmeters and ammeters are obtainable with the scales and working parts visible to the lecturer from behind and also to the class in front. Similar visible working apparatus is easily constructed and the overhead projection of numerous demonstration experiments using simple devices made in the laboratory workshop has been described by Alyea[2].

SETTING UP APPARATUS FOR DEMONSTRATION

For setting up apparatus for demonstration purposes, some lecturers prefer to submit a list of items to the technician and this should be checked carefully to see that no necessary items have been overlooked.

The setting up of material on the lecture bench is analogous to the setting of a stage. The apparatus must be arranged so that the vital parts are facing towards, and are nearest to, the audience. Arranged in this way it will appear to the class as if they themselves are carrying out the experiment. The analogy can be carried a little further, in that the apparatus not immediately needed should be placed, as inconspicuously as possible, on one side of the bench. If necessary, the apparatus not required immediately may be covered up, so as not to distract the attention of the class from the experiment in progress.

Assembled apparatus should be well-constructed and well arranged in the way that the lecturer would wish his students to erect it in the laboratory. Although glass assemblies should be firmly supported, as a rule only one clamp is required for this purpose provided it is correctly placed at the centre of gravity. A single clamp avoids strain in the apparatus and does not cloak the experiment. The weight of the apparatus must be over the centre of the foot of the retort stand.

Condensers and similar apparatus, if glass-to-rubber joints are involved, should be tested for leaks. All necessary safety precautions should be taken and may be purposely accentuated so as to be readily observed by the students.

Corks must be well-fitting and should not protrude too far into the neck of the apparatus or obstruct temperature readings.

Thermometers should extend the correct distance into necks of flasks, and in distillation experiments the bulb should be opposite the side-arm outlet.

All glass apparatus must be clean and dry. Liquid levels in pneumatic troughs and boiling flasks should be at the correct height and the stems of thistle funnels in gas generators should extend below the level of the liquid. The bends in glass tubing should be perfectly rounded so that the tube is not collapsed.

The wiring for electrical experiments should be short and easy to follow and coloured wires may be used to distinguish one from another. Voltmeters and ammeters with a single face should be arranged with their scales facing the class.

Projection apparatus and sound amplifiers must be tested and arranged before the class arrives. If slides are to be projected these should be arranged in order of projection. Wires leading from projectors must be neatly coiled away so as not to constitute a tripping hazard.

PREPARATION ROOM

The preparation room should be equipped and serviced as efficiently as any laboratory and should incorporate some light workshop facilities. It must also be spacious to comfortably house all the materials required for preparation work and the repeat experiments which should be left set up and stored in boxes or, if bulky, under dust covers. The use of numbered storage bins, in which apparatus is kept and checked by a card file system, has been described by Nicholson[3]. A list of contents and a short summary of the experimental details should be attached to the bins. Electrical experiments for repetitive use may be erected on frames. The disadvantage of the extra space required for the storage of set-up experiments is amply compensated by the time saved in their assembly.

If a book is compiled over the years containing full details of the various experiments, it will in time acquire an enormous value.

LECTURE THEATRE

The lecture theatre is the heart of any institution and pumps life into its various activities. Quite apart from normal class lectures, it is the centre for the extraordinary scientific meetings which can mean so much to a department. It should therefore be comfortable and efficient and its equipment must be of the best.

The best situation for the theatre is close to the main entrance. This allows easy access at times when the department is normally closed down. It should also be close to refreshment and cloakroom facilities.

The hall must be well provided with artificial lighting and with extra lighting above the blackboard and lecture bench area. The controls for the lights need to be in a position close to the lecture bench and double switching should be arranged so that the lights may also be controlled by the projectionist. Ceiling lights should be accessible for the replacement of tubes or globes.

Ventilation is extremely important and there should be no draughts. Air conditioning is best, but in any event ventilation must be adequate when the dark blinds are drawn.

The theatre should also have good acoustic properties and the advice of an expert should be sought in this connection.

The lecture theatre seating arrangements should be such that the seats are elevated by shallow steps in the floor. The height of the steps should be consistent with a good sight angle from each row of seats so that each person has a good view of the lecture bench. The width of the hall, too, should allow this. The seats themselves should be comfortable and those of the tip-up variety are best since they save space and allow easy access along the rows. They must be of solid construction, however, as they are more easily damaged than the solid wooden type of fixed seating. All seats must be accessible from a door at the rear or side of the hall to avoid the distraction of the lecturer and the embarrassment of the latecomer.

At the windows dark blinds must be provided and should be mechanically controlled so that the room may be darkened quickly. Modern methods allow this to be done by push button control.

In physics lecture rooms access should ideally be provided into the ceiling above the lecture bench and a trapdoor or other means of suspending objects over the lecture bench provided.

LECTURE BENCH

The size of the lecture bench must be in accordance with the large-scale apparatus necessary for purposes of demonstration. The services to the bench differ according to the technological subject for which they are intended. The services may include the following: electricity supply, a.c. or d.c., gas, water, vacuum, and in some cases steam and compressed air. Whichever services are used the outlets should not take up too much bench space and should be well distributed along the bench. The fittings should not interfere with the view of the audience, or cast shadows when projection equipment is used. Water fittings may cause an obstruction but a pattern is available which folds down into the sink.

It is desirable that the services have frontal controls situated on the lecturer's side of the bench. Gas cocks and recessed gas outlets may also be on the lecturer's side of the bench but in both cases a sufficient bench overhang should be allowed.

Provided a preparation room is adjacent, large cupboards in

lecture benches are unnecessary. A small cupboard and a few drawers, in which to store the many small items which may be required during a lecture, are useful. The side of the lecture bench nearest the audience may be glass-fronted above the normal kicking level and this can be used as a specimen display cabinet.

A fume cupboard is necessary for chemistry lecture demonstrations. These are difficult to situate in a position where the audience can easily see what is going on inside the cupboard. To overcome this difficulty a fume hood may be used on the lecture bench itself and the fumes extracted from it through a duct in the bench top.

OPEN-DAY EXHIBITIONS

It is the practice of educational establishments occasionally to invite the public to see the institution and the work done by the students. Open-day exhibitions may be regarded by the staff as an unnecessary interruption of their normal duties, or may on the other hand be entered into in a spirit of enthusiasm. This spirit, which ensures the success of the exhibition, is the right one and the time spent on preparing the exhibits will be found to be in itself instructive and rewarding.

The benefits to be gained from the open day, by all concerned, are numerous. The institution itself benefits in prestige by capturing the imagination of prospective students, and by showing the general public the facilities it has to offer.

PRELIMINARY ORGANISATION

Organisation for the open day necessitates a preliminary meeting of all the laboratory and teaching staff. Only the broad outlines of the departmental effort are established at this meeting and various responsibilities allocated to sections within the department. A further meeting should then be arranged. This procedure allows time for all those taking part to assess their equipment and the possibilities of assistance, and to crystallise their ideas.

At a subsequent meeting the various proposed exhibits and experiments should be discussed in detail to avoid any overlapping of ideas. The various sections may find, too, at this stage, that they are able to assist each other by the loan of apparatus and materials.

To ensure a co-ordinated effort and continuity of presentation one person must be made responsible for the general arrangements and requirements.

GENERAL REQUIREMENTS

Many of the materials required are of a similar nature and range in variety from drawing pins to bouquets. It is expeditious and economical if some items are purchased in bulk and distributed as required.

Invitation cards

The printing and addressing of invitation cards may be dealt with by the general office. It should be requested that all invitations be acknowledged and this allows an estimate of the possible number of visitors to be made so that adequate refreshment and other facilities can be provided.

Invitations to schools

Local schools should be invited to send their students to see the exhibition. These attendances should preferably be made at convenient times during the day, leaving the evening free for adult visitors.

Legends

Irrespective of size, legends on the notices for exhibits should be lettered in the same style and these are more likely to be satisfactory if one competent artist is given the task of preparing them. The various exhibiting sections should be advised as to whom they should contact for the preparation of their cards and should be required to submit lists of legends and indicate the size of cards they require. The size of the cards should be compatible with the size of the equipment described.

Direction notices

Since the visitors will be in unfamiliar surroundings, plenty of direction notices should be displayed. It is also advisable to recruit a number of students to act as guides and assist guests. Special visitors should be met at the entrance by members of the staff. The guides and demonstrators should wear some form of identification, the nature of which should be stated in the programme.

Programmes

Although a general programme may be issued, it is also advisable to make available to visitors a departmental one which gives full details and the locations of the departmental exhibits. The locations should be indicated by room number and by the number of the stand if applicable. The name of persons responsible for the exhibits in the various sections should also be given as should the times of the commencement of non-continuous items such as films, glass-blowing, and other demonstrations.

Condition of the department

The general condition of the department cannot escape notice and all normal laboratory facilities and apparatus must bear close inspection. Some of the main items of everyday equipment may even be given prominence by attaching notices to them to explain their use.

NATURE OF EXHIBITS

All exhibits must be safe, easily visible, and capable of sustaining interest. They should in general be simple and easily understood. A large proportion should be 'live' exhibits or working models and whenever possible experiments should be spectacular and thus of more interest to visitors than immobile equipment or displays. Most of the experiments should be continuous or easily repeated and conducted experiments should not take too long to demonstrate.

Within reason, the visitors should be invited to operate certain exhibits for themselves and there should be some exhibits of this kind. On the other hand, some exhibits actually need to be protected from the attention of visitors and on these a 'do not touch' card should be displayed.

SUGGESTED EXHIBITS

It would be well-nigh impossible to list the numerous possible exhibits which could be shown. It is to be hoped that the following few suggestions will serve to illustrate the interesting possibilities of the open-day exhibition.

Films

Films may be borrowed free of charge from various companies and from certain film libraries. The film-show absorbs a suitable proportion of the visitors and prevents overcrowding in the laboratories. What is more important, it allows an opportunity for a short relaxation from the tiring round of exhibits. A limited time should be allocated to each showing and it is preferable to show three short films of varied interest rather than one long one. If two projectors are available both should be used as this saves time in changing the reels.

Visual aids

Many visitors are interested in visual aids from the point of view of either teaching or home use. A working display of projectors, such as filmstrip and micro-projectors, lanterns, epidiascopes, and similar equipment, is itself an absorbing exhibition and is made more so if the workings of the projectors are verbally explained.

Glassblowing

The interest which glassblowing techniques hold for everybody need not be elaborated. It is sufficient to point out that in fairness to the operator the demonstration should take place at specified intervals.

Liquid air

Under safe conditions and proper supervision, liquid-air demonstrations leave nothing to be desired in the way of entertainment. The display of freezing processes and the hammering to pieces of grapes and rubber tubing are in themselves of sufficient interest to make visitors feel that their visit was worthwhile.

Teaching demonstrations

A demonstration on the lecture bench can also provide unparalleled entertainment, but a suitable subject incorporating a number of spectacular experiments should be chosen.

Students' experiment

A pleasant diversion from the more serious displays can be provided by allowing students to set up their own 'rag' experiment. This kind of experiment, which may introduce alchemic experimental methods and usually consists of a complicated mass of meaningless apparatus which leads nowhere, may also have an amazing end product. In performing the experiment every safety regulation may (in a harmless way) be contravened and an enjoyable time is had by all.

Novel displays

No open day is complete without the inclusion of a number of novel experiments. Physical apparatus lends itself particularly well to this kind of exhibit. Although the experiments listed as examples have a physical or chemical flavour, it requires but little imagination to think of similar ones applicable to other science subjects and many are described in various textbooks.

1. The everflowing water tap.
2. The growth of crystals.
3. A coral garden.
4. Freezing of mercury by evaporation.
5. The colour of flames (analysis).
6. Chemical flowers.
7. The electrolysis of water.
8. The solubility of ammonia.
9. The conductivity of electrolytes.
10. A lead tree.
11. The preparation of ozone.
12. The purification of mercury.

DISPLAY OF EQUIPMENT, SAMPLES, AND SPECIMENS

By the erection of suitable apparatus various examples of the day-to-day processes employed by students in the laboratories may be illustrated. The experiments may be set up as immobile exhibits with suitable descriptions while others such as reflux or continuous distillation processes may be left to work continuously.

Prepared specimens which have been mounted by students may also be shown and in this field the biological subjects have the advantage. Similarly, engineering and other departments can show finished pieces of work of particular merit which have been made by students.

The sets of apparatus issued to students' lockers and other items of general apparatus, such as magnetic stirrers, are of particular interest to parents and may be of special interest to visiting teachers.

MORE ADVANCED EXPERIMENTS

To cater for those visitors who are interested in the higher work of the institution, more advanced experiments and equipment should be shown. Demonstrators are needed to explain this kind of exhibit, which might, for example, involve spectrographic, crystallographic, photomicrographic, or other specialised techniques.

In addition to the departmental exhibits scientific firms and local industries may also be asked to contribute material, and in some cases to participate in the exhibition.

EXHIBITIONS ARRANGED BY CHEMICAL SOCIETIES

Exhibitions and conversaziones arranged by scientific societies need special organisation which, owing to the diverse nature of the exhibitors, is not without its difficulties. The exhibitors are usually not directly connected with one another and some may be completely unknown to the organiser. Scientific societies, too, often make a practice of regularly changing the site of their exhibitions, which may be held at a different scientific institution each year.

The duties of the organiser probably include the organising of a trade show and the various scientific firms invited must be written to requesting their participation. The firms should be carefully selected according to the type of apparatus in which they specialise, so that the variety of equipment on show will be interesting to scientists and technicians working in different scientific fields. When replies have been received from the firms the accepting parties should be again written to and thanked for their co-operation. These firms, and society member exhibitors, are then sent a questionnaire form which they are asked to complete by indicating the amount of bench space and the number of gas, water, electrical, and other service points they will require, plus any special requirements.

When all the information has been obtained, the organiser is in a position to allocate the bench space, and for this purpose he must have a first-hand knowledge of all the facilities and space available.

It is as well, at this stage, to make a rough plan of the area, including the layout of the benches and the position of the service outlets. The various exhibitors are allocated space in accordance with their needs and the available facilities. It is usually found that the number of service outlets, particularly electrical sockets, is insufficient and temporary distribution methods will have to be arranged.

When everyone's needs have been met as far as possible, the exhibitors are informed of the provisions which have been made for them and are given a number which refers to their allocated space. On the day of the exhibition the benches are similarly numbered.

The compilation and dispatch of the invitations, the arrangements for speakers, refreshments, cloakroom facilities, and extra technical assistance, represent a few more of the problems which confront the organiser, but with the co-operation of colleagues employed in the building in which the exhibition is to be held the conversazione is assured of success.

REFERENCES

1 Bacon, E. K., 'Displaying Chemical Exhibits', *J. chem. Educ.*, **15**, 219 (1938)
2 Alyea, H. N., 'Tips in General Chemistry Tested Overhead Projection Series', *J. chem. Educ.*, **39**, 12 (1962)
3 Nicholson, D. G., 'Storage of Lecture Demonstration Equipment', *J. chem. Educ.*, **32**, 138 (1955)

11
Organisation of the work and training of laboratory technicians

ENGAGEMENT OF STAFF

INTERVIEWING

The successful training of laboratory technicians depends mainly upon the quality of the trainees. The manner and accuracy of selection of potential technicians are therefore very important. The responsibility for the engagement of new technical staff rests usually with the laboratory supervisor, who should be fully acquainted with the method of interviewing persons seeking employment. The methods are based on the intelligence of the interviewer, the exercise of his skill in dealing with his fellow beings, and a sound knowledge of the work for which the candidate is applying.

A successful interviewer must have a kind and understanding manner in order to gain the respect and confidence of a person being interviewed. Much depends upon whether or not the interviewer has himself had a wide experience and the extent to which he is aware of his own thoughts about life. He should not allow his judgement to be swayed by preferences and his findings must be based on an objective and methodical assessment of the applicant. The interviewer must appear relaxed so that candidates are not given the impression that they are encroaching on his time.

Sincerity gives the interviewer a great advantage provided he behaves naturally and does not endeavour to cloak his true identity in artificial cordiality. A sincere approach is certainly fairer to the candidate, with whom he may eventually have to work. If the interviewer lacks specialised training this may be largely overcome by adequate preparation and by reading literature on the subject. The successful interviewer must also be acquainted with information concerning present-day educational and training systems, the various types of schools, apprenticeships, training schemes, school leaving examinations, and so on. This is particularly important at the present time when the educational system is undergoing vast changes.

The lack of suitable applicants may present a serious handicap to the interviewer in his task and the number of candidates should be sufficient to allow a reasonable choice and their previous education and training should be such that they are fitted for the job.

The chief purpose of the interview is to observe the candidate's appearance and conduct and by questioning him to find out his capacity for making a success of the job. It also affords an opportunity to assess him personally and to obtain a knowledge of his achievements. It can also be estimated whether or not he is sufficiently intelligent and whether his personality is suitable for the job.

More time is available for such observations if factual information is gathered together beforehand. For this purpose a well-designed application form provides a wealth of information. The application form should tell the interviewer the nature of the candidate's present employment, his qualifications, previous employment, interests, education, marital status, and reasons for wanting the job. Should a large number of applications be received, a suitable short list of candidates can be prepared. It is equally helpful to candidates if they are sent a descriptive leaflet giving details of the conditions of employment.

The suitability of candidates is assessed by considering the following factors:

1. Circumstances (financial, social, geographical).
2. Physical characteristics (health, smartness, bearing, general appearance).
3. Attainments (work, school, games, hobbies).
4. General intelligence.
5. Special aptitudes (e.g. mechanical dexterity, literary ability).
6. Interests (intellectual, practical, social).
7. Disposition (attitude to himself, his work, others).

Circumstances

In fairness to the candidate and to the interviewer, it must be ascertained whether or not the candidate can afford the job. It is not unusual to find candidates whose keenness obscures their financial judgement. The interviewer must be sure that the salary offered is adequate to meet the candidate's travelling and other expenses and is sufficient to keep him free from worry while in the job. The hours of work when added to his travelling and study time must allow him sufficient opportunity to mix socially with the other laboratory staffs and his background must also be suitable for this.

Physical

The candidate must be physically capable of the work he will be required to carry out and he must not be likely to suffer ill effects from it. It is quite reasonable to suppose for instance that the work in the laboratory may require freedom from colour blindness, a condition from which a particular candidate may suffer. Similarly, skin complaints such as dermatitis may well be aggravated by laboratory work.

Attainments and aptitude

It is possible that although the applicant may exhibit outstanding intelligence he has not had the level of education necessary in the job.

Similarly, although the applicant may have excellent aptitudes they may serve him no useful purpose in his work. An excellence in music may have no bearing on a successful scientific career.

Interests

Social interests, such as membership of a youth movement, often serve as a guide to a person's qualities and usually indicate a reliable person and one used to team work. The candidate's intellectual and practical interests may also reflect his interest in the field in which he is seeking employment.

Disposition

The person selected must have a disposition which suits the work. A strong self-interest would be of no help in an appointment which calls for a very social person and the possible effects upon his fellow employees in this connection would be well worth consideration. The interviewer should also assess the candidate's attitude to the work and his capabilities for supervising others should this be required of him.

Arranging the interview

It is extremely important that the interviews be properly arranged. When the applications are received, acknowledgements should be sent directly to the candidates. A short list of applicants should be prepared and due allowance made for those who may later drop out. Suitable times of appointment should be notified to the candidates, which should be carefully calculated to allow at least thirty minutes to each person. This allows time for writing up notes about them immediately they have been seen and before the next person is called in.

Once the timetable is decided upon it must be very rigidly adhered to, so as to ensure that candidates are not kept waiting and that each has a similar length of time allocated to him.

On arrival the candidates should be received and accommodated in a cheerful waiting room in which some reading matter should be provided. The room in which the interview is conducted should also be cheerful and well lit and should have reasonable seating accommodation. Complete privacy should be ensured and steps taken to prevent any interruption.

As he enters the room the visitor should be greeted by name and the employment he seeks verified, which avoids any possibility of error. The interviewer should also make his name and position known.

Before entertaining the applicant, the written information which he has provided must be assimilated, for this gives the impression that the person's history and background are well known to the interviewer and allows any matter not clearly explained by the candidate on the form to be adequately cleared up. A few general questions not calling for lengthy replies allow the candidate time to settle down and enable the interviewer to establish an easy relationship with him. This encourages the applicant to talk freely.

The applicant should be encouraged to talk and the conversation should be guided by remarks from the interviewer into the channels which allow him to gain information. One way of doing this is to ask short leading questions which allow plenty of scope for a reply. The applicant should not be inundated with questions and it is better to ask a few and be patient in allowing adequate time for a reply. The questions asked should not savour of suspicion or be provocative in their nature. It is well to avoid matters affecting an applicant's personal life, and any such details important to the selection should be sought with care and tact. The candidate's ambitions are important and he should be asked his future aims in life.

At the conclusion of the interview, the applicant should be invited to ask questions concerning the job and be told when he might expect to learn the result of the interview. Any confidences given to the interviewer must be fully respected.

Notes taken during the interview should be discreetly concealed from the candidate and this saves him the embarrassment of having to avoid glancing at them. The notes made should be brief and should not be written when a candidate is speaking of his private life. It allows the candidate more time for a considered reply if the interviewer asks questions while he is making notes of the previous reply.

Expert interviewers resort to a rating scale which assists in assessing the candidate and marks are awarded under certain headings. Examples of more general headings are appearance, bearing, clarity, knowledge of the job, alertness, and attitude towards others.

From the observations and notes made at the interview the suitability of the candidate can now be accurately gauged. In the actual summing up, however, the greatest care must be exercised in determining the degree of honesty of the candidate. If his past record has been properly ascertained, it is not difficult to see how he is likely to proceed in the future.

The interviewer should remember that few people do themselves justice in an interview and modesty and awkwardness are important factors affecting behaviour. When achievements are compared, the respective ages of the candidates should be borne in mind and due distinction drawn between brilliance and perseverance. Although either of these could be important to the work, one without the other might not be enough.

When it is difficult to make a decision about a candidate it is well to imagine him in some responsible aspect of his future employment. Impartiality is absolutely necessary and candidates should be judged solely in terms of the vacancy which they will be required to fill.

Any difficulties experienced in obtaining replies to advertisements should not be allowed to influence the interviewer in his choice so that he engages persons having little chance of success in the job. Such persons should never be accepted for such a choice would, in the long run, prove detrimental to the working of the whole department. It is better to be patient, choose wisely, and be happy afterwards.

ORGANISATION OF LABORATORY WORK

The way the work of the technical staff is organised depends upon the size and nature of the particular institution. Larger institutions are normally composed of departments within each of which a responsible technician is in charge of the work of the technical staff. In larger institutions those departments which employ suitable numbers of technicians have a laboratory superintendent in charge, and in this case the establishment for personnel may be such that chief technicians assume charge of the various major sections within a department. Other grades of technical staff are senior technicians, technicians, and junior technicians. In smaller institutions the organisation of the work of the technical staff may be the responsibility of a chief technician or senior technician. The seniority of the appointment is usually governed by the number of technical staff allocated to the institution by the employing authority.

The organisation of the work within the departmental sections such as the specialised laboratories, photographic darkroom, glass-blowing room, and engineering workshops, should be left to the responsible laboratory technician. The laboratory superintendent in charge should, however, be well acquainted with the abilities and experience of the technicians in the various sections, so that he may confidently allocate the work of the department. The rate of progress of the work within the sections should be carefully assessed to ensure it constitutes a satisfactory output in relation to the work of the department as a whole. Efficient organisation requires fair but firm discipline on the part of the laboratory superintendent, who should also have a good knowledge of the various personalities of his staff and be able to appreciate their difficulties. He should manage the introduction of new people to the department in such a way that any injustice to existing staff is avoided.

DISTRIBUTION OF WORK

Any requests by various experimental groups within the department for work to be done should be submitted to the laboratory superintendent, so as to maintain a fair distribution of work. This system also permits advice to be given to groups when particular parts of the work which they require to be done may be better carried out by sections other than that which they have nominated. Priorities can also be assessed and unnecessary personal contact and monopolisation or sidetracking of individuals in particular sections may be avoided. A further advantage of the system is that it allows the supervisory technician to advise the departmental head on claims placed by sections for new equipment or machinery. Whenever work is given to sections, the manner of its execution and its allocation to individuals should be left to the responsible technician in charge of the section.

In every department certain general routine duties are necessary for its smooth running. In fairness to all concerned these should be allocated by a rota system, notice of which should be exhibited on the staff notice board. Any general duties of a special nature appertaining to individual sections should be allocated by the responsible technician to his own staff.

Cleaning duties

A rota of cleaning duties should be published for the benefit of the cleaning staff. This also indicates to other staff the times at which their rooms and laboratories will be attended and precautions for the safety of the cleaning staff may be taken. On engagement all cleaning staff should have a short training under supervision in the laboratories.

ORGANISATION OF TRAINING

Training facilities for laboratory technicians may be provided (a) on the job or (b) in technical colleges.

To organise successfully a comprehensive scheme for training on the job, sufficient numbers of trainees are necessary. It follows, therefore, that organised schemes of training are generally restricted to establishments which employ large numbers of technical staff. In many industrial firms and in scientific establishments of national status internal training schemes are widely employed. Such schemes are mainly devoted to teaching the trainees the organisation and the special work relating to the establishment. The trainees are normally also required to attend the courses in laboratory technicians' work held at a technical college to prepare them for qualification by examination.

With all due respect to the attendant difficulties of applying internal training schemes, there has in the past been little attention paid towards organising these in educational institutions. In most establishments technical staff were encouraged to attend academic classes held within the institution. So those who may have been happy to have been permanently engaged on technical laboratory work left the service to pursue academic qualifications. In recent years the enormous turnover in technical staffs employed in educational establishments led some educational authorities to recognise this situation, and junior technicians are now encouraged by them to attend part-time day courses in laboratory technicians' work at technical colleges.

In spite of this advancement, the need for on-the-job training is still very important. This is particularly necessary for new entrants, and wherever possible new staff should be given organised instruction, even though it may need to be restricted to a departmental basis.

It is important that the new entrant be introduced to the particular work of the establishment through the medium of a well-informed instructor. To establish new staff in this way, rather than by merely turning them loose into a laboratory, is beneficial both to the individual and to the establishment.

TRAINING JUNIOR TECHNICIANS

Reception of new arrivals

One function of the technician in charge of the department is to receive new technical staff. He should give them all necessary instructions and make arrangements with the appropriate office staff with regard to matters affecting their employment. The history and geography of the establishment, and the relationship between the different departments, should also be explained. Information should also be given in respect of the refectory and such social facilities as may be available. The new entrant should also be introduced to existing staff, in particular to those responsible for his subsequent advancement. The courses of study available to him should be explained in detail and any arrangements which have been made for him to enrol for these. By the attention given to these and other matters the entrant is assured that the institution has an interest in his welfare. Finally, the newcomer should be handed over to a senior person from whom he will receive his first instruction.

Responsibilities of the instructor

Since young persons beginning their laboratory career do so at a transitional stage in their life, the senior staff given the task of instructing them bear a heavy responsibility. The trainee is undergoing rapid physical and mental changes which are further complicated by a change from school and home life to a completely new environment. He may therefore adopt a rebellious and aggressive outlook, which is psychologically explained as being due to a feeling of insecurity. Physical changes, such as change in weight and height, also cause glandular upsets which result in lethargy and clumsiness. Because of these changes, the adolescent, as he rapidly develops, is in need of help and guidance from his seniors to direct him along the path upon which his scientific abilities lie. Particular attention, too, must be paid to his moral and social development. Persons responsible for the training of junior technicians are in some respects at an advantage, since young persons tend towards hero worship. By his example and fairness, the instructor may therefore easily win the loyalty and affection of the trainee. By poor handling he may, of course, equally well lose it. A little kindness on his part will go a long way with young people, who are often in need of kind words, to soften the sting of failure or to stamp in the mark of success. Kind but firm treatment is required from the instructor to bring out the best in his students.

Duties of the instructor

Since the abilities of the junior technician are latent, the first duty of the instructor is to develop them. This may be done by the sincerity, conviction, and enthusiasm of the adult leader. Good instructors are not necessarily those with the longest experience, or those most expert at the job. Some people are happiest only when they are teaching others, and it is more likely that this type of person will in fact make the best instructor. The selection of the instructors is therefore in itself of great importance and they should be chosen for their willingness and readiness to impart their knowledge.

Objects of training

The main object of training is to impart facts and to make the trainee more efficient in his work. His habit of thought should be developed, as should his sense of loyalty and responsibility. A strong sense of moral responsibility is particularly important in scientific laboratories, where expensive apparatus and prolonged experimental work may be easily ruined by the lack of it. In the initial stages of the training, it should be the object of the instructor to acquaint the junior staff with all the dangerous aspects of the work and the precautions which should be taken to prevent accidents.

Conditions for learning

Instruction is more readily absorbed if the observer is comfortable and relaxed. If practical work is to be done seated then the seating provided should be comfortable. Similarly, other conditions such as lighting and ventilation should also be adequate.

Method of instruction

Good instruction involves first obtaining the full co-operation of the trainees, and to increase the relevance of the instruction they should be made fully aware of its importance. All the methods used should aim at making the instruction as vivid as possible and linking it in some way with the trainees' previous knowledge. In the later stages of instruction this may be accomplished by recalling previous work done.

At all times a high standard of cleanliness, tidiness, and care in working must be insisted upon. This helps to maintain a high standard of technique and prevent accidents.

When a full explanation of the work has been given and the person in charge is fully satisfied that the trainee knows what is required of him, then the trainee should be given an opportunity to carry out the exercise himself. It is important that the trainee be kept active in this way and any tendency on the part of the instructor to monopolise the exercise himself should be avoided. When the student is seen to be proceeding along the right lines, he should be left alone to practise the technique.

Although methods of repetition may be applied to increase the beginner's skill, these may be made more interesting by variations in the nature of the task and the tools used for its execution.

Whenever practicable, items of equipment should be dismantled to demonstrate their construction and workings. To arouse his curiosity and to spur him to extra learning, the trainee should be allowed to assist in dismantling the apparatus. Only through the experience of handling the equipment can he make progress. Competition between the trainees should also be encouraged, provided that due care is taken to see that no individual is allowed to feel superior or inferior as a result.

The trainees should from time to time be made aware of the progress they are making and occasionally a few words of praise should be given to encourage them.

Throughout the training, the various instructors should assess the performance of the individual junior technicians and the assessments should be correlated and used as a guide for the eventual placement of the trainee in the department.

Training for senior staff

The further education of senior technical staff should not be neglected. To maintain the full efficiency of its staff, the employing authority should give every encouragement to senior persons to attend specialised or refresher courses which enable them to keep abreast of new developments and techniques.

PROFESSIONAL ORGANISATIONS FOR TECHNICAL STAFFS

Although the work of all technicians in the various branches of science has certain similarities, the duties tend to resolve themselves into three main categories. Accordingly, technicians may be eligible for membership of the Institute of Science Technology, the Institute of Medical Laboratory Technology, or the Institute of Animal Technicians. These three professional bodies, established to meet the academic requirements of technicians, have as their common objective a desire to provide suitable training, examinations, and qualifications for technical staffs. In certain instances the work of some technicians brings them into contact with more than one branch of science and it is not uncommon to find such persons qualifying for membership of more than one of the above-mentioned institutes.

The functions of these three institutes are very similar and each has set up its own branches which hold regular meetings and scientific discussions and organise visits to places of interest. Each organisation further stimulates the interest of its members by publishing a journal devoted to scientific articles.

The training of the members belonging to these three bodies is largely gained in the establishment in which they are employed and by attending part-time day and evening classes conducted by local education authorities.

INSTITUTE OF SCIENCE TECHNOLOGY

The formation of the Institute of Science Technology satisfies a need for an organisation to advance the professional standing and efficiency of science laboratory technicians.

The Institute, formed out of the membership of the Science Technologists Association, has built up a world-wide membership. In 1959 the Institute of Science Technology and the City and Guilds of London Institute announced that they jointly proposed to rationalise their examination and qualifications for science laboratory technicians and that one series of examinations would be conducted by the City and Guilds on behalf of both Institutes.

Members of the Institute are to be found in all branches of science but most of them are employed in educational laboratories, schools, technical colleges, colleges of advanced technology, and universities, while others are engaged in industrial and National Health Service laboratories.

Student membership

Student membership of the Institute may be awarded to applicants who are normally not less than sixteen years of age and who are in scientific employment approved by Council and who:

1. possess four GCE 'O' level passes or four CSE Grade I passes, two of which must be scientific subjects and one a test of the English language; *or*
2. possess two subjects at credit level and one pass from the G* course; *or*
3. possess a qualification of equivalent standard approved by Council; *or*
4. have gained entry to an approved technician course of study.

Ordinary membership

Ordinary membership of the Institute may be awarded to applicants who hold one of the following qualifications and who have at least *three* years' experience in scientific employment approved by Council:

IST/CGLI Ordinary Certificate
IST Certificate
Ordinary National Certificate in an appropriate subject
Ordinary National Diploma in an appropriate subject
* CGLI Mechanical Engineering Technicians' Certificate, Part II
* CGLI Electrical Technicians' Certificate, Part II
* CGLI Chemical Technicians' Ordinary Certificate
* Biological Technicians' Ordinary Certificate (Bristol Technical College under special relationship with the CGLI)
* CGLI Photographic Technicians' Certificate (345)
* CGLI Radio and Television Servicing Final Certificate
Diploma for Technicians of the North Cheshire Central College of Further Education (under special arrangement with the Institute of Science Technology)

or

Another examination approved by Council acting on the assessment of the Examinations Board

Associate membership

Associate membership of the Institute may be awarded to applicants who hold one of the following qualifications and who have at least *five* years' experience in scientific employment approved by Council:

IST/CGLI Advanced Certificate
IST Diploma
Higher National Certificate in an appropriate subject
Higher National Diploma in an appropriate subject
* CGLI Mechanical Engineering Technicians' Certificate, Part III
* CGLI Electrical Technicians' Certificate, Part II, with two endorsement subjects
* CGLI Chemical Technicians' Advanced Certificate
* Biological Technicians' Advanced Certificate (Bristol Technical College under special relationship with the CGLI)

or

Another examination approved by Council acting on the assessment of the Examinations Board

Fellowship

Fellowships may be awarded to Associate members of not less than two years' standing who are not less than twenty-five years of age.

The Council of the Institute when making such elections take into consideration the professional skill, knowledge, length of service, and administrative ability of the applicant. An Associate member may support his application by:

1. Taking a special examination.
2. Presenting a thesis or dissertation in accordance with the regulations laid down by Council.

(Note: Qualifications marked with an asterisk will be reviewed when the revised IST/CGLI examinations become operative.)

Examinations

The existing courses leading to the examinations of the City and Guilds of London Institute in conjunction with the Institute of Science Technology consist of an ordinary certificate course and an advanced certificate course.

Science Laboratory Technicians Ordinary Certificate

The City and Guilds of London Institute is responsible for the organisation and conduct of the examinations held in approximately fifty centres in the U.K. and some overseas countries. The age for admission for the examination is normally not less than sixteen years. The course is a part-time one of three years' duration or the equivalent in, for instance, block release schemes.

The subjects taken for the examination are Science and Laboratory Methods and Equipment which provide a very broad and firm foundation on which technicians can build before proceeding to the final examination.

Science Laboratory Technicians Advanced Certificate

At the advanced level, Laboratory Procedure and Administration must be taken by all candidates. The student also has a choice of Laboratory Techniques which include the following: Physics, Chemistry, Biology, Geology, and General Laboratory Techniques.

In both the certificate and the advanced certificate examinations, first- or second-class certificates, in accordance with the standard achieved, may be awarded to successful candidates.

Note that since the Government has embarked on a policy of

expansion of technical education the rationalisation of the examinations of the two Institutes is of singular importance. The enormous increases in the number of technologists will undoubtedly result in even further increases in the number of technicians who will be required in industry, in research, and in educational institutions.

INSTITUTE OF MEDICAL LABORATORY TECHNOLOGY

The parent body of the Institute of Medical Laboratory Technology was the Pathological and Bacteriological Laboratory Assistants' Association formed in 1912. The Association expanded its activities and in 1921 was successful in introducing a scheme of examination certification for Medical Laboratory Technicians. Arising out of the Association the Institute was incorporated under the Companies Act in 1942 and since that time its membership has grown rapidly until today it embraces over 10 000 technicians.

In many countries throughout the world, members of the Institute are employed in hospital, public health, blood transfusion, university, and research laboratories. Their prime function is the control and treatment of disease and this work, which demands both skill and devotion, also involves considerable responsibility.

INSTITUTE OF ANIMAL TECHNICIANS

The Animal Technicians Association was founded in 1950 as the result of the recommendations of a committee set up to consider the training of technicians in animal husbandry and animal techniques and under the guidance of the elected Council of the Association the Institute of Animal Technicians was formed in 1965.

Technical duties

Animal Technicians are devoted to animal care in all its aspects and contribute in great measure to the success of animal experimentation in laboratories in many parts of the world. The breeding of animals and the care of experimental animal houses and the provision of skilled technical assistants are now recognised as important functions in the furtherance of biological research.

FUTURE OF LABORATORY TECHNICIANS

Although enormous strides have been made in the last twenty years in the production of highly trained technical staffs much still remains to be accomplished. The raising of the school leaving age coupled with greater opportunities for young people at the universities may tend to accentuate rather than relieve the present shortage of technicians. The expansion of schools, colleges, universities, and industry is itself promoting innumerable vacancies which are becoming progressively more difficult to fill. In addition, the extra funds now available to higher educational institutions as a result of the Robbins Committee report, and the consequent reorganisation of practical courses, have created even more careers for technicians.

The young person entering laboratory service today as a student technician in a higher educational establishment can expect to embark immediately on a training scheme of high standard. Governing bodies and local education authorities not only encourage on-the-job training but also arrange that the student performs duties of a varied nature, which means that he is introduced to the widest possible spectrum of scientific work. The City and Guilds courses, too, are purposely geared to this end. In addition local education authorities offer day release up to the age of twenty-one for courses of study which lead to higher qualifications.

The present shortage of technicians has also resulted in quicker promotion, and at present student technicians can expect promotion to technician grade at the age of about twenty-one. This is subject, however, to the attainment of certain qualifications such as the intermediate certificate of the City and Guilds of London Institute in Laboratory Technicians' work. Possession of the advanced certificate of the City and Guilds or its equivalent normally results in promotion to a senior grade at the age of twenty-four or twenty-five. Further advancement depends on additional experience and specialist qualifications.

Today's technicians can therefore look forward to a continuous programme of learning and advancement in step with the complexities of their work arising from the increasing use of instrumentation. The technician must not only be capable of operating new equipment, but is expected to know its inner workings and the correct maintenance procedures.

In spite of the undoubted advantages of day-release opportunities allowed by local educational authorities and others, it is desirable that expanded facilities for training technicians should be provided which are tailored to suit the general trend of advancement in the country. It has already been suggested that technical work could be made more attractive by overcoming the necessity for school-leavers to take up unskilled appointments for the dubious advantage of attending part-time day and evening classes. Such classes, in view of other attractions, may now prove to be insufficient incentive for young persons to follow their chosen vocation. It is thought by some that the solution is to provide full-time pre-employment courses which would provide young aspirants with a chance equal to that of persons pursuing academic careers. It would also provide the technician with a theoretical and practical laboratory education before he takes up a position in laboratory service, and this would also benefit the employer.

Whatever the answer, it is certain that technicians themselves are fully aware of the situation, and through the medium of their own Institutes will play a great part in bringing about the desired changes.

The future for technicians is assured and a highly interesting professional occupation is now offered to young persons who desire to pursue a technical career in laboratory service.

12

Elements of experimental procedure

It is comparatively easy to conduct an experiment following predetermined patterns and produce an end result. Nowadays technicians are also interested in, and are required to understand the scientific reasoning involved in, any experiment they may be asked to perform; indeed in many cases they are required to formulate the experiment and interpret its findings entirely to the satisfaction of a senior scientist.

For this reason, I have included a chapter on this subject.

The enquiring mind will ask:

1. What is the experiment set up to do?
2. What techniques are available in the literature?
3. What equipment is best suited to the experiment?
4. Is it to be a comparative or an absolute experiment?
5. What controls are necessary?
6. What statistical treatment of the data found will be necessary?

It is well to remember at the outset that all observations made during experimental work are subject to experimental error, no matter how carefully the experiment is carried out.

Having decided what the experiment is set up to do, the very first action of the technician should be to consult the appropriate literature to ascertain which techniques he may use to obtain the best results. The same source may well indicate appropriate equipment, but he should take due care to inquire whether development has taken place in the field of instrumentation which would make a modern piece of equipment more suitable, e.g. automatic recording devices which remove human error but which themselves must be checked from time to time.

In the preliminary stages of any investigation, it is absolutely essential to retain room for manoeuvre, since the early experiments may well suggest a more fruitful line of approach. It is bad technique to have to complete an experiment, or series of experiments, to a plan which is seen to be deficient after the first experiment.

The best plan, determined from literature search, should be adopted as a 'pilot experiment', and before proceeding with the ultimate design the results of the 'pilot experiment' should be carefully considered, not only by the experimenter but through consultation with his senior colleagues.

Results obtained from an experiment must be reproducible in any part of the world, and to this end nothing may be taken for granted and only reasonable assumptions made. The usual test for these assumptions is the reproducibility of the experiment.

Assuming the pilot experiment or a modification of it has been successful, it may then be necessary to select a design for the whole project. A simple arithmetical mean of the results may be adequate in a simple set of experiments, but more likely it will be necessary to consider uniformity, chance characteristics, and variability, and here statistical design can greatly enhance the value of an experiment.

Further consideration of basic statistical procedures is given later in this chapter.

EXPERIMENTAL RECORDS

The ultimate success of an experiment may depend upon the accuracy and detail with which records are kept. It is no use whatsoever to record an experimental procedure on pieces of paper which are easily lost or destroyed. It is absolutely essential to maintain notebooks which may be referred to, if necessary, years after the experiment has been performed. A good experimenter must, it cannot be overemphasised, keep clearly written, concise notes of all his work and record therein all the relevant details. Observations should be immediately recorded as the experiment proceeds, no matter how insignificant they may seem at the time.

Recently in the teaching of science it has become common practice to encourage learning from the experiment and to promote original thought. The experimenter is taught to think for himself and to refer to books, journals, and original papers for his information. The standard format for recording laboratory experiments has therefore also been discouraged. Nevertheless, the experimenter should maintain his notes in such a way as to present a logical sequence in the method of recording.

The title of the experiment should leave no doubt as to the type of work being done and the record of the experiment should enable later workers clearly to understand what the objects of the experiment were and what it was that the experimenter set out to accomplish. The methods used to accomplish the results or modifications resulting from a 'pilot experiment' should also be concisely recorded.

It is not always essential to sketch the apparatus used, as a good photograph is usually better, especially if the work is to be published later.

The recorded results of an experiment should be meticulously accurate and checked at least once. It is sometimes the practice to record results on special pro formas but to avoid confusion a note should be made in the notebook to this effect. Even if this procedure is customary, it is a good plan to record in the permanent notebook the salient features of the experimental results since loose pro formae may be lost or destroyed and unless a general picture of experimental findings is kept much valuable time and effort may have been wasted.

A common error in records is an attempt to indicate a greater accuracy than can be obtained experimentally; thus, if temperatures are read to 0.1 °C and the average of many readings is taken it is little use recording this to so many decimal places greater than the original observations.

In presenting data gathered by experimentation a simple procedure may be to prepare some diagrams of the results. The commonest form of description diagram is a graph plotting values found experimentally against a standard constant, or correlation graphs where one uncontrolled factor is plotted against another (e.g. the increase in personal income with increasing economic growth of a country). Results might also be shown as a 'frequency' graph or the number of times an event of any particular magnitude occurs (e.g. how many times a marksman hits a particular target in a given period).

It should be clearly remembered that such graphs prove little if anything but will draw attention to obvious relationships which may suggest either further experimental work to prove a particular relationship or a certain statistical treatment of the data.

BASIC STATISTICAL METHODS

Generally speaking, the same test carried out under identical conditions by the same experimenter on the same chemical tends to give the same result. It is obvious that individual experiments may not produce precisely the same result but the frequency of the same, or nearly the same, result does tend to become obvious.

The commonest statistical parameter with which all experimenters will be familiar is the *average* or *mean*. This is defined as the sum of all observations divided by the number of observations.

As an example, which illustrates various other statistical treatments, we can consider the following data on the melting-point of a certain chemical:

First Run	Second Run
1. 36.4 °C	1. 35.3 °C
2. 38.9 °C	2. 34.2 °C
3. 42.4 °C	3. 38.4 °C
4. 39.4 °C	4. 32.2 °C
5. 40.7 °C	5. 39.4 °C

The average or mean melting-point over both runs is given by the sum of all observations (377.3 °C) divided by the number of observations (10), which equals 37.73 °C. Note that although the result is shown to two decimal places, statistically this ought to be approximated to the nearest decimal point, i.e. 37.7 °C, in reporting the findings.

Another technique which describes the results more clearly when the mean is unduly affected by very large or very small outlying observations is the *median*. This is the centre value of a series of observations when these observations are ranged in order from lowest value to highest. Since there is not a great scatter around the mean it is not of great value in the example given here.

The *mode* is the maximum point on the curve of frequency distribution, i.e. that measurement which occurs most frequently.

These simple arithmetical processes, however, tell us little of the variation within the ten results, although in the example chosen this is clearly not likely to be great. A method which gives us more precise information on this random type of variation is the *standard deviation* (s.d.). This is computed as the square root of the arithmetic average of the squares of the difference between the observations and the mean.

A useful formula for this calculation is

$$\text{s.d.} = \sqrt{\left(\frac{\Sigma x^2}{n} - \bar{x}^2\right)}$$

where s.d. is the standard deviation, Σx^2 is the sum of the squares of each observation, $\bar{x}$ is the arithmetic mean of all observations, and n is the number of observations.

In this connection Davies[1], in whose book an unbiased estimate of the standard deviation is defined using $n - 1$ instead of n, should also be consulted. This may be important when a small number of readings is involved. But using our formulae for the example given the sum would become

$$\text{s.d.} \sqrt{\left(\frac{14324.26}{10}\right) - 1423.553}$$

$$= \sqrt{8.873}$$

$$= 2.98 \text{ (approx.)}$$

We now have more precise information on our raw data, i.e. that the average over all observations is 37.73 °C and that the spread of observed data around this mean is 2.98 °C. With such a small spread it is probable that the true result is fairly near the mean.

As mentioned earlier in the chapter, all experimental work is plagued by errors, human, instrumental, or otherwise, and it is essential that the errors, be they inherent in the technique or not, can be computed. The concept of *standard error* considers how much variation may be expected to occur merely by chance in the various characteristics of samples drawn equally randomly from one and the same population. In this example the word population means the same chemical compound but it may equally mean the same age-group, same race, etc.

Consequently, although we now have the mean and the standard deviation of our observations we may wish to know what this chance variation, *the standard error of the mean* (s.e.m.), may be and this is given as

$$\text{s.e.m.} = \frac{\text{s.d. of observations in the sample}}{\sqrt{(\text{number of observations})}}$$

In the worked example this is

$$\frac{2.98}{3.16} = 0.94$$

If you wish to go further with this subject, there are several excellent statistical textbooks now available which present various techniques in readable form for the less mathematically minded. These indicate the various tests of significance, correlation, coefficients, variance analysis, and many other valuable statistical means of assessing the value of an experiment or series of experiments.

REFERENCE

1 Davies, O. L. (ed.), *Statistical Methods in Research and Production*, 3rd edn, Oliver and Boyd, London (1949)

13
Technical literature

Although many institutional and university scientific libraries are geared to supplying or finding academic scientific information, technical information is not usually so readily available. As a consequence the laboratory technical worker has to be more or less self-sufficient and may have to spend a considerable amount of time in tracking down suppliers of a particular piece of routine apparatus and then comparing the products of the various manufacturers for cost and efficiency. Any system which cuts down this search time should be welcome as more time is available for evaluation.

Some technicians' time is taken up with building apparatus for specific experiments, and where such apparatus is not available commercially it is possible that time and effort may be saved by looking through published material relating to similar pieces of apparatus constructed by the authors.

LIBRARY SERVICES

The first step to be taken by the enquirer after technical literature is to approach the library within his own organisation or place of employment. If the information is not to be found on the library shelves, or in the archives, the librarian may well direct him to one of the following sources.

1. *Public, municipal, or county libraries*. In Britain these cover the very widest range of subjects and publications, and while a particular branch library may not possess the material required on its own shelves the exchange system practised ensures its expeditious production from another branch.

2. *University or technical college libraries.* Here an approach may best be made by the organisation's librarian to the librarian of the university or college.
3. *Libraries of co-operative research associations and development associations.* There are many of these associations, which have much specialist information which they are pleased, generally speaking, to pass on to customers or other interested persons. Examples of these associations are the Lead Development Association, 34 Berkeley Square, London W1, and the Natural Rubber Producers Research Association, 48–56 Tewin Road, Welwyn Garden City, Hertfordshire.
4. *National Lending Library for Science and Technology, Boston Spa, Yorkshire.* This library normally lends documents by post to institutions, not to individuals.
5. *National Reference Library of Science and Invention, 25 Southampton Buildings, Chancery Lane, London WC2.* This is open to the public during office hours but although it provides photocopies it does not lend documents.

Much of the above information is to be found in the booklet entitled *Technical Services for Industry*, issued by the Department of Scientific and Industrial Research, State House, High Holborn, London WC1. This carries the addresses and telephone numbers of the D.S.I.R. Headquarters, Branch Offices, and the Regional Technical Information Centres, also information on the D.S.I.R. National Lending Library for Science and Technology (see above) and the Research Stations. A great deal of information is given in this publication about the many research associations mentioned previously, and apart from addresses and telephone numbers it contains information concerning the scope of the associations' work, their library facilities, enquiry services, and other information.

Aslib—the Association of Special Libraries and Information Bureaux—3 Belgrave Square, London SW1, issues a two-volume Directory described as 'a guide to sources of information in Great Britain and Ireland'. Volume 1 contains the Classified Index, Subject Index, Name Index and appendices. Volume 2 has a Directory of Libraries and Information Services and a Regional Index.

The Organisation for Economic Co-operation and Development issues a *Guide to European Sources of Technical Information*. The alphabetical index at the back of this Directory indicates the coverage of a wide range of scientific, technical, and industrial subjects, and there is a list of National Technical Information Liaison Offices.

A very comprehensive volume is the *British Technology Index*, issued by the Library Association, Chaucer House, Malet Place,

London WC1. It is a subject guide to the major articles published in 400 British technical journals during the year, and comprises about 30 000 entries arranged in a single alphabetical sequence of subject headings, together with supporting references. It also contains a list of journals, with addresses, ranging from *Associated Electrical Industries Engineering*, through *Journal of Science Technology* and *Ultrasonics* to *World's Paper Trade Review*.

Possession of these four publications enables the laboratory technical worker, even if he has access to no library facilities at all, at least to find an address to which he can send his enquiries knowing he will be provided either with a reply to his query or another recommended address to which a successful application probably may be made.

PROFESSIONAL TECHNICAL ORGANISATIONS

It is obviously to the advantage of the laboratory worker to become a member of his or her relevant professional association or institute, since the benefits of such membership are likely to include a regular supply of the organisation's publications, such as journals, bulletins, and gazettes. These publications carry scientific and technical papers relevant to the profession which they serve; they may also carry theses or dissertations submitted to the organisation in support of applications for higher membership or qualification.

It is often possible for non-member individuals or places of employment to receive these professional journals on subscription. For a comprehensive list of organisations consult the British Technology Index, mentioned above.

STANDARD WORKS

TEXTBOOKS

It is impossible to list all the textbooks a laboratory technical worker might find useful, but some of the organisations to be found in the publications already mentioned are examining bodies, or run tuition schemes, and in consequence are likely to have available lists of recommended textbooks. The City and Guilds of London Institute, 76 Portland Place, London W1, is such a body, which examines candidates often in conjunction with another organisation. Application should be made to the relevant organisation for an up-to-date copy of such a list.

Fortunately, an increasing number of textbooks are being written by practising laboratory technicians; this means the author is writing from his own experience and not merely copying out formulae and methods from previous publications.

Textbooks which have been published for some little time soon become known, and the library services mentioned above can usually supply lists of them to supplement those supplied by examining bodies. What the technician requires is early notification of new publications, and he can best ensure this by regular study of publishers' advertisements in the various journals related to his particular speciality, and those with a more general field. In addition, of course, many journals carry a book review section which supplies independent assessments of the newest books. The regular purchaser of textbooks may find it an advantage to have his name entered on the publishers' mailing lists, which are likely to include the types of books he is interested in. Booksellers are only too pleased to keep the reader abreast of the latest publications and some institutes have a medical, scientific, and technical lending library with books available for loan or purchase.

REFERENCE BOOKS

Textbooks are, of course, reference books, but reference books are not necessarily textbooks. Books of reference which are not textbooks fall roughly into two groups, those to which the laboratory worker must have ready access in the laboratory and those which need not be so readily accessible, and may be available in some central library to all the staff. Into this latter category falls the monumental *McGraw-Hill Encyclopedia of Science and Technology*. Published in 1960, this is an international reference work in fifteen volumes, including an index. It is most comprehensive, and since its publication yearbooks have been issued as supplements to the encyclopedia which keep the reader up to date on the previous year's advances. No one library can be expected to keep many reference books of this type and the reader is restricted to volumes held by the libraries to which he has access; possibilities of loans of this type of book are limited.

The other group of reference books includes those books of constants typified by the *Handbook of Chemistry and Physics* published by Chemical Rubber Publishing Company, 2310 Superior Avenue, N.E., Cleveland, Ohio, U.S.A. This book describes itself as 'a ready reference book of chemical and physical data'. In its almost 3000 pages are a very considerable number of items of value to the laboratory worker. It is frequently to be found, in a well-thumbed state, on the laboratory shelves.

Another large reference volume is the Merck Index of Chemicals and Drugs, published by Merck and Company Incorporated, Rahway, New Jersey, U.S.A. It is an encyclopedia for chemists, pharmacists, physicians, and members of allied professions. The first edition was published in 1889, and the current (7th) edition (1960) has a Cross Index of Names with about 30 000 entries. The monographs, which include the chemical and physical characteristics of the material being described, and its medical and other uses, number almost 3400. The appendices carry a very large number of physical and chemical constants which are most useful to the laboratory worker.

Specialised dictionaries are useful books of reference, and are available from a number of different publishers. Penguin Books have produced dictionaries of science, biology, civil engineering, and psychology amongst others. A recent addition is the series of scientific and technical dictionaries published by George Newnes Ltd. This series covers aeronautical engineering, astronautics, ceramics, civil engineering, dyeing and textile printing, electronics, mathematics, mining, plastics, and printing, papermaking and bookbinding, each in separate volumes. The entries tend to be longer and more detailed than in some general dictionaries, and trade names are included.

Many co-operative research associations and development associations issue booklets, often free on application, relating to the particular industry which they exist to serve. These booklets consist, more or less, of factual information relating to their subject, and so can be considered as reference books. A similar service is offered by laboratory supply firms of all kinds, and in many cases the catalogues themselves contain information such as conversion tables and other lists of constants which make them most useful.

BRITISH STANDARDS

The technician often has cause to refer to important sources of technical information known as British Standards, issued by the British Standards Institution. By making available this information, the British Standards Institution is able to co-ordinate the efforts of producers and users and to standardise many products, including laboratory equipment. Thus, laboratory technicians are assured of the dimensions, quality, and performance of the apparatus they use.

British Standards are produced as the result of requests made to the Institution by authoritative bodies and the subsequent efforts of the Institution's appointed technical committees. The indispensable information made available to technicians in this way, and use of standard apparatus, permit greater accuracy, promote safer working, and assist laboratory techniques.

British Standards may be purchased from the Institution by post or at the sales counter at the Institution's Mayfair offices at 2 Park Street, London W1.

COMMERCIAL PRODUCTS

As mentioned at the beginning of this chapter, it is very necessary for the laboratory technician to be able to locate rapidly the source of supply of any item required in his laboratory, be it furniture, clothing, apparatus, chemicals, or equipment of any sort. Obviously, then, he should build up a library of catalogues from any firms which he knows are able to fulfil his requirements. Such catalogues are issued by the larger laboratory suppliers and are very comprehensive. They list a considerable proportion of those items most likely to be needed in the majority of laboratories and include many items made or supplied by firms other than those within their own group. The catalogues of the smaller firms must not be ignored, however; in many cases such firms specialise in particular apparatus, such as educational, clinical, or electronic, and can offer a quicker and better service in those items; others specialise in imported apparatus and so can fill the gaps in domestic suppliers' lists. It is advisable to make written application to these firms to be placed on their mailing lists, and once one's name is entered new editions of catalogues should automatically be dispatched.

Careful perusal of all the catalogues in one's possession may still not give a source of supply for the piece of apparatus one requires. In this case the next step to be taken is to request one or more of one's regular suppliers to track down the instrument. Most suppliers are well equipped to do this as they have experts in many fields on their staff and large numbers of contacts outside the firm. If this fails to bring a satisfactory result, either the firm or the customer may make use of the Enquiry Service of the Scientific Instrument Manufacturers Association of Great Britain, 20 Queen Anne Street, London W1. S.I.M.A. can usually answer questions such as which firms supply a particular type of scientific or technical instrument, or alternatively whether such a piece of equipment is commercially available and if so its whereabouts. Of course, not each and every maker of laboratory apparatus is a member of S.I.M.A.

The advice to get one's name on to the mailing lists of those firms in which one is interested applies also to those requiring chemicals and chemical supply firms are pleased to send their catalogues to genuine enquirers. In addition, the Association of British Chemical Manufacturers, Cecil Chambers, 86 Strand, London WC2, issues a biennial guide to their members' products entitled *British Chemicals and their Manufacturers*; this lists almost 250 members with their addresses and telephone numbers, and indicates the groups of chemicals which they manufacture. Details of affiliated associations are given, and between 10 000 and 12 000 chemicals are listed. In addition, a long list of proprietary and trade names is given, and a number of proprietary and trade marks.

In order to keep abreast of the latest items available from the supply firms, the technician cannot do better than set aside some time, at regular intervals, to peruse the advertisements in the journals previously mentioned. Although, if his name is on the firms' mailing lists, he will eventually learn of additions to and deletions from their catalogues, his first intimation of additions is likely to be from advertisements.

Two publications in particular can assist in the search for established equipment and for new developments. The first is the *Journal of Scientific Instruments*, published by the Institute of Physics and the Physical Society, 47 Belgrave Square, London SW1. This contains numbers of articles detailing new or improved pieces of apparatus, developments, or methods of evaluating equipment.

The second is *Laboratory Equipment Digest*, published by Gerard Mann Limited, 225c Balham High Road, London SW17. This is circulated free to a large number of senior laboratory executives each month; a charge is made to overseas subscribers. Apart from editorial matter, such as news of the trade, the journal carries some sixty pages of varied advertisements which keep the reader informed of the latest developments in laboratory apparatus, materials, and techniques, for research, industry, education, and medicine.

FILMS

Although this chapter is entitled *Technical Literature*, aids other than the printed word must surely be included in a review of sources of technical information.

Visual aids, in the form of black and white or colour slides, photographs, silent films, and sound films have been used in education for a considerable time. The position has now been reached when films may be loaned or hired on almost any subject whatsoever, and very often the only cost to the borrower is the return postage after use.

Since the total number of sources of 16-mm films of all kinds in the United Kingdom is nearly a thousand, it is impractical to list them all here, but the following are some examples of suppliers of technical, scientific, and educational films on free loan and on hire, and suggestions as to where comprehensive lists may be found.

Abbott Laboratories Limited (Technical Service Department), Queenborough, Kent, have available films on such subjects as anaesthetics, cytology, nursing gynaecology, etc. British Film Institute (Distribution Department), 81 Dean Street, London W1, has over 1500 titles on a variety of subjects, including the sciences, industry, medicine, child psychology, and many others. Petroleum Films Bureau, 4 Brook Street, London W1, apart from supplying many films on the various aspects of the oil industry, also has a large number of films on such subjects as engineering, science, agriculture, biology, geology, etc. The Royal Institute of Chemistry, 30 Russell Square, London W1, has about 1200 films available from various sources on pure and applied sciences.

Each of the above sources of films, plus some 800 others, is to be found in the monthly journal *Film User* incorporating *Industrial Screen*, published on the first of each month from Davis House, 69 High Street, Croydon, Surrey, postal address P.O. Box 109, Croydon, Surrey. Each January issue carries an index of all the films released, and reviewed in *Film User*, during the previous twelve months. Later in the year an article entitled 'Factual Films from A to Z' lists an alphabetical directory of firms and other organisations which have 16-mm prints to lend or hire. The following month's issue carries an article headed 'Where to Find It', which contains a subject index which enables the reader to track down the source of supply of films on the subject or subjects he is interested in. Once the source, or probable source, is known, the reader will now be able to write for a specific film, for a catalogue of films or for further information.

Another British publication which lists, among many others, technical and scientific films is the weekly *Amateur Cine World* published by Fountain Press Limited, 46 Chancery Lane, London WC2.

From Germany comes the *Integral Catalogue of Scientific Films*, published by the Institut für den Wissenschaftlichen Film, 72 Nonnenstieg, 34 Göttingen, Germany; almost all disciplines in science and technology are covered by the films, some modern and others more than thirty years old. Included in this publication are the films available from the *Encyclopaedia Cinematographica*, which covers a similar range of subjects to the Institut films. All these films may be purchased or hired from the Institut für Wissenschaftlichen Film.

The publishing firm of Macmillan and Company Limited has recently added to its educational aids two series of single-concept films. One series is known as *Macmillan Cineloops* and the other as *Eothen Cinettes*; these are loops of film, running for four or five minutes, on such subjects as languages, physics, nursing, and other scientific and technical subjects, and are intended to be used during lessons in the Technicolour 800/E2 back projector to illustrate particular points. Details may be obtained from the Education Sales Manager at Macmillan's.

It is probable that single-concept films will become available from some other sources, such as certain film libraries.

Appendix 1

Table A1 COMMON NAMES OF CHEMICALS*

Common name	Chemical name	Formula
Alum	Potassium aluminium sulphate	$K_2SO_4.Al_2(SO_4)_3.24H_2O$
Alumina	Aluminium oxide	Al_2O_3
Alumino-ferric	Crude aluminium sulphate containing iron sulphate	$Al_2(SO_4).xH_2O$
Ammonia liquor	Ammonium hydroxide	NH_4OH
Aqua fortis	Nitric acid	HNO_3
Aqua regia	Nitric acid + Hydrochloric acid	$HNO_3 + 3HCl$
Arsenic	Arsenious oxide	As_2O_3
Aspirin	Acetylsalicylic acid	$o\text{-}C_6H_4(O.CO.CH_3)COOH$
Baking soda	Sodium bicarbonate	$NaHCO_3$
Barytes	Barium sulphate	$BaSO_4$
Blanc fixe	Barium sulphate	$BaSO_4$
Bleaching powder	Calcium chlorohypochlorite	$CaOCl_2$
Blue vitriol	Copper sulphate	$CuSO_4.5H_2O$
Boracic acid	Boric acid	H_3BO_3
Borax	Sodium tetraborate	$Na_2B_4O_7.10H_2O$
Burnt lime	Calcium oxide	CaO
Calomel	Mercurous chloride	Hg_2Cl_2
Cane sugar	Sucrose	$C_{12}H_{22}O_{11}$
Carbolic acid	Phenol	C_6H_5OH
Carbonic acid	Carbon dioxide	CO_2
Carborundum	Silicon carbide	SiC
Caustic potash	Potassium hydroxide	KOH
Caustic soda	Sodium hydroxide	$NaOH$
Chalk	Calcium carbonate	$CaCO_3$
Chile saltpetre	Sodium nitrate	$NaNO_3$
Chloride of lime	Calcium chlorohypochlorite	$CaOCl_2$
Chrome alum	Potassium chromium sulphate	$K_2SO_4Cr_2(SO_4)_3.24H_2O$
Copperas	Ferrous sulphate	$FeSO_4.7H_2O$
Corrosive sublimate	Mercuric chloride	$HgCl_2$

* Reproduced from *Useful Technical Data*, 1964, by permission of African Explosives and Chemical Industries Ltd.

Table A1 COMMON NAMES OF CHEMICALS—*continued*

Common name	Chemical name	Formula
Corundum	Aluminium oxide	Al_2O_3
Cream of tartar	Potassium hydrogen tartrate	$KHC_4H_4O_6$
Dextrose	Glucose	$C_6H_{12}O_6$
Eau de Javelle	Sodium hypochlorite solution	$NaOCl$
Epsom salts	Magnesium sulphate	$MgSO_4.7H_2O$
Essence of mirbane	Nitrobenzene	$C_6H_5NO_2$
Fluorspar	Calcium fluoride	CaF_2
Formalin	37% Aqueous solution of formaldehyde plus 7% methanol	$HCHO$
French chalk (talc)	Hydrated magnesium silicate	$Mg_3Si_4O_{11}.H_2O$
Fruit sugar	Fructose	$C_6H_{12}O_6$
Fuller's earth	Hydrated silicates of magnesium and aluminium	
Fulminating mercury	Mercuric fulminate	$HgC_2O_2N_2$
Glauber's salt	Sodium sulphate	$Na_2SO_4.10H_2O$
Grape sugar	Glucose	$C_6H_{12}O_6$
Green vitriol	Ferrous sulphate	$FeSO_4.7H_2O$
Gypsum	Calcium sulphate	$CaSO_4.2H_2O$
Hypo	Sodium thiosulphate	$Na_2S_2O_3.5H_2O$
Laughing gas	Nitrous oxide	N_2O
Lime	Calcium oxide	CaO
Litharge	Lead monoxide	PbO
Lithopone	Zinc sulphide + Barium sulphate	$ZnS + BaSO_4$
Lunar caustic	Silver nitrate	$AgNO_3$
Magnesia	Magnesium oxide	MgO
Magnesite	Magnesium carbonate	$MgCO_3$
Marble	Calcium carbonate	$CaCO_3$
Marsh gas	Methane	CH_4
Methanol	Methyl alcohol	CH_3OH
Milk of lime	Calcium hydroxide	$Ca(OH)_2$
Milk of magnesia	Magnesium hydroxide	$Mg(OH)_2$
Milk sugar	Lactose	$C_{12}H_{22}O_{11}$
Muriate of ammonia	Ammonium chloride	NH_4Cl
Muriatic acid	Hydrochloric acid	HCl
Nitre cake	Anhydrous sodium bisulphate	$NaHSO_4$
Nitro lime	Calcium cyanamide	$CaCN_2$
Oil of wintergreen	Methyl salicylate	$o\text{-}C_6H_4(OH)COOCH_3$
Oleum	Fuming sulphuric acid	$H_2SO_4 + SO_3$
Paris green	Copper acetoarsenite	$Cu(C_2H_3O_2)_2.3CuAs_2O_4$
Pearl ash	Potassium carbonate	K_2CO_3
Picric acid	Sym-trinitrophenol	$C_6H_2(NO_2)_3OH$
Plaster of Paris	Calcium sulphate	$CaSO_4.\frac{1}{2}H_2O$
Precipitated chalk	Calcium carbonate	$CaCO_3$
Prussian blue	Ferric ferrocyanide	$Fe_4(Fe(CN)_6)_3$
Prussic acid	Hydrocyanic acid	HCN
Quicklime	Calcium oxide	CaO
Quicksilver	Mercury	Hg
Quinol	Hydroquinone	$p\text{-}C_6H_4(OH)_2$

Table A1 COMMON NAMES OF CHEMICALS—*continued*

Common name	Chemical name	Formula
Red lead	Lead tetroxide	Pb_3O_4
Rochelle salt	Potassium sodium tartrate	$KNaC_4H_4O_6.4H_2O$
Rouge	Ferric oxide	Fe_2O_3
Saccharin	Benzoic sulphimide	$C_6H_4SO_2NHCO$
Salammoniac	Ammonium chloride	NH_4Cl
Salt	Sodium chloride	$NaCl$
Salt cake	Impure sodium sulphate	Na_2SO_4
Saltpetre	Potassium nitrate	KNO_3
Scheele's green	Copper hydrogen arsenite	$CuHAsO_3$
Slaked lime	Calcium hydroxide	$Ca(OH)_2$
Soda	Sodium carbonate	$Na_2CO_3.10H_2O$
Soda ash	Anhydrous sodium carbonate	Na_2CO_3
Sodium hyposulphite	Sodium thiosulphate	$Na_2S_2O_3.5H_2O$
Spirits of salts	Hydrochloric acid	HCl
Spirits of wine	Ethyl alcohol 94/96%	C_2H_5OH
Tartar emetic	Potassium antimonyl tartrate	$K(SbO)C_4H_4O_6.\frac{1}{2}H_2O$
TNT	2,4,6-Trinitrotoluene	$C_6H_2(CH_3)(NO_2)_3$
Verdigris	Basic copper acetate	$2Cu(C_2H_3O_2)_2 + CuO$
Vitriol	Sulphuric acid	H_2SO_4
Washing soda	Sodium carbonate	$Na_2CO_3.10H_2O$
Water glass	Aqueous solution of sodium silicates	$Na_2O.xSiO_2$
White arsenic	Arsenious oxide	As_2O_3
White lead	Basic lead carbonate	$2PbCO_3 + Pb(OH)_2$
White vitriol	Zinc sulphate	$ZnSO_4.7H_2O$
Whiting	Calcium carbonate	$CaCO_3$
Wood alcohol	Methyl alcohol	CH_3OH

Appendix 2

Table A2 USUAL CONCENTRATIONS OF SOME ACIDS

	% by weight	Density $D^{20°}_{4°}$	Baumé† degrees	Normality‡
Acetic acid glacial 96%	96	1.06	8	17
Acetic acid glacial 99–100%	99–100	1.06	8	18
Acetic acid diluted	30	1.04	5.4	5
Acetic anhydride	90	1.07	10	—
Formic acid	98–100	1.22	26	26
Hydrochloric acid	25	1.12	16	8
Hydrochloric acid (1.16)	32	1.16	20	10
Hydrochloric acid (1.18)	36	1.18	22	12
Hydrochloric acid fuming	38	1.19	23	12.5
Nitric acid	25	1.15	18.6	5
Nitric acid (1.40)	65	1.40	41	14
Nitric acid fuming (abt 1.50)	abt 99	1.51	49	21
Phosphoric acid	25	1.15	19	9
o-Phosphoric acid (1.710)	85	1.71	59	45
o-Phosphoric acid (1.750)	89	1.75	62	48
Sulfuric acid (1.84)	95–97	1.84	66	36
Sulfuric acid diluted	16	1.11	14	4
Sulfuric acid fuming (abt 65% SO_3)		1.99	72	—

* Courtesy of E. Merck, Darmstadt.

† According to formula $145 - \dfrac{145}{\text{Density}} = °\text{Bé (Density at } 20°/4°)$

‡ Approximate value.

Bibliography

Chapter 1 Designing the laboratory

Adams, C. S., 'University or College Laboratory', *Ind. Engng Chem.*, **39**, 457 (1947)
Bailer, J. C., Jun., 'The General Chemistry Laboratory', *J. chem. Educ.*, **24**, 327 (1947)
Beach, D. M., 'A Large Industrial Research Laboratory', *Ind. Engng Chem.*, **39**, 448 (1947)
Black, H. R., 'Metal Laboratory Furniture Installations', *Lab. Pract.*, **12**, 999 (1963)
Brookfield, K. J. and Pimblett, A., 'Fibreglass as a Constructional Material in the Laboratory', *Lab. Pract.*, **12**, 990 (1963)
Brown, L. D., 'The Design of Laboratories for Work with Radioisotopes and Other Radiation Sources', *Lab. Equip. Dig.*, **8**, No. 8, 45 (1970)
Cairns, R. W., 'Selection of Laboratory Location', *Ind. Engng Chem.*, **39**, 440 (1947)
Case, L. O., 'Laboratories for Physical Chemistry', *J. chem. Educ.*, **24**, 338 (1947)
Building Research Station, 'Estimating Daylight in Buildings', *Building Research Station Digest*, 2nd ser., Nos. 41 (1963) and 42 (1964), H.M.S.O., London
Coleman, H. S. (ed.), *Laboratory Design*, National Research Council of the U.S.A., Reinhold, New York (1951)
Darby, G. M., Roberts, E. J., and Grothe, J. D., 'Process Engineering Research Laboratories', *Ind. Engng Chem.*, **39**, 453 (1947)
Davies, R. L., 'The Design of Research Laboratories', *Lab. Pract.*, **6**, 405 (1957)
Department of Scientific and Industrial Research, Lighting Committee of the Building Research Board, *The Lighting of Buildings*, H.M.S.O., London (1944)
'Designing for Science, Oxford School Development Project', *Building Bulletin* No. 39, H.M.S.O., London (1967)
Dobson, E. W., 'Laboratory Benches', *Lab. Pract.*, **12**, 830 and 908 (1963)
Dobson, E. W., 'Laboratory Services', *Lab. Pract.*, **12**, 994 (1963)
Eastman Kodak, *Darkroom Design and Construction*, Publication No. K.13, 2nd edn, Eastman Kodak Co., Rochester, N.Y. (1958)
Fisher, *Manual of Laboratory Safety*, Bulletin F.S.201, Fisher Scientific Co., New York (n.d.)
Gawen, D., 'Laboratory Prefabricated Services', *Lab. Pract.*, **12**, 1091 (1963)
Hall, G. R., 'The Design and Management of the Radiochemical Laboratory', *Lab. Pract.*, **12**, 249 (1963)
'Harris College, Preston', *Building Bulletin* No. 29, H.M.S.O., London (1966)

Hurd, C. D., 'The Organic Laboratory', *J. chem. Educ.*, **24**, 333 (1947)
Joliffe, G. O., 'Automation in the Lecture Room', *Lab. Pract.*, **12**, 1086 (1963); **13**, 120 (1964); **13**, 838 (1964)
Kay, H. D., 'The Design of Laboratories: Some General Considerations', *Lab. Pract.*, **12**, 818 (1963)
Kodak, *Darkroom Design and Construction*, Data Sheet K13, Kodak, London (1958)
Kruk-Schuster, A., 'Safety Aspects in Laboratory Design', *Lab. Pract.*, **12**, 835 (1963)
Lewis, H. F. (ed.), *Laboratory Planning for Chemistry and Chemical Engineers*, Chapman & Hall, London (1962)
Lovett, A. B. E., 'Safety and the Technician', *S.T.A. Bull.*, **4**, No. 4, 2 (1955)
Mann, S., 'Laboratory Flooring', *Lab. Equip. Dig.*, **8**, No. 9, 60 (1970)
Marvin, G. G., 'The Analytical Laboratory', *J. chem. Educ.*, **24**, 329 (1947)
Munce, J. F., 'Furnishings and Fittings', *Lab. Pract.*, **12**, 905 (1963)
Munce, J. F., 'Future Laboratory Developments', *Lab. Pract.*, **12**, 1073 (1963)
Munce, J. F., *Laboratory Planning*, Butterworths, London (1962)
Munce, J. F., 'Services: Planning and Layout', *Lab. Pract.*, **12**, 984 (1963)
Munce, J. F., 'Special Laboratories', *Lab. Pract.*, **12**, 1070 (1963)
Munce, J. F., 'Structure and Fabric', *Lab. Pract.*, **12**, 902 (1963)
Munce, J. F., 'The Architect's Role in Laboratory Planning', *Lab. Pract.*, **12**, 823 (1963)
Munce, J. F., 'The Architectural Significance of Radioactivity in the Laboratory', *Lab. Pract.*, **12**, 349 (1963)
Munce, J. F., 'The Planning of Various Types of Laboratories', *Lab. Pract.*, **12**, 826 (1963)
Nuffield Foundation, Division of Architectural Studies, *The Design of Research Laboratories*, Oxford University Press, London (1961)
Packard, R. J., 'Technical Photography', *Lab. Pract.*, **3**, 110 (1954)
Palmer, R. R., and Rice, W. M., *Modern Physics Buildings*, Reinhold, New York (1961)
Rassweiler, C. F., 'New Ideas from Industrial Laboratory Design', *J. chem. Educ.*, **24**, 346 (1947)
Royal Institute of British Architects, *Teaching Laboratories, Report of a Symposium on Design of Teaching Laboratories in Universities and Colleges of Advanced Technology*, R.I.B.A., London (1958)
Royal Institute of Chemistry, 'Report of a Symposium on Laboratory Layout and Construction', *R. Inst. Chem. Report* No. 6 (1949)
Savage, Sir G., *The Planning and Equipment of School Science Blocks*, Murray, London (1964)
Schramm, W., *Chemistry and Biology Laboratories*, Pergamon, London (1965)
Smith, P. C., 'Design for Facilities for Research', *Ind. Engng Chem.*, **39**, 445 (1947)
Weber, H. C., 'Design of Laboratories for Chemical Engineering Instruction', *J. chem. Educ.*, **24**, 341 (1947)
Young, R. R., and Harrington, P. J., 'Design and Construction of Laboratories', *R. Inst. Chem. Lecture Ser.* No. 3 (1962)

Chapter 2 Installation of laboratory equipment

Attree, V. H., 'The Use of Silicones for Viscous Damping of Galvanometer Mountings', *J. scient. Instrum.*, **25**, 423 (1948)

Baker, S. C., 'A Laboratory Galvanometer Stand', *J. scient. Instrum.*, **29**, 299 (1952)
Barbour, R., *Glassblowing for Laboratory Technicians*, Pergamon, London (1968)
Boos, R. N., 'The Microanalytical Laboratory', in *Laboratory Design*, Reinhold, New York (1951)
Coates, G. E., and Coates, J. F., 'A Simple Anti-vibration Galvanometer Support', *J. scient. Instrum.*, **22**, 153 (1945)
Frost, J. A., 'Laboratory Glassblowing Workshops', *Lab. Pract.*, **12**, 1091 (1963)
Hilger, *A Specification for a Spectrochemical Laboratory*, Catalogue Adam Hilger, Ltd.
Wild-Barfield, *Installation, Working and Maintenance Instructions for Horizontal Rectangular Muffle Furnaces*, Wild-Barfield Electric Furnaces Ltd., Watford, Herts

Chapter 3   Stores management

British Transport Commission, *Conditions of Carriage, British Road Services*, London
British Railways Board, *Dangerous Goods by Freight Train and by Passenger Train or Similar Service. List of Dangerous Goods and Conditions of Acceptance*, London (1966)
British Railways Board, *Packing Regulations for Goods (Other than Dangerous Goods) to be Carried by the Board's Services* (1971)
Commissioners of Customs and Excise, *Conditions Relating to the Receipt and Use of Duty Free Alcohol*, London
Edwards, J. A., 'Laboratory Management', *Chem. Age, Lond.*, **68**, 955 (1953)
Hartshorn, C., *Systematic Bookkeeping for South African Students*, 4th edn, Juta, Capetown (1933)
Haskins, A. L., 'Operating an Agricultural Biochemistry Stockroom', *J. chem. Educ.*, **27**, 391 (1950)
Hiscocks, E. S., *Laboratory Administration*, Macmillan, London (1956)
Holtzinger, A. H., 'The Operation of a Multiple Stockroom', *J. chem. Educ.*, **30**, 512 (1953)
Jackson, J. H., *Elements of Accounting*, 3rd edn, McGraw-Hill, New York (1952)
Jobling, *Tubing Sizes, Pyrex Laboratory Glass*, James A. Jobling Co. Ltd, Sunderland (1962)
Loytty, O. M., 'Purchase and Stocking of Laboratory Glassware', *J. chem. Educ.*, **27**, 393 (1950)
McKenzie, D. H., *Fundamentals of Accounting*, Macmillan, New York (1947)
Melinsky, B., *Industrial Storeskeeping Manual*, Chilton, U.S.A. (1956)
Sutcliffe, A., *School Laboratory Management*, 2nd edn, Murray, London (1950)

Chapter 4   Preparation and storage of reagents

African Explosives and Chemical Industries, *Useful Technical Data*, African Explosives and Chemical Industries Ltd, Johannesburg (1964)
Baker, F. J., Silverton, R. E., and Luckcock, E. D., *An Introduction to Medical Laboratory Technology*, 4th edn, Butterworths, London (1966)
Baker, J. R., *Cytological Techniques*, 4th edn, Methuen, London (1960)
Belcher, R., and Wilson, C. L., *Qualitative Inorganic Microanalysis*, Longmans, Green, London (1964)
Belcher, R., and Wilson, C. L., *New Methods in Analytical Chemistry*, Chapman & Hall, London (1955)

BIBLIOGRAPHY

British Drug Houses, *Analar Standards for Laboratory Chemicals*, 5th edn, British Drug Houses and Hopkin and Williams, London (1957)

Clayden, E. C., *Practical Section Cutting and Staining*, 4th edn, Churchill, London (1962)

Daldy, F. G., 'Apparatus for Storing Standard Solutions', *Sch. Sci. Rev.*, **17**, 460 (1935)

Diamond, P. S., and Denham, R. F., *Laboratory Techniques in Chemistry and Biochemistry*, 2nd edn, Butterworths, London (1973)

Duddington, C. L., *Practical Microscopy*, Pitman, London (1960)

Field, F. B., 'Dirty Solutions', *Sch. Sci. Rev.*, **17**, 459 (1935)

Gabb, M. H., and Latchem, W. E., *A Handbook of Laboratory Solutions*, Deutsch, London (1967)

Gatenby, J. B., *Biological Laboratory Technique*, 4th edn, Churchill, London (1937)

Gatenby, J. B., and Painter, T. S., *The Microtomist's Vade-mecum*, 10th edn, Churchill, London (1937)

Gregory, T. E., 'A Convenient Method of Making Solutions', *Sch. Sci. Rev.*, **36**, 90 (1954)

Gurr, E., *A Practical Manual of Medical and Biological Staining Techniques*, 2nd edn, Hill, London (1956)

Gurr, G. T., *Biological Staining Methods*, 5th edn, Gurr, London (1967)

Holness, H., *Inorganic Qualitative Analysis, Semi-micro Methods*, Pitman, London (1954)

Levine, M., *Introduction to Laboratory Technique in Bacteriology*, Macmillan, New York (1947)

Linstead, R. P., and Weedon, B. C. L., *A Guide to Qualitative Organic Chemical Analysis*, Butterworths, London (1956)

McLean, R., and Cook, W., *Plant Science Formula*, 2nd edn, Macmillan, London (1958)

McLean, R. C., and Ivimey-Cook, W. R., *Textbook of Practical Botany*, Longmans, Green, London (1952)

Mahoney, R., *Laboratory Techniques in Zoology*, 2nd edn, Butterworths, London (1973)

Mann, F. G., and Saunders, B. C., *Practical Organic Chemistry*, 4th edn, Longmans, Green, London (1960)

Merck, *Complexometric Assay Methods with Titriplex*, 3rd edn, E. Merck AG, Darmstadt (n.d.)

Middleton, H., *Systematic Qualitative Organic Analysis*, 2nd edn, Arnold, London (1943)

Pantin, C. F. A., *Microscopical Technique for Zoologists*, Cambridge University Press, London (1946)

Parr, N. L. (ed.), *Laboratory Handbook*, Newnes, London (1963)

Peacock, H. A., *Elementary Microtechnique*, 3rd edn, Arnold, London (1966)

Purvis, M. J., Collier, D. C., and Walls, D., *Laboratory Techniques in Botany*, Butterworths, London (1964)

Siegfried, *Titrations with Complexone*, B. Siegfried AG, Zofingen (n.d.)

Sharpey-Schafer, Sir Edward, *The Essentials of Histology*, 13th edn, Longmans, Green, London (1934)

Smith, F. J., and Jones, E., *A Scheme of Qualitative Organic Analysis*, Blackie, London (1948)

Stevens, W. I., 'Standardization of Volumetric Solutions', *Lab. Pract.*, **12**, 564 (1963)

Sudborough, J. J., and James, T. C., *Practical Organic Chemistry*, Blackie, London (1909)

Sutcliffe, A., *School Laboratory Management*, 2nd edn, Murray, London (1950)
Van Nieuwenburg, C. J., and Gilles, J., *Reagents for Qualitative Inorganic Analysis*, Elsevier, Brussels (1948)
Vogel, A. I., *Text Book of Quantitative Inorganic Analysis*, 3rd edn, Longmans, Green, London (1961)
Vogel, A. I., *Practical Organic Chemistry*, Longmans, Green, London (1948)
Vogel, A. I., *Text Book of Qualitative Chemical Analysis*, Longmans, Green, London (1947)
Weast, R. C. (ed.), *Handbook of Chemistry and Physics*, 53rd edn, Chemical Rubber Publishing Co., Cleveland, Ohio (1972)

Chapter 5   Laboratory inspection and maintenance

Bangaru, P., 'Cleaning of Laboratory Glassware and Apparatus', *J. Sci. Technol.*, **12**, No. 3, 127 (1966)
Barer, R., *Lecture Notes on the Use of the Microscope*, Blackwell, Oxford (1953)
Bruce, J., and Harper, H., *Practical Chemistry*, 5th edn, Macmillan, London (1945)
Foner, A. H., 'The Use and Care of Platinum Laboratory Apparatus', *Lab. Pract.*, **14**, 944 (1965)
Green, G., 'Planned Maintenance', *Lab. Pract.*, **6**, 277 and 333 (1957)
Hare, R., and O'Donoghue, P. N. (eds), *The Design and Function of Laboratory Animal Houses*, Laboratory Animals Ltd, London (1968)
Lewin, S., 'Cleaning Methods', *Lab. Pract.*, **6**, 213 (1957)
Jones, B. E., *Electric Accumulators*, Cassell, London (1954)
Short, D. J., and Woodnott, D. P., (eds), *The A.T.A. Manual of Laboratory Animal Practice and Techniques*, Crosby Lockwood, London (1963)
Steere, N. V. (ed.), *Handbook of Laboratory Safety*, Chemical Rubber Publishing Co., Cleveland, Ohio (1967)
Sutcliffe, A., *School Laboratory Management*, 2nd edn, Murray, London (1950)
Vogel, A. I., *Practical Organic Chemistry*, Longmans, Green, London (1948)
Vogel, A. I., *Quantitative Analysis*, Longmans, Green, London (1948)
Wild-Barfield, *Installation, Working and Maintenance Instructions for Horizontal Rectangular Muffle Furnaces*, Wild-Barfield Electric Furnaces Ltd, Watford, Herts.

Chapter 6   Safety in laboratory and workshop

Association of British Chemical Manufacturers, *Safety and Management—A Guide for the Chemical Industry*, London (1964)
Association of Universities of the British Commonwealth, *Code of Practice for Protection of Persons Exposed to Ionizing Radiations in University Laboratories*, London (1961)
Barnes, D. E., and Taylor, D., *Radiation Hazards and Protection*, 2nd edn, Newnes, London (1963)
Bournsell, J. C., *Safety Techniques for Radioactive Tracers*, Cambridge University Press, London (1958)
British Chemical Industry Safety Council, *Protection of the Eyes*, London (1963)
British Oxygen, *Safety in the Use of Compressed Gases*, pamphlet issued by British Oxygen Co., London (n.d.)
British Titan, *Laboratory Safety Manual*, British Titan Products Co. Ltd, London
Brookes, V. J., and Jacobs, M. B., *Poisons*, 2nd edn, Van Nostrand, New York (1958)

Church, F. W., 'Industrial Toxicology', *J. chem. Educ.*, **26**, 309 (1949)
Department of Scientific and Industrial Research, *Safety Measures in Chemical Laboratories*, 3rd edn, H.M.S.O., London (1964)
Dunlop, *Accident Prevention Manual*, Dunlop Rubber Co. Ltd, London (n.d.)
Dunlop, *Safety in the Laboratory*, 2nd edn, Dunlop Rubber Co. Ltd, London (n.d.)
Ellington, O. C., 'Safety Arrangements in Laboratories Handling Explosives and Kindred Materials', *R. Inst. Chem Report* No. 4, 35 (1949)
Factories Act 1961. H.M.S.O., London
Fawcett, H. H., 'Flammable Liquids Refrigerated', *Chem. Engng News*, **27**, 2102 (1949)
Fawcett, H. H., 'Safety Checks for Chemical Laboratories', *J. chem. Educ.*, **24**, 296 (1947)
Fawcett, H. H., 'Misdirected Curiosity Produces Fatal Results', *J. chem. Educ.*, **24**, 457 (1947)
Fawcett, H. H., 'Supplementing the Chemical Curriculum with Safety Education', *J. chem. Educ.*, **26**, 108 (1949)
Fisher, *Manual of Laboratory Safety*, Bulletin F.S. 201, Fisher Scientific Co., New York (n.d.)
Fisher, *Laboratory Emergency Chart*, Fisher Scientific Co., New York (n.d.)
Freeman, W. I., *Protective Clothing and Devices*, United Trade Press, London (1962)
Geigy, *Safety in the Laboratory*, The Geigy Company Ltd, Manchester (n.d.)
Gillette, *Safety in the Laboratory*, Gillette Industries Ltd, London (1962)
Glazebrook, A. L., and Montgomery, J. B., 'High Pressure Laboratory', *Ind. Engng Chem.*, **41**, 2368 (1949)
Glasstone, S., *Source Book on Atomic Energy*, Macmillan, London (1952)
Grey, C. H. (ed.), *Laboratory Handbook of Toxic Agents*, R. Inst. Chem., London (1960)
Guy, K., *Laboratory First Aid*, Macmillan, London (1965)
Hiscocks, E. S., *Laboratory Administration*, Macmillan, London (1956)
Imperial Chemical Industries Ltd, *Safety in the Laboratory*, London (1961)
Imperial College, *Code of Practice Against Radiation Hazards*, Imperial College of Science and Technology, University of London (1962)
International Atomic Energy Authority, *Safe Handling of Radio Isotopes*, Vienna (1958)
Lawson Helme, E. T., *Electric Light and Power Installations*, Cassell, London (1954)
Lovett, A. B. E., 'Safety and the Technician', *S.T.A. Bull.*, **4**, No. 4, 2 (1955)
Mahley, H. S., 'Operating High Pressure Equipment', *Chem. Engng News*, **27**, 3860 (1949)
Manufacturing Chemists' Association, *Guide for Safety in the Chemical Laboratory*, 2nd edn, General Safety Committee of the Manufacturing Chemists' Association, Inc., Van Nostrand, New York (1954)
May and Baker, *Safety in the Chemical Laboratory*, May and Baker Ltd, London (1965)
Medical Research Council, *Safety Precautions in Laboratories*, Medical Research Council, London (1960)
Miller, H. C., 'Safe Handling of Fluorine Chemicals', *Chem. Engng News*, **27**, 3854 (1949)
Ministry of Education, *Safety Precautions in Schools*, Ministry of Education Pamphlet No. 13, H.M.S.O., London (1948)
National Union of Public Employees, *Legal Aid and Union Benefits*, National Union of Public Employees Handbook No. 1, London (1954)
Parsons, O. H., *Guide to the Industrial Injuries Act*, L.R.D. Publications Ltd, London (1961)
Schenk, G., *The Book of Poisons*, Weidenfeld & Nicolson, London (1956)

Royal Institute of Chemistry, 'Report of a Conference on the Origins and Prevention of Laboratory Accidents', *R. Inst. Chem. Report* No. 4 (1949)
Siebe Gorman, *Personal Protection*, Siebe Gorman & Co. Ltd, London (1965)
Young, E. G., 'Are You Guilty', *J. chem. Educ.*, **26**, 258 (1949)

Chapter 7 Special needs of teaching laboratories

Bedow, N. E., 'A Simple Method for Recovering Iodine', *S.T.A. Bull.*, **2**, 8 (1952)
Cannon, W. A., 'An Automatic Mercury Vacuum Still', *J. chem. Educ.*, **28**, 272 (1951)
Dunning, W. G., 'The Maximum Use of the Chemistry Laboratory', *J. chem. Educ.*, **25**, 457 (1948)
Edwards, J. A., 'Laboratory Management', *Chem. Age, Lond.*, **68**, 955 (1953)
Gaddis, S., 'Semi-micro Equipment for the Elementary Lab.', *J. chem. Educ.*, **19**, 530 (1942)
Guy, K., 'An Inexpensive Hot Air Dryer', *J. Sci. Technol.*, **2**, No. 1, 29 (1956)
Guy, K., 'A Method for Removing Stuck Stopcocks', *S.T.A. Bull.*, **4**, No. 1, 7 (1954)
Haskins, A. L., 'Operating an Agricultural Biochemistry Stockroom', *J. chem. Educ.*, **27**, 391 (1950)
Jelinek, H. N., Huber, C. F., and Astle, M. J., 'A Continuous Mercury Still', *J. chem. Educ.*, **26**, 597 (1949)
Joncich, M. J., Alley, C. A., and Kowaka, M., 'Apparatus for the Triple Distillation of Mercury', *J. chem. Educ.*, **33**, 607 (1956)
Loytty, O. M., 'Purchase and Stocking of Laboratory Glassware', *J. chem. Educ.*, **27**, 393 (1950)
McNevin, W. M., 'Administrative and Teaching Problems of Large Classes in Quantitative Analysis', *J. chem. Educ.*, **25**, 589 (1948)
Newbury, N. F., *The Teaching of Chemistry*, Heinemann, London (1934)
Smith, A., and Hall, E. H., *The Teaching of Chemistry and Physics in the Secondary School*, Longmans, Green, New York (1907)
Sutcliffe, A., *School Laboratory Management*, 2nd edn, London, Murray (1950)

Chapter 8 Optical projection

Ansley, A. J., *An Introduction to Laboratory Technique*, 2nd edn, Macmillan, London (1952)
Atkinson, N. J., *Practical Projection for Teachers*, London, Current Affairs Ltd (1948)
Briggs, G. A., *Sound Reproduction*, 2nd edn, Wharfdale Wireless Works, Bradford (1950)
Campbell, F. W., Law, T. A., Morris, L. F., and Sinclair, A. T., *Sound Film Projection*, 4th edn, Newnes, London (1951)
De Kieffer, R., and Cochran, L. W., *Manual of Audio-visual Techniques*, Prentice-Hall, Englewood Cliffs, N.J. (1955)
Eastman Kodak, *Foundation for Effective Audio-visual Projection*, pamphlet issued by Eastman Kodak Co., Rochester, N.Y. (n.d.)
Frayne, J. G., *Elements of Sound Recording*, Wiley, New York (1949)
Gopsill, G. H., *Projection in Schools*, Current Affairs Ltd, London (1953)
Haas, K. B., and Packer, H. B., *Preparation and Use of Audio Visual Aids*, 3rd edn, Prentice-Hall, New York (1955)
Jenkins, N., *How to Project*, 2nd edn, Focal Press, London (1951)

Kidd, M. K., and Long, C. W., *Filmstrip and Slide Projection*, Focal Press, London (1949)
Kodak, *Projection for Teaching Purposes*, data sheet ED-1, Kodak, London (n.d.)
Leitz, *Large Lecture Hall Projectors*, pamphlet issued by Ernst Leitz GmbH, Wetzlar, W. Germany (n.d.)
McKnown, H. C., and Roberts, A. B., *Audio Visual Aids to Instruction*, McGraw-Hill, New York (1949)
Park-Winder, W. E., *Microprojection in Science Teaching*, Crosby Lockwood, London (1951)
Rank, *Bell & Howell 640 Recording Techniques*, Rank Precision Industries, London (n.d.)
Sands, L. B., *Audio-visual Procedures in Teaching*, Ronald Press, New York (1956)

Chapter 9   Laboratory records and technical information

Berwick Sayers W. C., *Manual of Classification*, 3rd edn, Grafton, London (1955)
Collison, R. L., *Indexes and Indexing*, Benn, London (1953)
Eastman Kodak, *Copying*, Kodak data book, 5th edn, Eastman Kodak Co., Rochester, N.Y. (1962)
Elliott, G. M. (ed.), *Film and Education*, Philosophical Library, New York (1948)
Ilford, *Document Copying with Ilford Materials*, 2nd edn, Ilford, London (1957)
Kahn, G., and Yerian, T., *Progressive Filing*, 6th edn, McGraw-Hill, New York (1955)
Kinder, J. S., *Audio Visual Materials and Techniques*, 2nd edn, American Book Co., New York (1959)
Kodak, *Contact Document Copying*, Kodak Ltd, London (n.d.)
Kodak, *Photographic Copying of Documents*, Kodak data sheet D.C.1, London (1955)
Kodak, *Storage and Care of Kodak Colour Film*, Kodak pamphlet No. E.30, Rochester, N.Y. (n.d.)
Library of Congress, *Rules for descriptive cataloguing in the Library of Congress. Motion Pictures and Filmstrips*, Washington (1952)
Macdonald and Evans, 'Reproduction Processes, Pt. I, Duplicators', and 'Pt. II, Photocopying and Microfilming', Manual of Modern Business Equipment, Macdonald and Evans, London (1956)
Odell, M. K., and Strong, E. P., *Records, Management and Filing Operations*, McGraw-Hill, New York (1947)
Ralph, R. G., *The Library in Education*, Turnstile Press, London (1949)
Rufsvold, M. I., *Audio-Visual School Library Service*, American Library Association, Chicago (1949)
Sands, L. B., *Audio Visual Procedures in Teaching*, Ronald Press, New York (1956)
Schell, M., 'Filing and Classification Methods of Photographic and Illustrative Material', *J. biol. Photogr. Ass.*, **19**, No. 3, 129, and No. 4, 158 (1951)
Stickney, E. P., and Scherer, H., 'Developing an A-V Program in a Small College Library', *Library Journal*, **84**, 2457 (1959)
Weeks, B. M., *How to File and Index*, Ronald Press, New York (1946)

Chapter 10   Organisation of demonstrations and exhibitions

Alyea, H. N., 'Tested Demonstrations in General Chemistry', *J. chem. Educ.*, **32**, 28 (1955)
Bacon, E. K., 'Displaying Chemical Exhibits', *J. chem. Educ.*, **15**, 219 (1938)

Baker, H. W., and Phares L. I., 'A Chemistry Exhibit', *J. chem. Educ.*, **9**, 501 (1932)
Curtis, W. C., 'Project Teaching in High School Chemistry', *J. chem. Educ.*, **18**, 293 (1941)
Briggs, D. B., 'The Preparation and Exhibition of Crystals', *Sch. Sci. Rev.*, **10**, 35 (1928–9)
Day, J. E., 'The Modern Lecture Hall and Preparation Room', *J. chem. Educ.*, **6**, 1887 (1929)
Eastman Kodak, *Effective Lecture Slides*, pamphlet issued by Eastman Kodak Co., Rochester, N.Y. (n.d.)
Eastman Kodak, *Producing Slides and Film Strips*, 3rd edn, Eastman Kodak Co., Rochester, N.Y. (1963)
Hutton, K., 'Lantern Slides for Writing On', *Sch. Sci. Rev.*, **35**, 126 (1953–4)
Jolly, W. L., 'Transparencies by Photographic Reversal', *J. chem. Educ.*, **39**, 83 (1962)
Lawson, D. F., 'The Effective Preparation of Graphs, Illustrations and Kymographs for the Purpose of Reproducing by Photographic Means', *Lab. Pract.*, **6**, 700 (1957)
Edwards, J. J., and Edwards, M. J., *Medical Museum Technology*, Oxford University Press, London (1959)
Fowles, G., *Lecture Experiments in Chemistry*, Bell, London (1937)
Institute of Physics, *Design of Physics Research Laboratories*, Institute of Physics, Chapman & Hall, London (1959)
Meyers, D. K., 'Element of the Week', *J. chem. Educ.*, **27**, 82 (1950)
Newbury, N. F., *The Teaching of Chemistry*, Heinemann, London (1934)
Nicholson, D. G., 'Storage of Lecture Demonstration Equipment', *J. chem. Educ.*, **32**, 138 (1955)
Nokes, M. C., *Demonstrations in Modern Physics*, Heinemann, London (1952)
Panush, L., 'Science Exhibits', *J. chem. Educ.*, **28**, 25 (1951)
Payne, V. F., 'The Lecture Demonstration and Individual Laboratory Methods Compared', *J. chem. Educ.*, **9**, 1277 (1932)
Rakestraw, N. W., 'Functions and Limitations of Lecture Demonstration', *J. chem. Educ.*, **6**, 1882 (1929)
Reed, R. D., 'High School Chemistry Demonstrations', *J. chem. Educ.*, **6**, 1905 (1929)
Robertson, G. R., 'Design of a Chemistry Lecture Room', *J. chem. Educ.*, **36**, 197 (1959)
Rosenfield, J., 'Making Lantern Slides for Immediate Projection', *J. chem. Educ.*, **35**, 314 (1958)
Shaw, W. H. R., and Aronson, J. N., 'The Preparation of Inexpensive Lantern Slides', *J. chem. Educ.*, **40**, 483 (1963)
Southern, J. A., 'Technique for Preparing Lantern Slides', *J. chem. Educ.*, **34**, 324 (1957)
Spence, T. F., *Teaching and Display Techniques in Anatomy and Zoology*, Pergamon, London (1967)
Sutton, R. M., *Demonstration Experiments in Physics*, McGraw-Hill, New York (1938)

Chapter 11   Organisation of the work and training of laboratory technicians

Better opportunities in technical education, H.M.S.O., London (1963)
Boyer, R. M., 'Service Personnel for University Chemistry Departments', *J. chem. Educ.*, **33**, 387 (1956)
Hiscocks, E. S., *Laboratory Administration*, Macmillan, London (1956)

Institute of Animal Technicians, *Introductory Booklet*, London (n.d.)
Institute of Science Technology, *Introductory Booklet*, London (n.d.)
Lindley, P., and Johnson, R. M., 'Notes for Beginners in Laboratories', *Lab. Pract.*, **6**, 581 (1957)
Mills, H. R., *Teaching and Training: Techniques for Instructors*, Macmillan, London (1966)
Central Youth Employment Executive, *Choice of Careers No. 57—The Medical Laboratory Technician*, H.M.S.O., London (1960)
Pfiffner, J. M., *Supervision of Personnel*, Prentice-Hall, New York (1951)
Starks, S. L., 'Training in Industry', *J. chem. Educ.*, **21**, 285 (1944)
Taylor, M. H., 'Qualifications for Laboratory Technicians', *Bull. Inst. Sci. Technol.*, **10**, No. 9, 8 (1965)
Topping, J., 'The Education of Technicians and Technologists', *J. Sci. Technol*, **6**, No. 1, 7 (1960)
Vernon, P. E., *The Training and Teaching of Adult Workers*, University of London Press, London (1943)
War Office, *Successful Instruction*, Training Publication, London (1951)

Chapter 12   Elements of experimental procedure

Baird, D. C., *Much Simpler but Still Very Good Experimentation*, Prentice-Hall, New York (1962)
Beveridge, W. I. B., *The Art of Scientific Investigation*, Heinemann, London (1951)
Cox, D. R., *The Planning of Experiments*, Wiley, New York (1958)
Davies, O. L., *Statistical Methods in Research and Production*, 3rd edn, Oliver and Boyd, London (1958)
Freedman, P., *Principles of Scientific Research*, Macdonald, London (1949)
Lewin, S., 'Variations and Errors in Experimental Investigation', *Lab. Pract.*, **10**, 99, 151, 361, 474, 556, and 636 (1961)
Moroney, M. J., *Facts from Figures*, Pelican, London (1964)
Parr, N. L. (ed.), *Laboratory Handbook*, Newnes, London (1963)
Reichmann, W. J., *Use and Abuse of Statistics*, Pelican, London (1961)
Wilson, E. B., *An Introduction to Scientific Research*, Oliver and Boyd, London (1952)

Chapter 13   Technical literature

Aslib Directory, 3 Belgrave Square, London, SW1
British Chemicals and Their Manufacturers, The Association of British Chemical Manufacturers, 86 Strand, London, WC2
'British Standards in the Laboratory' Pt. 1, *Lab. Pract.*, **4**, 251 (1955)
British Technology Index, Library Association, Chaucer House, Malet Place, London, WC1
Carey, R. J. P., *Finding and Using Technical Information*, Camelot, London (1966)
Edwards, J. A., *Laboratory Management and Techniques*, Butterworths, London (1960)
Guide to European Sources of Technical Information, Organisation for Economic Co-operation and Development. Château de la Muette, 2 Rue Andre Pascal, Paris 16ème

Handbook of Chemistry and Physics, 53rd edn, Chemical Rubber Publishing Co., Cleveland, Ohio (1972)
Merck Index of Chemicals and Drugs, Merck and Co., Inc., Rahway, N.J. (1960)
Parr, N. L. (ed.), *Laboratory Handbook*, Newnes, London (1963)
Technical Services for Industry, Department of Scientific and Industrial Research, State House, High Holborn, London, WC1

Index

Accidents, 83, 84, 245, 340 (*see also under* Safety)
  from glass, 237
  notification, 335
  radiation, 290
Accounting, petty cash, 159–165
Accounting, stores, 115
Accumulator acid, 209
Accumulators, 209–212
  alkaline, 211
  charging, 210
  cleaning, 211
  maintenance, 209
  storage, 211
Acetylene
  cylinders, 272
  explosion hazard, 260
  valves, 272
Acid(s)
  burns, 317
  concentrations, 504
  safety measures, 242
  stores, safety measures, 265
Agar media, 191
Air
  conditioning, 50
  filtration, 50
  receivers, 333
Alcohol, purchase, 121
Alkali burns, 317
α-particles, 279
Ammonia stores, safety measures, 266
Amplifiers, 407
Ampoules, safety measures, 243
Animals
  dispatch, 134, 135

Animals *continued*
  handling techniques, 295
Antidotes, 320
Apparatus, 498
  cleaning, 223–235
  loans, 115
  maintenance, 214
  old and redundant, 130
  preparation room, 167
  repairs, 115
  storage, 148
    shelves, 112
Artificial respiration, 308–309
Asbestos blankets, 254
Ashless papers, 154
Aslib, 493
Asphalt mastic, 44
Asphalt tiles, 44
Asphyxia, 307
Atmosphere, 83
Autoclaves, safety measures, 263

Balance room, 95
Balance supports, 96
Balances, 168, 348
  cleaning, 208
  installation, 95
  maintenance, 207
Barometers, 99–100
B.C.F. fire extinguishers, 253
Bench(es), 2–19
  cleanliness, 339
  comparative price levels, 16
  design, 15
  dimensions, 2

Bench(es) *continued*
  fixed, 15
  glassblowing, 102
  height, 2, 85
  maintenance, 213
  services, 13, 15, 19
  space between, 5
  teaching laboratories, 344
  tops, 6
    fabricated, 11
    glass, 10
    glazed earthenware tiles, 10
    linoleum, 9
    maintenance, 213–214
    metal, 7
    plastics, 11
    quarry tiles, 10
    rubber, 9
    slate, 9
    soapstone, 11
    soft asbestos, 10
    supporting of, 16–19
    timber, 6
  types, 2
  understructure, 13
  unit-constructed, 15
  width, 3, 85
β-particles, 279
Biological hazards, safety measures, 294–296
Biological reagents, 183–196
Blackboards, 212
Boilers, 333
Bottles
  reagent, 222, 349
  residues, 352
Brightness, 72
British Standards, 30, 65, 496–497
British Technology Index, 493, 494
Bromine burns, 318
Burettes, 215–216
Burners, 102, 256
Burns, 315–318

Camera copying, 444
Canada balsam bottles, 351
Carbon dioxide cylinders, 199
Carbon dioxide fire extinguishers, 253
Carbon removal, 225
Card index, 423, 431, 432
Carrier, responsibility of, 132
Carriers, safety measures, 265
Catch pots, 27–28

Ceilings, 198
Celastoid, 370
Celluloid, postal regulations, 138
Chemical reagents. *See* Reagents
Chemical receivers, 26
Chemicals
  classes, 147
  common names, 501–503
  disposal, 267–270
  explosion hazard, 260
  hazardous, 84, 266
  inorganic, 147
  organic, 147
  preservation and storage, 146
  storage, 266–267
    shelves for, 111
  suppliers, 498
  teaching laboratory, 351
Chlorine cylinders, 273
Cine film projector, 391–396
  mechanism, 393
  procedure, 399–401
Cinematographic materials, packing, 138
City and Guilds of London Institute, 482, 484
Claims, 133
Clamps, maintenance, 214
Clean air, 36
Cleaning, 475
  accumulators, 211
  apparatus, 223–235
  balances, 208
  glassware, 223–231
  microscope slides, 230
  plastics laboratory ware, 234
  platinum ware, 233–234
  porcelain, 231
  silica ware, 231
  syringes, 230
Cleanliness, 83
Coat racks, 87, 90
Cocks. *See* Outlets, service
Colour, of walls and furniture, 78
Compressed air, 30, 36
  valves, 42
Compressors, 36–37
Concrete floors, 43
Condensers, 363
Conditions of sale, 120
Conical joints, 150
Consignment notes for dangerous goods, 137
Containers
  empty, return, 130

Containers *continued*
  isotope, 280
  labelling, 267
  of radioactive materials, 289
  safety measures, 266
Control valves, 30
Copying processes, document, 434–451
Cork tiles, 45
Corks, boring, 239
  storage, 152
Credit note, 145
Customs and Excise, 126
Customs procedure, 128
Cyanide poisoning, 327–338

Damage, notification, 134
Dangerous Drugs Act, 123
Dangerous goods, 133
  consignment notes, 137
  rail transport, 136
Darkroom, 88–93
  basic requirements, 88
  decoration, 91
  fire precautions, 91
  floor surfaces, 92
  furniture, 89
  heating, 91
  lighting, 88–89
  plumbing, 90
  sinks, 92
  storage units, 92
  ventilation, 91
Daylight
  factor, 80
  measurement, 80
Debit note, 145
Decoration, 78, 197–198
  darkroom, 91
Delivery note, 144
Demonstrations, 452, 455–462
  teaching, 465
Demulcents, 320, 321
Desiccators, 221
Design, of laboratory. *See* Laboratory design
Dewar vessels, 104
Dewey Decimal Classification, 424, 432
Diascopic projection, 361
Diazo process, 381, 450
Direct positive process, 447
Dispatch
  advice, 144
  of goods, 130–143

Disposal, of chemicals, 267–270
Distillation, 249
Distilled water, 342
Document copying and reproduction processes, 434–451
Documentation, stores, 144
Doors, 83, 85
Dosemeter, 282
Drainage, 198
  teaching laboratory, 342
  (*see also* Waste pipes)
Drip cups, 22
Dry box, 287
Duplicators
  gelatine, 434
  spirit, 438
  stencil, 435, 436
Dust, 333
  explosion hazard, 259
  extraction, 84

Earth connections, 277
Effluent, 23, 26, 27, 30
  dilution, 27
Electric furnaces, maintenance, 204
Electric lighting. *See* Lighting
Electric shock, 278, 307
Electrical equipment and components, storage, 158
Electrical fittings, inspection, 199
Electrical plant, installation, 276
  safety precautions, 277
Electrical supply, 30, 32, 105
  for cine projectors, 395
  high-tension, 273, 275
  overloading, 276
  safety measures, 273–278
Electrical wiring, experimental set-ups, 276
  installation, 276
Electronic stencils, 436
Emetics, 319, 320, 324
Employment
  of women, 335
  of young persons, 335
Enclosures, types, 274
Epidiascope, 374
Episcope, 374
Episcopic projection, 362
Equipment, 498
  heavy, 103–104
  installation, 94–108
  safety measures, 241

Ether peroxides, explosion hazard, 261
Exhibitions, 452–468
 by chemical societies, 467–468
 open-day, 462–467
 permanent and semi-permanent, 452–455
Experimental procedure, 486–491
Experiments
 continuous, 340
 pilot, 487
Explosions
 causes, 262
 precautions against, 258–264
Explosives, 333 (*see also* Explosions)
Extraction, 249
Eye injuries, 318
Eye protection, 242, 299
Eyeshields, 244

Factories Act, 328–335
Fanlights, 83
Fans, 48–55
Filing
 films, 432
 special materials, 427
 technical information, 429
Filing cabinets, 418
Filing control, 426
Filing equipment, 418
 visible, 421
Filing systems, 416
 basic arrangements, 422
 geographic, 426
 lateral, 421
 memory, 426
 subject, 425
 suspended method, 418
Film badges, 282
Films, 431, 499–500
 packing, 138
 sizes, 392
 sound, 393
 storage, 432
Filmstrip projector, 382–384
Filmstrips, 384–386, 432
 storing and filing, 431
Filter flasks, damaged, 218
Filter papers
 characteristics, 155–157
 storage, 153–157
Filter pumps, 35
Filtration, 153

Fire
 causes, 251
 extinguishers, 199, 251
  types, 252–255
 gas cylinder, 273
 hazards, 84, 252
 means of escape, 333
 precautions, 248–257
  darkroom, 91
  inflammable solvents, 248–250
  general, 256–257
 prevention, 251
Firefighting, 255
 equipment, 85
First aid, 237, 245, 301–338
Fixatives, 183
Floor surfaces, 42–45, 87
 darkroom, 92
 workshop, 297
Focal length, 363
Folders, 417
Fortin barometer, 99
Freezing mixtures, 193
Fuel gas, explosion hazard, 260
Fume cupboards, 51–53, 84, 86, 244, 246, 248, 263, 353
 balanced extract, 59
 centre back exhaust, 59
 compensated extract, 59
 direct extract, 59
 ducting, 64
 extraction systems, 53–65
  central, 56
  ejector, 55
  individual, 56
  positive, 53
 framings and panels, 67
 general design, 65–70
 inspection, 200
 joints and jointing compounds, 65
 Kjeldahl, 63
 multi-level extract, 57
 normal purpose, 57
 radioactive work, 293–294
 sashes, 69
 services, 70–72
 single top exhaust, 57
 special purpose, 59
 walk-in, 70
 washed extract, 63
 working top, 68
Fume hoods, 84
Fumes, dangerous, 246, 333
 poisoning by, 322

INDEX

Furniture, 85
  colour, 78
  darkroom, 89
  maintenance, 212
  preparation room, 166
  radioactive work, 294
  safety measures, 241
  teaching laboratories, 344–348

γ-rays, 279
Gas appliances, 256
Gas burners, 102, 256
Gas cylinders, 86, 246
  colour code, 270
  fires, 273
  laboratory use, 272
  safety measures, 270
  storage, 270
  testing for leaks, 270
  transportation, 271
  valves and fittings, 271
Gas holders, water-sealed, 333
Gas supply, 30, 33, 87, 102
  hazards, 255–256
  outlets, 40
  valves, 42
Gas taps, 40, 199
Gases
  explosion hazard, 259
  inflammable, 86, 333
  poisoning by, 322
  toxic, 244
Geiger–Müller tube, 283
Gelatine duplicators, 434
Glass
  accidents, 237
  broken, disposal, 240
  carrying, 239
  cutting, 237
  dangers from, 240
  heating, 239
Glass annealing, ovens, 103
Glass apparatus containing mercury, 230
Glass tubing, 221
  and rod, storage, 151
  bending, 238
  boring corks for, 239
Glassblowing, 465
  bench design, 102
  equipment, 100–103
Glassware
  cleaning, 223–231

Glassware *continued*
  for vacuum or pressure work, 240
  graduated, fillings in, renewal of, 222
  hard deposits, 226
  issue, 239
  maintenance, 215
  sintered, 228
  stained, 226
  storage, 149
  volumetric, 226
Glassworking lathes, 102
Glazed porcelain, marking, 194
Gold adhesion filter, 359
Graduated glassware, fillings in, renewal of, 222
Grease removal, 225
Greases, high-vacuum, 158
Grinding, 84
Groundglass joints, 150

Haemorrhage, 313–314
Handles, 18
Health, Factories Act, 332, 334
Heating
  darkroom, 91
  of low-boiling-point solvents, 249
  workshop, 297
Heavy equipment, 103–104
Humidity, 50
Hydrofluoric acid burns, 317
Hydrogen, explosive properties, 259
Hydrogen cyanide, 327
Hydrogen peroxide, explosion hazard, 261
Hydrogen sulphide, explosion hazard, 260
  safety measures, 244
Hyvac pump, 358

Illuminating sources for optical projectors, 364–368
Illumination, 72, 74, 79
  depreciation, 77
Illustrative material, 428
Imperial Standard wire gauge, 123
Import Duties Act (1958), 126
Import licence, 127
Imprest system, 165
Index guides, 417
Indicator papers, 158
Industrial diseases, notification, 335

INDEX

Infection
  of wounds, 311
  prevention, 295
Inflammable liquids, storage, 266
Inflammable materials, 334
  packing, 138
  safety measures, 265
  storage, 87, 250
Inflammable solvents, 248–250
  explosion hazard, 258
Information, sources of, 493, 514–515
Injuries and their treatment, 306
Ink stain remover, 194
Inspection, 197–201
Inspectors, of factories, 237, 330
Installation, of laboratory equipment, 94–108
Institute of Animal Technicians, 479, 483
Institute of Medical Laboratory Technology, 479, 483
Institute of Science Technology, 479, 480
Instruments, damage, 340
Interviewing, 469–474
Invoice, 144
Iodine, recovery, 353

Joints
  conical, 150
  seized, 217
  spherical, 150

Kew barometers, 100
Kjeldahl fume cupboard, 63

Labelling
  of bottles, 349
  of containers, 267
  of packages containing dangerous materials, 137
Laboratory design, 1–93
Laboratory equipment. *See* Equipment
Laboratory regulations, 246, 337, 341
Laboratory technicians
  future, 484
  organisation and work, 469–485
Laboratory work, organisation, 474–479
Lamphouse, 364
Lamps, for optical projectors, 364–368
Lantern plates, 413
Lantern slides, 409–414
  storage, 430

Lathes, glassworking, 102
Lecture demonstrations, 455
Lecture experiments, 455
Lecture rooms, inspection, 200
Lecture theatre, 460
Lenses, 363
  maintenance, 205–206
Library services, 492
Light
  artificial, 297
  measurement, 72
Lighting, 72–83 (*see also* Illumination)
  darkroom, 88
  diffused, 77
  direct, 77
  efficiency, 74
    of fittings, 75
  electric, 73–74
  fittings, height of, 79
  indirect, 77
  natural, 80–83, 297
  semi-direct and semi-indirect, 77
  stores, 113
  teaching laboratory, 341
  values, 79
  workshop, 296
Linoleum, 42
Liquid air
  demonstrations, 465
  explosion hazard, 259
  plant, 104
Literature, technical, 492
Litmus papers, 158
Loans, of apparatus, 115
Lockers, 346

Machinery
  maintenance, 300
  safety measures, 300
Magnetic recorders, 406
Magnetic stripe, 404
  recording on, 407
Maintenance, 201–223
  machinery, 300
  personal equipment, 338
Manesty pattern stills, 105
Marking, of glazed porcelain, 194
Measuring cylinders, 219
Mercury, 84, 247–248
  contaminated, 354–359
  glass apparatus containing, 230
  purification of, 354–359
Mercury barometer, 99

INDEX

Mercury collector, 248
Mercury vapour, 84
Metalware
  maintenance, 223
  storage, 158
Microcopying, 445
Microfilming, 445
Microphones, 405
Micro-projection, 386–391
  with compound microscope, 386–389
Micro-projector, 386
Microscope lenses, maintenance, 205
Microscope slides, cleaning, 230
Microscopes
  maintenance, 205
  optical projection, 386 ff
Microtome knives, sharpening, 207
Mounting media, 190

Nife accumulators, 211
Nuffield Teaching Project, 344

Offset lithography, 441–443
Oil removal, 225
Open-day exhibitions, 462–467
Optical components, 363
Optical projection, 361–415
  acoustics, 398
  arrangements, 396
  cine film, 391–396
    procedure, 399–401
  darkening arrangements, 399
  diascopic, 361
  episcopic, 361, 362
  filmstrip, 382–386
  lighting, 398
  micro-projection, 386–391
    with compound microscope, 386–389
  overhead, 377–380
  projectors, 372–380
  reflector, 368
  screen(s), 368–372, 380
    position, 396
  seating, 396
  sound recording and reproduction, 401–409
  ventilation, 398
Order form, 144
Outlets, service, 37–42
Ovens
  glass annealing, 103

Ovens *continued*
  maintenance, 205
  Overcrowding, 84
Oxalic acid, 246
Oxygen
  cylinders, 101
  supply, 101
  valves, 272

Packing
  faulty, 135
  inflammable materials, 138
  regulations, 136, 137
  special substances, 138–140
Paint, packing, 140, 142
Paper, storage, 153
Perchloric acid
  explosion hazard, 263
  safety measures, 244
Perishable merchandise, 135
Peroxides, explosion hazard, 261
Personnel facilities, 245
Pestles, maintenance, 214
Petty cash accounts, 159–165
Pharmacy and Poisons Act (1933), 121, 319
Phenol burns, 318
Phosphorus burns, 317
Photocopying processes, non-optical, 446
Photographic materials, packing, 138
Photographic stencils, 436
Photographic transfer process, 448
Photostat photographic copying machine, 444
Pipettes, safety measures, 243, 295
Piping materials, 36
Plastics apparatus, storage, 152
Plastics laboratory ware, cleaning, 234
Plastics varnish floor surfaces, 45
Platinum
  purchase, 123
  recovery, 354
  sheet, weight per square centimetre, 124
  ware, cleaning, 233–234
  wire, weight per metre, 122
Plumbing
  darkroom, 90
  radioactive work, 294
Plumbing lines, 18
Poisoning, 319–328
  cyanide, 327–328

INDEX

Poisoning *continued*
  symptoms, 321
  treatment, 322, 324
Poisons, 148
  classification, 321
  purchase, 121
  safety measures, 246
Porcelain
  cleaning, 231
  glazed, marking, 194
Postal regulations, 137–143
Preparation room, 166
  apparatus, 167
  furniture, 166
  services, 166
Pressure vessels, safety measures, 263
Professional organisations, 479–483
Projector(s), 372–380
  cine film, 391–396
    mechanism, 393
    procedure, 399–401
  filmstrip, 382–386
  illuminating sources, 364–368
  lamps, 364–368
  micro-, 386
  overhead, 377–380
  position, 397
Protective clothing, 242, 244, 300, 338
Protective equipment, 241
Prussic acid, 327
Pumps
  filter, 35
  rotary vacuum, maintenance, 201–204
  water filter, 199, 343
  water jet, 35
Purchasing
  of alcohol, 121
  bulk, 126
  of materials, 118–126
    overseas, 126
    special, 120
  of platinum, 123
  of poisons, 121
P.V.C. tiles, 44

Quality, of materials, 120
Quarry tiles, 44
Quickfit joints, 150

Rad, 279
Radiation
  accidents, 290

Radiation *continued*
  active areas, 288
  air sampling, 285
  contamination, measurement, 284
  decontamination, 289–290
  dose, 279
  dry box, 287
  effects, 281
  maximum permissible dose, 281
  measurement, 281
  monitor, 283
  quantity, 280
  safety measures, 278–294
  sealed and unsealed sources, 280
  shielding, 285–287
  smear tests, 284
  types of, 279
  unit, 279
Radioactive materials, 85
  containers, 289
  methods of working, 288
  storage, 86, 291–292
  transportation in buildings, 292
Radioactive waste disposal, 290–291
Radioactive work
  furniture, 294
  laboratory design, 292
  plumbing, 294
Radioisotopes, containers, 280
Rail transport, 133, 135
Razors, maintenance, 206
Reagent bottles
  maintenance, 222
  teaching laboratory, 349
Reagent shelves, 3–5
Reagents
  biological, 183–196
  chemical, 170–183
  distribution, 168
  preparation and storage, 166–196
  quality, 168
  special, 192
Records, 145, 416
  experimental, 487
Recovery, of residues, 352–359
Reference books, 495
Reflectance value
  metals, 75
  room interior surfaces, 79
Reflex copying, 382
Reflex process, 446
Releasing solutions, 223
Rem, 279
Rep, 279

Repairs, 197
  to apparatus, 115
Reproduction processes, document, 434–451
Residues, recovery, 352–359
Residues bottles, 352
Road transport, 131
Röntgen, 279
Roof ventilators, 47
Rotary vacuum pumps, maintenance, 201–204
Rubber bungs, 239
  storage, 152
Rubber teats, and policemen, storage, 152
Rubber tubing, storage, 151
Rubbish box units, 18

'S' bend traps, 28
Safety apparatus, inspection, 199
Safety devices, inspection, 201
Safety measures, 83–87, 236–336
  acid stores, 265
  acids, 242
  ammonia stores, 266
  ampoules, 243
  autoclaves, 263
  biological hazards, 294–296
  carriers, 265
  chemical stores, 264
  electricity, 273–278
  equipment, 241
  Factories Act, 332, 334
  furniture, 241
  gas cylinders, 270
  hazardous operations, 257
  inflammable materials, 265
  machinery, 300
  perchloric acid, 244
  pipettes, 243, 295
  poisons, 246
  pressure vessels, 263
  radiation, 278–294
  shelving, 265
  sodium, 243, 257
  spectographic equipment, 274
  sulphuretted hydrogen, 244
  teaching laboratories, 337
  trolleys, 265
  workshop, 200, 296–301
Safety officer, 237
Safety personnel, radiation, 288
Samples, postal regulations, 141
Sand, 254
Scintillation counter, 283
Scintillation probe, 283
Screens, for optical projection, 368–372, 380, 396
Sealing compounds, for high-vacuum work, 160
Seized joints, 217
Separating funnels, 219
Service boxes, 18
Service(s), 29–42
  on benches, 19
  colour code, 30
  fume cupboard, 70–72
  lecture bench, 461
  misuse, 338
  outlets, 37–42
  piping, materials recommended, 36
  preparation room, 166
  teaching laboratory, 342
Shelving
  for storage of apparatus, 112
  for storage of chemicals, 111
  reagent, 3–5
  safety measures, 265
Shock, 306, 315
  electrical, 278, 307
Silica materials, transparent, 233
Silica ware, cleaning, 231
Silver, recovery, 352
Silvering solution, 195
Sinks, 19
  darkroom, 92
  fitting, 20
  glazed fireclay, 20
  location, 19
  materials, 20
  metal or metal-lined, 21
  outlets, 22
  polythene, 21
  porcelain on metal, 21
  for washing glassware, 22
Sintered glassware, 228
Slides, 409–414, 430
Soap solution, 195
Soda acid or water–CO_2 fire extinguishers, 252
Sodium, safety measures, 243, 257
Solutions, standard, 168
  strength of, 169
Solvents
  inflammable, 248–250
    explosion hazard, 258
  low-boiling-point, heating, 249
  recovery, 353

Sound head, 404
Sound recording and reproduction, 401–409
Sources of information, 493, 514–515
Spectrographic equipment, 106–107
 safety measures, 274
Spectrographic laboratory, 106
Spherical joints, 150
Spirit duplicators, 438
Staff, engagement of, 469
Stained glassware, 226
Stains, 185
Standard deviation, 489–490
Standard error, 490
Standard solutions, 168
Statistical methods, 488–491
Statutory Rules and Orders, 133
Steam supply, 30, 34
 outlets, 40
Steam valves, 40
Stencil duplicators, 435, 436
Stencils
 electronic, 436
 photographic, 436
Stills, 105, 358
 maintenance, 204
Stock, recording, 145
Stockbook, 116
Storage, 146–158
 of apparatus, 148
 of chemicals, 266–267
 of films, 432
 of filmstrips, 431
 of inflammable materials, 250
 of radioactive materials, 291–292
Storage units, darkroom, 92
Storekeeper, function, 114
Stores, 86, 109–165
 accounting, 115
 acid, safety measures, 265
 ammonia, safety measures, 266
 central, 109
 chemical, safety measures, 264
 design, 110
 dispensing, 110
 documentation, 144
 economy, 115
 function of, 114
 general rules, 117
 goods, dispatch, 130–143
 inspection, 200
 layout, 117
 lighting, 113

Stores *continued*
 main, 110
  design, 110
 purchasing, 118–126
 size, 117
 stock, recording of, 145
 teaching laboratory, 348
 types, 109–110
Students' experiment, 466
Sulphuretted hydrogen. *See* Hydrogen sulphide
Sulphuric acid, 242–243
Syringes, cleaning, 230

Tables, maintenance, 213
Tape recorders, 406
Taps, 37, 38, 40, 199
Tar removal, 225
Teaching demonstrations, 465
Teaching laboratories, 337–360
 benches, 344
 chemicals, 351
 discipline, 337–341
 furniture, 344
 reagent laboratory, 349
 stores, 348
Technical information, filing, 429
Technical literature, 492
Technical Services for Industry, 493
Terrazzo floors, 43
Theft, prevention, 339
Thermometers, repairs, 221
Tiles, 44–45
Tools, hand, 300
Toxic fumes, 246
Toxic gases, 244
Trade catalogues, 427
Training, 484
 organisation, 475–479
Transmission printing process, 447
Transparencies, preparation, 380–382
Trap units, 26, 28, 248
Treasury Direction, 127–129
Trolleys, safety measures, 265
Troughs, 19
 materials, 20
Tubing, glass, 221

Vacuum, 30, 34
Vacuum points, 101
Vacuum pumps, rotary, maintenance, 201–204

Vacuum taps, 40
Valves, 37
  control, 30
  for compressed air and other gases, 42, 87
  gas cylinders, 271
  oxygen, 272
  steam, 40
  water, 87
Vapours, 333
Varnish, packing, 140, 142
Ventilation, 45–50, 84, 245 (*see also* Fume cupboards)
  central systems, 48
  darkroom, 91
  for optical projection, 398
  local systems, 49
  mechanical, 47
  natural, 46
  teaching laboratory, 341
  workshop, 297
Ventilators, roof, 47
Vibration, 94–99
  isolated table, 99
Vitreosil, transparent, 233
Vitreosil ware, 231
Vitreous tiles, 44
Volatile liquids, storage, 266
Volumetric apparatus, 226

Walls, colour, 78
Waste containers, 339
Waste disposal, radioactive materials, 290–291
Waste pipes, 23
  cast-iron, 25
  chemical stoneware, 23

Waste pipes *continued*
  glass, 26
  lead, 25
  polythene, 24
  P.V.C., 25
Water filter pumps, 199, 343
Water jet pumps, 35
Water pressure, 343
Water supply, 30, 87
  cold, 32
  outlets, 38
Waxes, high-vacuum, 158
Welfare, Factories Act, 334
Windows and window frames, 198
  design, 80–82
  obstructions, 82
Women, employment, 335
Wood floors, 42
Workshop, 87
  floor surfaces, 297
  heating, 297
  inspection, 200
  lighting, 296
  safety measures, 296–301
  ventilation, 297
Wounds, 309
  infection, 311
  treatment, 312
  types, 311

Xerography, 439–441
Xerox copying equipment, 381
X-rays, 279

Young persons, employment, 336